PARVALBUMIN

A VOLUME IN MOLECULAR ANATOMY AND PHYSIOLOGY OF PROTEINS SERIES

Molecular Anatomy and Physiology of Proteins Series

Vladimir N. Uversky, Series Editor

Alpha -Lactalbumin
Eugene A. Permyakov
2005. ISBN: 1-59454-107-8

Creatine Kinase
Vladimir N. Uversky (Editor)
2006. ISBN: 1-59454-715-7

Neuronal Calcium Sensor Proteins
Pavel P. Philippov and Karl-Wilhelm Koch (Editors)
2006. ISBN: 1-59454-978-8

Human Stefins and Cystatins
Eva Žerovnik and Nataša Kopitar Jerala (Editors)
2006. ISBN: 1-60021-233-6

PARVALBUMIN

A VOLUME IN MOLECULAR ANATOMY AND PHYSIOLOGY OF PROTEINS SERIES

EUGENE A. PERMYAKOV

Nova Science Publishers, Inc.
New York

For permission to use material from this book please contact us:
Telephone 631-231-7269; Fax 631-231-8175
Web Site: http://www.novapublishers.com

LIBRARY OF CONGRESS CATALOGING-IN-PUBLICATION DATA

Permiakov, E. A. (Evgenii Anatolevich)
Parvalbumin / Eugene A. Permyakov.
p. cm. -- (Molecular anatomy and physiology of proteins series)
Includes index.
ISBN 13 978-1-60021-337-3
ISBN 10 1-60021-337-5
1. Calcium-binding proteins. I. Title.
QP552.C24P465 2006
572'.69--dc22 2006021204

Published by Nova Science Publishers, Inc. ✣ New York

CONTENTS

ACKNOWLEDGMENTS

This work was supported by the Program of the Russian Academy of Sciences "Molecular and Cellular Biology". I am indebted to Dr. Denesyuk who kindly helped me to draw molecular structures and to Dr. S. Permyakov and A. Bakunts who helped me in obtaining some experimental data for book illustrations.

INTRODUCTION

In 1955 Henrotte isolated from carp muscle extracts a protein of low molecular weight with an unusual ultraviolet absorption spectrum (the absence of tryptophan and tyrosine absorption and the presence of distinct phenylalanine absorption). It was only in 1965 that a full characterization of this type of protein was published (Hamoir & Konosu, 1965; Konosu *et al*., 1965). The characterization included amino acid analysis and precise estimation of the length of the polypeptide chain, which turned to be a little bit more then 100 residues. Several isotypes of this protein were isolated from the same muscle. It was concluded that they constitute a family of homologous proteins. The first amino acid sequence of this protein was determined in 1971 for the major component isolated from hake white muscle (Pechére *et a*l., 1971b). The general name of *parvalbumins* (small albumins) was proposed, designating at the same time, two of their main characteristics: low molecular weight and high solubility in water (Pechére, 1968).

Three-dimensional structure of a parvalbumin from carp muscle was first elucidated in 1972 (Nockolds *et al*., 1972; Kretsinger & Nockolds, 1973). This was the first X-ray structure of one of the proteins of a large superfamily (EF-hand proteins) of calcium binding proteins with the highest affinity to calcium ions. It turned out that calcium binding domains (EF-hand) of parvalbumin consisting of two helices and a calcium binding loop between them were found in all other members of this protein superfamily.

The calcium binding ability of parvalbumins was found early (Fleron, 1968) and later it was confirmed by Pechére *et al*. (1971a) and Benzonana *et al*. (1972), who first measured stoichiometry of the binding and binding constant. It was suggested that parvalbumins, concentration of which in some types of muscles reaches millimoles per liter, take part in calcium regulation of muscle contraction and relaxation.

For some time parvalbumins were thought to be muscle-specific proteins although trace amounts of parvalbumin were found in other tissues extracts especially in brain (Baron *et al*., 1975). Later, parvalbumin from brains of rat, cat and man was isolated and its biochemical and immunological properties were found to be indistinguishable from their muscle counterparts (Berchtold *et al*., 1982b).

The regulation of intracellular free calcium has become very important problem with the discovery that calcium can act as a secondary messenger in the transduction of the signals from diverse external effectors to various intracellular systems. Two transduction pathways involve an increase in intracellular calcium: one pathway is achieved by inositol (1,4,5)-triphosphate (IP_3) and causes the mobilization of internal calcium stores; the other pathway acts *via* stimulation of calcium channels to produce an influx of external calcium ions. Free Ca^{2+} concentration inside eucariotic cells is 100-200 nM and the work of calcium pumps generates a 10,000-fold Ca^{2+} concentration gradient between the extracellular space and the cytosol. Ca^{2+} controls almost everything cells do during the stages between origin and death from development to adulthood. Now it is clear that Ca^{2+} is extremely essential for cell life, but also, potentially dangerous for it: cytosolic Ca^{2+} overload causes the death of the cells because of the necrotic disintegration by permanently activated Ca^{2+}-dependent proteases (reviewed by Carafoli, 2004).

In nonexcitable cells (blood cells, hepatocytes, and so on) the slow inositol (1,4,5)-triphosphate-mediated pathway predominates (reviewed by Clapham, 1995). Two receptor classes, the G protein-coupled receptor and the receptor tyrosine kinase, release IP_3. The G protein-coupled receptors activate phospholipase Cβ, while receptor tyrosine kinases stimulate phospholipase Cγ to convert phosphatidylinositol (4,5)-bisphosphate into IP_3 and diacylglycerol. IP_3 serves as an intracellular second messenger and binds to the specialized IP_3 receptor on the endoplasmic reticular membrane and triggers the release of Ca^{2+} from the endoplasmic reticulum. Nonexcitable cells enhance Ca^{2+} entry also by hyperpolarization caused by open potassium channels. Ca^{2+} enters through specialized voltage-independent Ca^{2+}-selective channels triggered by second-messenger molecules.

In addition to the system existing in nonexcitable cells, excitable cells contain voltage-dependent Ca^{2+} channels (reviewed by Clapham, 1995). Specialized Ca^{2+} trigger proteins near the plasma membrane inner surface initiate various cell processes. In neurons Ca^{2+} entering through voltage-dependent Ca^{2+} channels directly activate ryanodine receptors to release Ca^{2+} from intracellular stores. In skeletal muscles dihydropiridine receptors on the surface of the plasma membrane and in T tubules are located near the sarcoplasmic ryanodine receptors. Conformational changes induced by voltage in the dihydropyridine receptor result in Ca^{2+} influx and perhaps directly modulate the ryanodine receptors to release Ca^{2+} from sarcoplasmic reticulum.

Calcium pumps transport Ca^{2+} ions into the endoplasmic (sarcoplasmic) reticulum or extracellular space at the cost of one to two ATP molecules per Ca^{2+} ion removed (reviewed by Clapham, 1995). Both endoplasmic reticulum (SERCA) and plasma membrane (PMCA) Ca^{2+} pumps are P type ATPases. The sarcoplasmic reticulum Ca^{2+} ATPase pumps are the products of three different genes, known as SERCA1, SERCA2, and SERCA3 which are expressed respectively in fast-twitch skeletal muscles, cardiac and slow-twitch skeletal muscles, and in nonmuscle tissues. Mitochondria accumulate Ca^{2+} at up to 0.5 mM levels in the mitochondria matrix due to a large electrochemical gradient created by hydrogen exchange. Mitochondrial Ca^{2+} uniporters are characterized by lower affinities for Ca^{2+} than SERCA pumps.

It is clear that intracellular free Ca^{2+} concentration is a result of a balance between Ca^{2+}-increasing, Ca^{2+}-decreasing and Ca^{2+}-buffering mechanisms. Ca^{2+} waves and oscillations are observed in cells; Ca^{2+} gradients within cells were proposed to explain various intracellular processes. However, our understanding of the specific mechanisms for these effects is still at an early stage.

It is of importance that during cell stimulation only a small portion of the released Ca^{2+} remains as free Ca^{2+}, most of it is bound to intracellular Ca^{2+} buffers. Intracellular Ca^{2+} buffers can be subdivided into molecular buffers, represented by the cytosolic calcium binding proteins and organelle buffers, mainly represented by mitochondria and endoplasmic reticulum. Ca^{2+} binding to high affinity binding sites often induces changes of the protein structure, thereby evoking some cellular responces.

Calcium-binding proteins of the EF-hand superfamily (see below), which can bind several ions of calcium per molecule with high affinity, play an important role in maintaining the intracellular free calcium concentration in the range that is required for normal cell functioning. The calcium-binding proteins seem to be involved also in a range of cell-specific functions, since they are present in different populations of cells. The diversity of distribution, ontological appearance, and coexistence of such calcium-binding proteins as parvalbumin, calbindin, calretinin, and S100 proteins in the brain and other tissues indicate that they can be involved in different cell-specific regulatory systems.

The EF-hand proteins can be subvivided into two groups: sensor (trigger) proteins and buffer (non-trigger) proteins. Upon calcium binding, the sensor proteins (such as calmodulin and troponin C) undergo a large conformational change and regulate a vast number of target proteins. The buffer proteins (such as parvalbumin and calbindin) take part in calcium homeostatic functions, such as uptake, transport and maintenance of proper calcium concentration in cell.

Parvalbumin is a small (M_r 12,000), acidic (pI 4-5), Ca^{2+}-binding protein of the EF-hand superfamily, that is very important from several points of view. First of all, this protein takes part in Ca^{2+} regulation of activity of some types of muscle cells, neurons and some other types of cells. At the same time, the exact physiological role of parvalbumin in some of these cells is not clear enough now. Second, parvalbumin has two high affinity Ca^{2+} binding sites and for this reason it is frequently used as a simple model Ca^{2+} binding protein. It is convenient for studies of effects of interactions between two calcium binding sites and is very useful for studies of calcium binding effects on interactions of the protein with another proteins, peptides, membranes and low molecular weight organic compounds, which frequently have physiological significance. Third, parvalbumin forms several partially folded intermediate states, which have been studied by many researchers interested in protein folding and stability problems. It is very attractive for studies of the properties and structure of intermediate molten globule-like states. The present book summarizes our knowledge on physical, chemical and physiological properties of parvalbumins.

Chapter 1

CALCIUM AND CALCIUM BINDING PROTEINS

CALCIUM AND OTHER PHYSIOLOGICALLY SIGNIFICANT METAL CATIONS

Six metals are present in human body in highest amounts: calcium (1700 g per 70 kg of body weight), potassium (250 g per 70 kg), sodium (70 g per 70 kg), magnesium (42 g per 70 kg), iron (5 g per 70 kg), and zinc (3 g per 70 kg). The content of another metals is less than 1 g per 70 kg of body weight.

Most biologically significant metals are located in the fourth period of the Periodic Table of Elements, only sodium and magnesium refer to the third period.

Electronic Structure of Ca^{2+}:

He $1s^2$
Ne $1s^2\ 2s^2\ 2p^6$
Ar $1s^2\ 2s^2\ 2p^6\ 3s^2\ 3p^6$
Ca $1s^2\ 2s^2\ 2p^6\ 3s^2\ 3p^6\ 4s^2$
Ca^{2+} $1s^2\ 2s^2\ 2p^6\ 3s^2\ 3p^6$ (K L $3s^2\ 3p^6$)

The ionization potential of the neutral calcium atom ($1s^2$, $2s^2\ 2p^6$, $3s^2\ 3p^6$, $4s^2$) for the first electron is 6.1132 eV, while the ionization potential for the second electron is 11.871 eV. Under normal geological or biological conditions the Ca^{1+} ion is never found. The filled octets of Mg^{2+}, of Ca^{2+}, and of Ba^{2+} ($2s^2\ 2p^6\ 3s^2\ 3p^6$, and $4s^2\ 4p^6$) have essentially no p character nor directional preference; they can be modeled as spheres of increasing atomic radius (0.65, 0.99, 1.35 Å, respectively) and decreasing charge density. A similar situation applies to Na^+ and to K^+ (1.00 and 1.52 Å, respectively).

In biological molecules, the four cations, Ca^{2+}, Mg^{2+}, Na^+ and K^+, almost always bond to oxygen, not nitrogen or sulfur. In contrast, divalent cations such as Zn^{2+}, and Mn^{2+} as well as the trivalent lanthanides have some directional preference and some covalent character in their bonding to electronegative ligands such as nitrogen, sulfur, or oxygen. Some physical properties of biologically significant metal cations are collected in Table 1.

Table 1. Properties of biologially significant bivalent metal cations.

	Ca	Mg	Mn	Zn
Ionic radius, Å	0.99	0.65	0.89	0.88
Intracellular concentration	0.1-1 μM	0.25-1 mM	μM ?	μM ?
Ligands	O	O	O, N	O, N, S
Coordinating amino acid residues	Asp, Glu	Asp, Glu	Asp, Glu, His	Asp, Glu, His, Cys
Coordination number	6-8	6	6	4-6
Interaction	electrostatic	electrostatic	partial covalent	partial covalent

In aqueous solution all these cations are hydrated. Usually the rate at which they react with another ligand(s) is determined by the rate of their own dehydration. This process in turn is usually controlled by the first water molecule leaving the first hydration sphere.

Na^+ and K^+ ions are distinguished primarily by their ionic radii and their hydration energies. These differences are reflected in many properties. The difference between magnesium and calcium ions is more essential. They differ in ionic radii and especially in hydration heats.

The equilibrium dissociation constant is the ratio of the dissociation (off) and association (on) rate constants k_{off} / k_{on} (s^{-1} /M^{-1} s^{-1}). The off rate for the Ca^{2+} and the Mg^{2+} ions in proteins are quite similar. The four decade higher affinity of many proteins for calcium in comparison with magnesium reflects the four decade higher dehydration rate and thus the four decade higher association rate constant k_{on}.

The atomic radius of Mg^{2+} is 0.65 Å; that of oxygen is 0.99 Å. Six oxygen atoms can surround the Mg^{2+} ion at the vertices of an octahedron to be in convenient van der Waals contact. Water dissociates slowly from hexa-aquo magnesium. Correspondingly, a protein must adjust its six oxygen ligands (if not provided by water) when binding magnesium to a near regular octahedron. The actual strength of a single Me-O bond is probably greater for magnesium than for calcium because of the greater charge density of Mg^{2+} *vs*. Ca^{2+}. Mg^{2+} is a "hard" ion and prefers "hard" ligands of low polarizability, with oxygen being the most preferred coordination atom, followed by nitrogen (reviewed by Dudev & Lim, 2003). Mg^{2+} binding sites in proteins contain at least one carboxylate ligand, which coordinates Mg^{2+} predominantly in a monodentate fashion. Among the noncharged ligands of Mg^{2+} in proteins, the side chains of Asn, Gln and backbone carbonyl groups are the most common, followed by Ser, Thr, His, and Tyr side chains.

In contrast, the atomic radius of Ca^{2+} is 0.99 Å and therefore it has enough surface area to accommodate 4-12 oxygen atoms in its primary coordination sphere, but coordination numbers 6-8 are most common. In a pentagonal bipyramid with five oxygens in the equatorial plane, the O-O distance is ~2.5 Å. This is a bit tight so the oxygens accommodate by buckling a bit out of the equatorial plane. The distance between the oxygens at the two vertices and the equatorial plane is 3.36 Å, well beyond van der Waals contact and can accommodate the buckling. This is the reason why one can see approximate pentagonal pyramid symmetry in high resolution protein structures. Ca^{2+}, like Mg^{2+}, prefers to bind to "hard" oxygen-containing ligands but, generally, with lower free energy gain (reviewed by Dudev & Lim, 2003). In order of decreasing prevalence, these ligands in proteins are carboxylates, carbonyls, water, and hydroxyl oxygen atoms. Compared to Mg^{2+}, Ca^{2+} shows a greater tendency toward bidentate carboxylate binding, but a lower affinity for water.

Unlike Mg^{2+} and Ca^{2+}, Zn^{2+} prefers "softer" ligands such as Cys and His, although it is also coordinated to Asp and Glu side chains (reviewed by Dudev & Lim, 2003; Christianson, 1991). Although in aqueous solution Zn^{2+} is octahedrically bound to six water molecules, in such proteins as Zn-finger proteins and enzymes Zn^{2+} is usually tertahedrally coordinated, but it can also adopt a 5- or 6-coordinate geometry.

Coordination numbers of metal cations observed in metalloproteins range from 3 to 8, with 6 as the most common value (reviewed by Babor *et al.*, 2005; Katz *et al.*, 1996). Mg^{2+} and Mn^{2+} have a main coordination number 6, whereas Zn^{2+} and Ca^{2+} show greater coordination flexibility, with values ranging from 4 to 6 and 6 to 8, respectively. The so called "hard" metals, Ca^{2+} and Mg^{2+}, are coordinated in proteins mainly by Asp and Glu side chains and very infrequently by His. The softer character of Mn^{2+} allows it to accept His ligands more often than Mg^{2+}, while Zn^{2+}, which is considerably softer (more polarizable), has the highest tendency to interact with His and Cys ligands.

Metal binding sites are usually located inside cavities and crevices of the protein structure. They are usually inaccessible to solvent, characterized by a low dielectric constant that enhances electrostatic metal-protein ligand interactions. The metal coordination number in a given complex is dictated by three factors: (1) the dielectric medium or solvent accessibility, (2) properties of the metal (mainly by its ability to accept charge from its ligands), and (3) the chemical characteristics of the ligands (reviewed by Dudev & Lim, 2003). "Border-line" metal cations, e.g., Zn^{2+}, in contrast to "hard" cations such as Mg^{2+}, can adopt different coordination geometries at a relatively low free energy cost.

Binding site selectivity appears to be anticorrelated with the natural abundance of the metal in living cells. Thus, Mg^{2+}-binding sites are not very specific for Mg^{2+}, which is the most abundant metal in the body fluids. These binding sites are weakly protected against other natural metal cofactors, in particular, Zn^{2+}. The type of metal ligands, their coordination mode and side chain interactions, as well as the overall charge and shape of the cavity govern the binding-site specificity (reviewed by Dudev & Lim, 2003).

Some metal binding sites are vulnerable to attacks by nonbiogenic "alien" cations such as Cd^{2+}, Hg^{2+}, and Pb^{2+}. Replacing the natural cofactor, e.g., Zn^{2+}, by a heavier metal may constitute one of the possible pathways for heavy-metal intoxication in living organisms.

Diffusion coefficient of Ca^{2+} ion in water calculated simply from its ionic radius is 1000 $\mu m^2/s$ and 800 $\mu m^2/s$ for hydrated Ca^{2+}. The occurrence of calcium binding proteins in water solution slows down the diffusion of Ca^{2+}. It was estimated that in a 300 μM calcium binding protein solution Ca^{2+} ion migrates no further than 0.1-0.5 μm lasting only about 50 μs before encountering a binding protein (assuming $k_{on}=10^8\ M^{-1}s^{-1}$) (Allbritton et al., 1992). Ca^{2+} diffusion coefficient also depends on the degree of saturation of Ca^{2+}-buffering proteins and varies between 15 and 65 $\mu m^2/s$.

The NMR active isotope ^{43}Ca is a quadrupolar (I = 7/2) nucleus that produces broad peaks when bound to a protein. Although much useful information can be obtained from studies of this nucleus, unfortunately it is usually impossible to resolve the signals of ^{43}Ca in individual calcium-binding sites in a protein with multiple binding sites. However, isomorphous replacement of the calcium ion with an ion with more favorable NMR properties often allows the resolution of NMR signals for individual sites and provides valuable insight into the coordination of the ion within the binding site from the observed shifts. ^{113}Cd, and to a lesser extent, ^{207}Pb, are two such metal nuclei (I = ½) that have been successfully used to characterize the Ca^{2+}-binding properties of a variety of metalloproteins (reviewed by Clarke & Fogel, 2002).

EF-Hand Calcium Binding Proteins

One of the most studied families of intracellular calcium binding proteins is the EF-hand protein family. The name "EF-hand" is connected with the name of their main calcium binding domain, which consists of two helices and a calcium binding loop between them (see below). This domain was first found in parvalbumin. Participants in virtually every Ca^{2+}-signaling pathway, the EF-hand proteins were the first intracellular Ca^{2+}-binding proteins identified and characterized. Parvalbumin, calmodulin, troponin·C, myosin light chains, calbindin and many other proteins belong to this family. Certain EF-hand family members, notably calmodulin and troponin C, function as Ca^{2+}-dependent regulatory proteins. Others, notably parvalbumin and calbindin, seem to serve as cytosolic Ca^{2+} buffers. By simply binding Ca^{2+}, the latter help to shape the Ca^{2+} signal, limiting its amplitude, extent of propagation, and/or duration.

Parvalbumin is supposed to play a role of a soluble relaxation factor in fast-twitch skeletal muscles, however it was isolated also from other tissues including the central nervous system, testis, kidney and several endocrine glands.

Chapter 2

ANATOMY OF PARVALBUMIN

AMINO ACID SEQUENCE AND GENE STRUCTURE OF PARVALBUMIN

Amino acid compositions of parvalbumins are characterized by the following features (reviewed by Permyakov, 1985, 1993):

1. High content of Asp (8 to 15 per molecule) and Glu (7 to 13) residues.
2. High content of Ala (9 to 23 per molecule, 29 in silver hake (*Merluccius Bilinearis*) parvalbumin (Zhang et al., 1990)), Leu (7 to 12), and Phe (8 to 10).
3. High content of Lys (10 to 16 per molecule).
4. Low content of Arg (1 to 3 per molecule), Met (0 to 3), Cys (0 to 3), Pro (0 to 3), Tyr (0 to 2), Trp (0 to 1), and His (0 to 3).
5. Acetylated N-terminus.

Two classifications of parvalbumins were proposed: one using Roman numerals is based on the electrophoretic mobility, the component I having the highest mobility at pH 8 and V the lowest one (Bhushana Rao *et al.*, 1969) and another one based on the isoelectric point (Pechére *et al.*, 1971c). Molecular mass of parvalbumin is close to 12 000; they are characterized by acidic isoelectric points ranging from 3.9 (carp) to 6.6 (lungfish), and high diffusion coefficient.

All parvalbumins have an acetylated N-terminal amino acid residue (Gerday & Bhushano Rao, 1970) except one isotype from coelacanth muscle (pI 5.44).

Table 1 contains amino acid sequences of 43 parvalbumins isolated from various sources. The EF-hand Ca^{2+}-binding motif (see below) consists of a 12-residue metal ion binding loop (marked black) flanked by short helices (marked gray). Positioned at the approximate vertices of an octahedron, the ligands are labeled +X, +Y, +Z, -Y, -X, and –Z (see below). The side chains of loop residues 1, 3, 5, 9, and 12 furnish the +X, +Y, +Z –X, and –Z ligands, respectively. The main chain carbonyl of loop residue 7 serves as the –Y ligand, and a water molecule often occupies the –X position. Coordination by the –Z carboxylate (generally glutamate) is bidentate for Ca^{2+} and monodentate for Mg^{2+}.

Table 2. Amino acid sequences of parvalbumins.

Ca^{2+}-binding loops are marked black; helices are marked grey.

1 – Oncomodulin (β-parvalbumin); *Cavia porcellus* (Guinea pig); Henzl *et al.*, 1997.
2 – Oncomodulin (β-parvalbumin); *Homo sapiens* (Human); Foehr *et al.*, 1993.
3 – Oncomodulin (β-parvalbumin); *Mus musculus* (Mouse); Banville *et al.*, 1992.
4 – Oncomodulin (β-parvalbumin); *Rattus norvegicus* (Rat); McManus *et al.*, 1983.
5 – α-Parvalbumin; *Amphiuma means* (Salamander, Two-toed amphiuma); Maeda *et al.*, 1984.
6 – α-Parvalbumin 4a; *Brachydanio rerio* (Zebrafish, Danio rerio); Hsiao *et al.*, 2002.
7 – α-Parvalbumin A1; *Cyprinus carpio* (Common carp); Coffee *et al.*, 1974.
8 – α-Parvalbumin; *Esox lucius* (Northern pike); Frankenne *et al.*, 1973.
9 – α-Parvalbumin; *Felis silvestris catus* (Cat); Hauer *et al.*, 1992.
10 – α-Parvalbumin; *Gerbillus sp.* (Gerbil); Hauer *et al.*, 1992.
11 – α-Parvalbumin; *Homo sapiens* (Human); Berchtold, 1989.
12 – α-Parvalbumin; *Latimeria chalumnae* (Latimeria, Coelacanth); Pechére *et al.*, 1978.
13 – α-Parvalbumin; *Macaca fuscata fuscata* (Japanese macaque); Hauer *et al.*, 1992.
14 – α-Parvalbumin; *Mus musculus* (Mouse); Zuehlke *et al.*, 1989.
15 – α-Parvalbumin; *Oryctolagus cuniculus* (Rabbit); Enfield *et al.*, 1975.
16 – α-Parvalbumin; *Raja clavata* (Thornback ray); Thatcher & Pechére, 1977.
17 – α-Parvalbumin pI 4.97; *Rana catesbeiana* (Bull frog); Sasaki *et al.*, 1990.
18 – α-Parvalbumin; *Rana esculenta* (Edible frog); Jauregui-Adell *et al.*, 1982.
19 – α-Parvalbumin; *Rattus norvegicus* (Rat); Berchtold *et al.*, 1982a.
20 – α-Parvalbumin; *Triakis semifasciata* (Leopard shark); Roquet *et al.*, 1992.
21 – β-Parvalbumin 1; *Salmo salar* (Atlantic salmon); Lindstroem *et al.*, 1996.
22 – β-Parvalbumin 2; *Salmo salar* (Atlantic salmon); Lindstroem *et al.*, 1996.
23 – β-Parvalbumin; *Amphiuma means* (Salamander, Two-toed amphiuma); Maeda *et al.*, 1984.
24 – β-Parvalbumin; *Boa constrictor* (Boa); Maeda *et al.*, 1984.
25 – β-Parvalbumin; *Brachydanio rerio* (Zebrafish); Xu *et al.*, 2000.
26 – β-Parvalbumin; *Cyprinus carpio* (Common carp); Coffee & Bradshaw, 1973.
27 – β-Parvalbumin; *Esox lucius* (Northern pike); Gerday, 1976.
28 – β-Parvalbumin; *Gadus callarias* (Baltic cod); Elsayed & Bennich, 1975.
29 – β-Parvalbumin; *Gadus morhua* (Atlantic cod); van Do *et al.*, 2001.
30 – β-Parvalbumin; *Graptemys geographica* (Map turtle); Maeda *et al.*, 1984.
31– β-Parvalbumin; *Latimeria chalumnae* (Latimeria, Coelacanth); Jauregui-Adell & Pechére, 1978.
32 – β-Parvalbumin V; *Leuciscus cephalus* (Chub); Gerday *et al.*, 1978.
33 – β-Parvalbumin; *Merluccius bilinearis* (Silver hake); Revett *et al.*, 1997.
34 – β-Parvalbumin; *Merluccius merluccius* (European hake); Capony *et al.*, 1973.
35 – β-Parvalbumin; *Merlangius merlangus* (Whiting); Joassin & Gerday, 1977.
36 – β-Parvalbumin; *Opsanus tau* (Oyster toadfish); Gerday *et al.*, 1989.
37 – β-Parvalbumin; *Rana esculenta* (Edible frog); Capony *et al.*, 1975.
38 – β-Parvalbumin; *Scomber japonicus* (Chub mackerel); Hamada *et al.*, 2002.
39 – β-Parvalbumin; *Theragra chalcogramma* (Alaska salmon); van Do *et al.*, 2001.
40 – β-Parvalbumin; *Xenopus laevis* (African clawed frog); Kay *et al.*, 1987.
41 – Parvalbumin, muscle; *Gallus gallus* (Chicken); Kuster *et al.*, 1991.
42– Parvalbumin, thymic (Avian thymic hormone) (ATH) (Thymus-specific antigen T1); *Gallus gallus* (Chicken); Brewer, 1990.
43 – Parvalbumin, thymic CPV3; *Gallus gallus* (Chicken); Hapak *et al.*, 1994.

#	01	02	03	04	05	06	07	08	09	10	11	12	13	14	15	16	17	18	19	20
1	S	I	T	D	V	L	S	A	D	D	I	A	A	A	L	Q	E	C	Q	D
2	S	I	T	D	V	L	S	A	D	D	I	A	A	A	L	Q	E	C	Q	D
3	S	I	T	D	I	L	S	A	D	D	I	A	A	A	L	Q	E	C	Q	D
4	S	I	T	D	I	L	S	A	E	D	I	A	A	A	L	Q	E	C	Q	D
5	S	M	T	D	V	I	P	E	A	D	I	N	K	A	I	H	A	F	K	A
6	A	M	K	N	L	L	K	D	D	D	I	K	K	A	L	D	Q	F	K	A
7	A	Y	G	G	I	L	N	D	A	D	I	T	A	A	L	E	A	C	X	A
8	A	K	D	L	L	K	A	D	D	I	K	K	A	L	D	A	V	K	A	E
9	S	M	T	D	L	L	G	A	E	D	I	K	K	A	V	E	A	F	T	A
10	S	M	T	D	L	L	S	A	E	D	I	K	K	A	I	G	A	F	A	A
11	S	M	T	D	L	L	N	A	E	D	I	K	K	A	V	G	A	F	S	A
12	T	K	K	M	S	E	I	L	K	A	E	D	I	D	K	A	L	N	T	F
13	S	M	T	D	L	L	N	A	E	D	I	K	K	A	V	G	A	F	S	A
14	S	M	T	D	V	L	S	A	E	D	I	K	K	A	I	G	A	F	A	A
15	A	M	T	E	L	L	N	A	E	D	I	K	K	A	I	G	A	F	A	A
16	S	S	K	I	T	S	I	L	N	P	A	D	I	T	K	A	L	E	Q	C
17	M	H	M	T	D	V	L	P	A	G	D	I	S	K	A	V	E	A	F	A
18	P	M	T	D	L	L	A	A	G	D	I	S	K	A	V	S	A	F	A	A
19	S	M	T	D	L	L	S	A	E	D	I	K	K	A	I	G	A	F	T	A
20	P	M	T	K	V	L	K	A	D	D	I	N	K	A	I	S	A	F	K	D
21	M	A	C	A	H	L	C	K	E	A	D	I	K	T	A	L	E	A	C	K
22	S	F	A	G	L	N	D	A	D	V	A	A	A	L	A	A	C	T	A	A
23	A	I	T	D	I	L	S	A	K	D	I	E	A	A	L	S	S	V	K	A
24	A	F	A	G	I	L	S	D	A	D	I	A	A	G	L	Q	S	C	Q	A
25	A	F	A	G	I	L	K	D	E	D	V	A	A	A	L	K	D	C	A	A
26	A	F	A	G	V	L	N	D	A	D	I	A	A	A	L	E	A	C	K	A
27	S	F	A	G	L	K	D	A	D	V	A	A	A	L	A	A	C	S	A	A
28	A	F	K	G	I	L	S	N	A	D	I	K	A	A	E	A	A	C	F	K
29	A	F	A	G	I	L	N	D	A	D	I	T	A	A	L	A	A	C	K	A
30	A	M	T	D	I	L	S	A	K	D	I	E	A	A	L	T	S	C	Q	A
31	A	V	A	K	L	L	A	A	A	D	V	T	A	A	L	E	G	C	K	A
32	A	F	G	L	K	E	A	D	I	T	A	A	L	E	A	C	K	A	A	D
33	A	F	S	G	I	L	A	D	A	D	V	A	A	A	L	K	A	C	E	A
34	A	F	A	G	I	L	A	D	A	D	I	T	A	A	L	A	A	C	K	A
35	A	F	A	G	I	L	A	D	A	D	C	A	A	A	V	K	A	C	E	A
36	S	F	A	G	I	L	S	D	A	D	I	D	A	A	L	A	A	C	Q	A
37	S	I	T	D	I	V	S	E	K	D	I	D	A	A	L	E	S	V	K	A
38	A	F	A	S	V	L	K	D	A	E	V	T	A	A	L	D	G	C	K	A
39	A	F	A	G	I	L	K	D	A	E	V	A	A	A	L	E	A	C	K	S
40	A	F	G	G	I	L	S	E	A	D	I	S	A	A	L	Q	N	C	Q	A
41	A	M	T	D	V	L	S	A	E	D	I	K	K	A	V	G	A	F	S	A
42	A	I	T	D	I	L	S	A	K	D	I	E	S	A	L	S	S	C	Q	A
43	M	S	L	T	D	I	L	S	P	S	D	I	A	A	A	L	R	D	C	Q

	21	22	23	24	25	26	27	28	29	30	31	32	33	34	35	36	37	38	39	40
1	P	D	T	F	E	P	Q	K	F	F	Q	T	S	G	L	S	K	M	S	A
2	P	D	T	F	E	P	Q	K	F	F	Q	T	S	G	L	S	K	M	S	A
3	P	D	T	F	E	P	Q	K	F	F	Q	T	S	G	L	S	K	M	S	A
4	P	D	T	F	E	P	Q	K	F	F	Q	T	S	G	L	S	K	M	S	A
5	G	E	A	F	D	F	K	K	F	V	H	L	L	G	L	N	K	R	S	P
6	A	D	S	F	D	H	K	K	F	F	D	V	V	G	L	K	A	L	S	A
7	X	D	S	F	N	A	K	S	F	F	A	K	V	G	L	S	A	K	T	P
8	G	S	F	N	H	K	K	F	F	A	L	V	G	L	K	A	M	S	A	N
9	V	D	S	F	D	Y	K	K	F	F	Q	M	V	G	L	K	K	K	S	P
10	A	D	S	F	D	H	K	K	F	F	Q	M	V	G	L	K	K	K	T	P
11	T	D	S	F	D	H	K	K	F	F	Q	M	V	G	L	K	K	K	S	A
12	K	E	A	G	S	F	D	H	H	K	F	F	N	L	V	G	L	K	G	K
13	I	D	S	F	D	H	K	K	F	F	Q	M	V	G	L	K	K	K	S	A
14	A	D	S	F	D	H	K	K	F	F	Q	M	V	G	L	K	K	K	N	P
15	A	E	S	F	D	H	K	K	F	F	Q	M	V	G	L	K	K	K	S	T
16	A	A	G	F	H	H	T	A	F	F	K	A	S	G	L	S	K	K	S	D
17	A	P	D	S	F	N	H	K	K	F	F	E	M	C	G	L	K	S	K	G
18	P	E	S	F	N	H	K	K	F	F	E	L	C	G	L	K	S	K	S	K
19	A	D	S	F	D	H	K	K	F	F	Q	M	V	G	L	K	K	K	S	A
20	P	G	T	F	D	Y	K	R	F	F	H	L	V	G	L	K	G	K	T	D
21	A	A	D	T	F	S	F	K	T	F	F	H	T	I	G	F	A	S	K	S
22	D	S	F	N	H	K	A	F	F	A	K	V	G	L	A	S	K	S	S	D
23	A	E	S	F	N	Y	K	T	F	F	T	K	C	G	L	A	G	K	P	T
24	A	D	S	F	S	C	K	T	F	F	A	K	S	G	L	H	S	K	S	K
25	A	D	S	F	N	Y	K	N	F	F	A	K	V	G	L	S	A	K	S	P
26	A	D	S	F	N	H	K	A	F	F	A	K	V	G	L	T	S	K	S	A
27	D	S	F	K	H	K	E	F	F	A	K	V	G	L	A	S	K	S	L	D
28	E	G	S	F	D	E	D	G	F	Y	A	K	V	G	L	D	A	F	S	A
29	E	G	S	F	D	H	K	A	F	F	T	K	V	G	L	A	A	K	S	P
30	A	D	S	F	N	Y	K	S	F	F	S	K	V	G	L	K	G	K	S	T
31	D	D	S	F	N	H	K	V	F	F	Q	K	T	G	L	A	K	K	S	N
32	S	F	N	H	K	A	F	F	A	K	V	G	M	S	A	K	S	A	G	D
33	A	D	S	F	N	Y	K	A	F	F	A	K	V	G	L	T	A	K	S	A
34	E	G	S	F	K	H	G	E	F	F	T	K	I	G	L	K	G	K	S	A
35	A	D	S	F	S	Y	K	A	F	F	A	K	C	G	L	S	G	K	S	A
36	A	E	S	F	K	H	K	E	F	F	A	K	V	G	L	S	A	K	T	P
37	A	G	S	F	N	Y	K	I	F	F	Q	K	V	G	L	A	G	K	S	A
38	A	G	S	F	D	H	K	K	F	F	K	A	C	G	L	S	G	K	S	T
39	A	G	S	F	D	H	T	K	F	F	K	S	C	G	L	A	G	K	S	S
40	A	D	S	F	N	F	K	T	F	F	A	Q	S	G	L	S	S	K	S	A
41	A	E	S	F	N	Y	K	K	F	F	E	M	V	G	L	K	K	K	S	P
42	A	D	S	F	N	Y	K	S	F	F	S	T	V	G	L	S	S	K	T	P
43	A	P	D	S	F	S	P	K	K	F	F	Q	I	S	G	M	S	K	K	S

	41	42	43	44	45	46	47	48	49	50	51	52	53	54	55	56	57	58	59	60
1	S	Q	V	K	D	V	F	R	F	I	D	N	D	Q	S	G	Y	L	D	E
2	N	Q	V	K	D	V	R	F	F	I	D	N	D	Q	S	G	Y	L	D	E
3	S	Q	L	K	D	I	F	Q	F	I	D	N	D	Q	S	G	Y	L	D	E
4	S	Q	V	K	D	I	F	R	F	I	D	N	D	Q	S	G	Y	L	D	G
5	A	D	V	T	K	A	F	H	I	L	D	K	D	R	S	G	Y	I	E	E
6	D	N	V	K	L	V	F	K	A	L	D	V	D	A	S	G	F	I	E	E
7	D	D	I	K	K	A	F	A	V	I	D	Q	D	K	S	G	F	I	E	E
8	D	V	K	K	V	F	K	A	I	D	A	D	A	S	G	F	I	E	E	E
9	D	D	I	K	K	V	F	H	I	L	D	K	D	K	S	G	F	I	E	E
10	D	D	V	K	K	V	F	H	I	L	D	K	D	K	S	G	F	I	E	E
11	D	D	V	K	K	V	F	H	M	L	D	K	D	K	S	G	F	I	E	E
12	P	D	D	T	L	K	E	V	F	G	I	L	D	Q	D	K	S	G	Y	I
13	D	D	V	K	K	V	F	H	I	L	D	K	D	K	S	G	F	I	E	E
14	D	E	V	K	K	V	F	H	I	L	D	K	D	K	S	G	F	I	E	E
15	E	D	V	K	K	V	F	H	I	L	D	K	D	K	S	G	F	I	E	E
16	A	E	L	A	E	I	F	N	V	L	D	G	D	Q	S	G	Y	I	E	V
17	P	D	V	M	K	Q	V	F	G	I	L	D	Q	D	R	S	G	F	I	E
18	E	I	M	Q	K	V	F	H	V	L	D	Q	D	Q	S	G	F	I	E	K
19	D	D	V	K	K	V	F	H	I	L	D	K	D	K	S	G	F	I	E	E
20	A	Q	V	K	E	V	F	E	I	L	D	K	D	Q	S	G	F	I	E	E
21	A	D	D	V	K	K	A	F	K	V	I	D	Q	D	A	S	G	F	I	E
22	D	V	K	K	A	F	Y	V	I	D	Q	D	K	S	G	F	I	E	E	D
23	D	Q	V	K	K	V	F	D	I	L	D	Q	D	K	S	G	Y	I	E	E
24	D	Q	L	T	K	V	F	G	V	I	D	R	D	K	S	G	Y	I	E	E
25	D	D	I	K	K	A	F	F	V	I	D	Q	D	K	S	G	F	I	E	E
26	D	D	V	K	K	A	F	A	I	I	D	Q	D	K	S	G	F	I	E	E
27	D	V	K	K	A	F	Y	V	I	D	Q	D	K	S	G	F	I	E	E	D
28	D	E	L	K	K	L	F	K	I	A	D	E	D	K	E	G	F	I	E	E
29	A	D	I	K	K	V	F	E	I	I	D	Q	D	K	S	D	F	V	E	E
30	D	Q	V	K	K	I	F	G	I	L	D	Q	D	K	S	G	F	I	E	E
31	E	E	L	E	A	I	F	K	I	L	D	Q	D	K	S	G	F	I	E	D
32	V	K	K	A	F	E	I	I	D	E	D	K	S	G	F	I	E	E	E	E
33	D	D	I	K	K	A	F	F	V	I	D	Q	D	K	S	G	F	I	E	E
34	A	D	I	K	K	V	F	G	I	I	D	Q	D	K	S	D	F	V	E	E
35	D	D	I	K	K	A	F	V	F	I	D	Q	D	K	S	G	F	I	E	E
36	D	V	I	K	K	A	F	Y	V	I	D	Q	D	K	S	G	F	I	E	E
37	A	D	A	K	K	V	F	E	I	L	D	R	D	K	S	G	F	I	E	Q
38	D	E	V	K	K	A	F	A	I	I	D	Q	D	K	S	G	F	I	E	E
39	D	D	V	K	K	A	F	G	I	I	D	Q	D	Q	S	D	F	I	E	E
40	D	D	V	K	N	V	F	A	I	L	D	Q	D	R	S	G	F	I	E	E
41	E	D	V	K	K	V	F	H	I	L	D	K	D	R	S	G	F	I	E	E
42	D	Q	I	K	K	V	F	G	I	L	D	Q	D	K	S	G	F	I	E	E
43	S	S	Q	L	K	E	I	F	R	I	L	D	N	D	Q	S	G	F	I	E

	61	62	63	64	65	66	67	68	69	70	71	72	73	74	75	76	77	78	79	80
1	E	E	L	K	F	F	L	Q	K	F	E	S	G	A	R	E	L	T	E	S
2	E	E	L	K	F	F	L	Q	K	F	E	S	G	A	R	E	L	T	E	S
3	D	E	L	K	Y	F	L	Q	R	F	Q	S	D	A	R	E	L	T	E	S
4	D	E	L	K	Y	F	L	Q	K	F	Q	S	D	A	R	E	L	T	E	S
5	E	E	L	Q	L	I	L	K	G	F	S	K	E	G	R	E	L	T	D	K
6	E	E	L	K	F	V	L	K	G	F	S	A	D	G	R	D	L	T	D	K
7	D	E	L	K	L	F	L	Q	N	F	S	A	G	A	R	A	L	T	D	A
8	E	L	K	F	V	L	K	S	F	A	A	D	G	R	D	L	T	D	A	E
9	D	E	L	G	F	I	L	K	G	F	Y	P	D	A	R	D	L	S	V	K
10	D	E	L	G	F	I	L	K	G	F	S	S	D	A	R	D	L	S	A	K
11	D	E	L	G	F	I	L	K	G	F	S	P	D	A	R	D	L	S	A	K
12	E	E	E	E	L	K	F	V	L	K	G	F	A	A	G	G	R	E	L	T
13	D	E	L	G	F	I	L	K	G	F	S	P	D	A	R	D	L	S	A	K
14	D	E	L	G	S	I	L	K	G	F	S	S	D	A	R	D	L	S	A	K
15	E	E	L	G	F	I	L	K	G	F	S	P	D	A	R	D	L	S	V	K
16	E	E	L	K	N	F	L	K	C	F	S	D	G	A	R	V	L	N	D	K
17	E	D	E	L	C	L	M	L	K	G	F	T	P	N	A	R	S	L	S	V
18	E	E	L	C	L	I	L	K	G	F	T	P	E	G	R	S	L	S	D	K
19	D	E	L	G	S	I	L	K	G	F	S	S	D	A	R	D	L	S	A	K
20	E	E	L	K	G	V	L	K	G	F	S	A	H	G	R	D	L	N	D	T
21	V	E	E	L	K	L	F	L	Q	N	F	C	P	K	A	R	E	L	T	D
22	E	L	K	L	F	L	Q	N	F	S	A	S	A	R	A	L	T	D	A	E
23	D	E	L	Q	L	F	L	K	N	F	C	S	S	A	R	S	L	S	N	A
24	D	E	L	K	K	F	L	Q	N	F	D	G	K	A	R	D	L	T	D	K
25	D	E	L	K	L	F	L	Q	N	F	S	A	G	A	R	A	L	T	D	A
26	D	E	L	K	L	F	L	Q	N	F	K	A	D	A	R	A	L	T	D	G
27	E	L	K	L	F	L	Q	N	F	S	P	S	A	R	A	L	T	D	A	E
28	D	E	L	K	L	F	L	I	A	F	A	A	D	L	R	A	L	T	D	A
29	D	E	L	K	L	F	L	Q	N	F	S	A	G	A	R	A	L	S	D	A
30	D	E	L	Q	L	F	L	Q	N	F	S	S	T	A	R	A	L	T	A	A
31	E	E	L	E	L	F	L	Q	N	F	S	A	G	A	R	T	L	T	K	T
32	L	K	L	F	L	Q	N	F	K	A	G	A	R	A	L	T	D	A	E	T
33	D	E	L	K	L	F	L	Q	V	F	S	A	G	A	R	A	L	T	D	A
34	D	E	L	K	L	F	L	Q	N	F	S	A	G	A	R	A	L	T	D	A
35	D	E	L	K	L	F	L	Q	V	F	K	A	G	A	R	A	L	T	D	A
36	D	E	L	K	L	F	L	Q	N	F	A	S	S	A	R	A	L	T	D	K
37	D	E	L	G	L	F	L	Q	N	F	R	A	A	A	R	V	L	S	D	A
38	E	E	L	K	L	F	L	Q	N	F	K	A	G	A	R	A	L	S	D	A
39	E	E	L	K	L	F	L	Q	N	F	S	A	S	A	R	A	L	S	D	A
40	E	E	L	K	L	F	L	Q	N	F	S	A	S	A	R	A	L	T	D	A
41	E	E	L	K	F	V	L	K	G	F	T	P	D	G	R	D	L	S	D	K
42	E	E	L	Q	L	F	L	K	N	F	S	S	S	A	R	V	L	T	S	A
43	E	D	E	L	K	Y	F	L	Q	R	F	E	C	G	A	R	V	L	T	A

	81	82	83	84	85	86	87	88	89	90	91	92	93	94	95	96	97	98	99	100
1	E	T	K	S	L	M	A	A	A	D	N	D	G	D	G	K	I	G	A	D
2	E	T	K	S	L	M	A	A	A	D	N	D	G	D	G	K	I	G	A	E
3	E	T	K	S	L	M	D	A	A	D	N	D	G	D	G	K	I	G	A	D
4	E	T	K	S	L	M	D	A	A	D	N	D	G	D	G	K	I	G	A	D
5	E	T	K	D	L	L	I	K	G	D	K	D	G	D	G	K	I	G	V	D
6	E	T	K	A	F	L	A	A	A	D	K	D	G	D	G	K	I	G	I	D
7	E	T	K	A	F	L	K	A	G	D	S	D	G	D	G	K	I	G	V	D
8	T	K	A	F	L	K	A	A	D	K	D	G	D	G	K	I	G	I	D	E
9	E	T	K	M	L	M	A	A	G	D	K	D	G	D	G	K	I	D	V	D
10	E	T	K	T	L	L	A	A	G	D	K	D	G	D	G	K	I	G	V	E
11	E	T	K	M	L	M	A	A	G	D	K	D	G	D	G	K	I	G	V	D
12	A	N	E	T	K	A	L	L	K	A	G	D	Q	D	G	D	D	K	I	G
13	E	T	K	T	L	M	A	A	G	D	K	D	G	D	G	K	I	G	V	D
14	E	T	K	T	L	L	A	A	G	D	K	D	G	D	G	K	I	G	V	E
15	E	T	K	T	L	M	A	A	G	D	K	D	G	D	G	K	I	G	A	D
16	E	T	S	N	F	L	A	A	G	D	S	D	G	D	H	K	I	G	V	D
17	K	E	T	T	A	L	L	A	A	G	D	K	D	G	D	G	K	I	G	M
18	E	T	T	A	L	L	A	A	G	D	K	D	G	D	G	K	I	G	V	D
19	E	T	K	T	L	M	A	A	G	D	K	D	G	D	G	K	I	G	V	E
20	E	T	K	A	L	L	A	A	G	D	S	D	H	D	G	K	I	G	A	D
21	A	E	T	K	A	F	L	K	A	G	D	A	D	G	D	G	M	I	G	I
22	T	K	A	F	L	A	D	G	D	K	D	G	D	G	M	I	G	V	D	E
23	E	T	K	A	F	L	F	A	G	D	S	D	G	D	G	K	I	G	V	D
24	E	T	A	E	F	L	K	E	G	D	T	D	G	D	G	K	I	G	V	E
25	E	T	K	A	F	L	S	A	G	D	S	D	G	D	G	K	I	G	V	D
26	E	T	K	T	F	L	K	A	G	D	S	D	G	D	G	K	I	G	V	D
27	T	K	A	F	L	A	D	G	D	K	D	G	D	G	M	I	G	V	D	E
28	E	T	K	A	F	L	K	A	G	D	S	D	G	D	G	K	I	G	V	D
29	E	T	K	V	F	L	K	A	G	D	S	D	G	D	G	K	I	G	V	D
30	E	T	K	A	F	M	A	A	G	D	T	D	G	D	G	K	I	G	V	D
31	E	T	E	T	F	L	K	A	G	D	S	D	G	D	G	K	I	G	V	D
32	K	I	F	L	K	A	G	D	A	D	G	D	G	K	I	G	I	D	E	F
33	E	T	K	A	F	L	K	A	G	D	S	D	G	D	G	A	I	G	V	D
34	E	T	A	T	F	L	K	A	G	D	S	D	G	D	G	K	I	G	V	E
35	E	T	K	A	F	L	K	A	G	D	S	D	G	D	G	A	I	G	V	E
36	E	T	E	T	F	L	K	A	G	D	S	D	G	D	G	K	I	G	I	D
37	E	T	S	A	F	L	K	A	G	D	S	D	G	D	G	K	I	G	V	E
38	E	T	K	A	F	L	K	A	G	D	S	D	G	D	G	K	I	G	I	D
39	E	T	K	A	F	L	K	A	G	D	S	D	G	D	G	K	I	G	V	D
40	E	T	K	A	F	L	A	A	G	D	S	D	G	D	G	K	I	G	V	E
41	E	T	K	A	L	L	A	A	G	D	K	D	G	D	G	K	I	G	A	D
42	E	T	K	A	F	L	A	A	G	D	T	D	G	D	G	K	I	G	V	E
43	S	E	T	K	T	F	L	A	A	A	D	H	D	G	D	G	K	I	G	A

	101	102	103	104	105	106	107	108	109	110	111	112	113
1	E	F	Q	E	M	V	H	S					
2	E	F	Q	E	M	V	H	S					
3	E	F	Q	E	M	V	H	S					
4	E	F	Q	E	M	V	H	S					
5	E	F	T	S	L	V	A	E	S				
6	E	F	E	A	L	V	H	E					
7	E	F	A	A	L	V	K	A					
8	F	E	T	L	V	H	E	A					
9	E	F	F	S	L	V	A	K	S				
10	E	F	S	T	L	V	S	E	S				
11	E	F	S	T	L	V	A	E	S				
12	V	D	E	F	T	N	L	V	K	A	A		
13	E	F	S	T	L	V	A	E	S				
14	E	F	S	T	L	V	A	E	S				
15	E	F	S	T	L	V	S	E	S				
16	E	F	K	S	M	A	K	M	T				
17	D	E	F	V	T	L	V	S	E	S			
18	E	F	V	T	L	V	S	E	S				
19	E	F	S	T	L	V	A	E	S				
20	E	F	A	K	M	V	A	Q	A				
21	D	E	F	A	V	L	V	K	Q				
22	F	A	A	M	I	K	G						
23	E	F	Q	A	L	V	R	S					
24	E	F	V	V	L	V	T	K	G				
25	E	F	A	L	L	V	K	A					
26	E	F	T	A	L	V	K	A					
27	F	A	A	M	I	K	A						
28	E	F	G	A	I	V	D	K	W	G	A	K	G
29	E	F	G	A	M	I	K	A					
30	E	F	Q	A	L	V	K	A					
31	E	F	Q	K	L	V	K	A					
32	A	A	L	V	K	A							
33	E	W	A	A	L	V	K	A					
34	E	F	A	A	M	V	K	G					
35	E	W	V	A	L	V	K	A					
36	E	F	A	D	L	V	K	E	A				
37	E	F	Q	A	L	V	K	A					
38	E	F	A	A	M	I	K	G					
39	E	F	A	A	M	V	K	A					
40	E	F	Q	S	L	V	K	P					
41	E	F	A	T	M	V	A	E	S				
42	E	F	Q	S	L	V	K	A					
43	E	E	F	Q	E	M	V	Q	S				

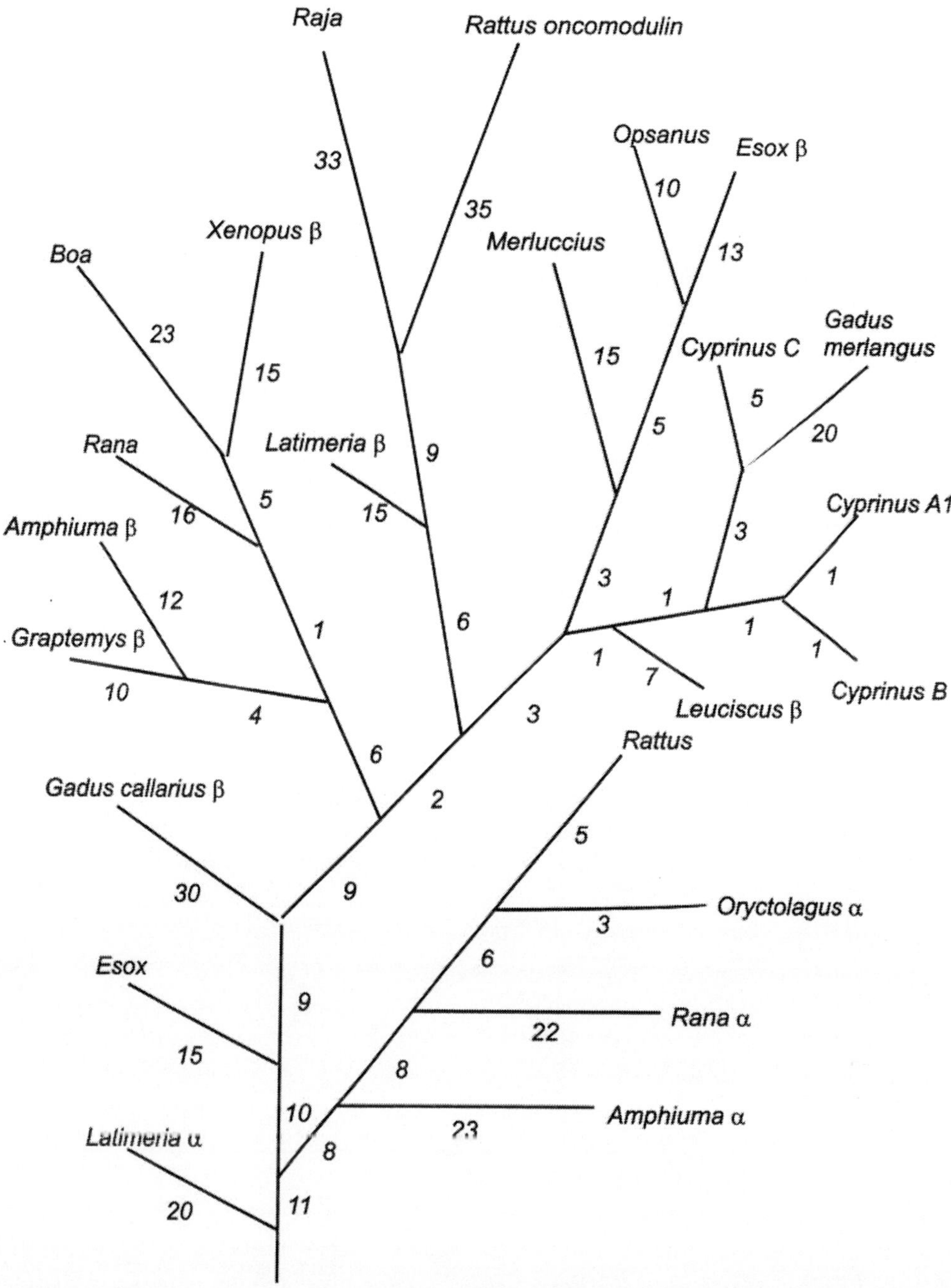

Figure 1. Relationships among members of parvalbumin subfamily (Moncrief et al., 1990). Numbers indicate branch length and represent relative amounts of divergence among amino acid sequences.

The relationships among 153 EF-hand (calcium-modulated) proteins of known amino acid sequence were studied in the work of Moncrief *et al.* (1990) using the method of maximum parsimony (a method in which the evolution of an ancestral amino acid sequence toward a contemporary one is supposed to always occur with the minimum number of nucleotide replacement). It turned out that these proteins can be grouped into 12 distinct subfamilies: calmodulin, troponin C, essential light chain of myosin, regulatory light chain of myosin, sarcoplasmic calcium binding protein, calpain, aequorin, *Stronglyocentrotus purpuratus* ectodermal protein, calbindin 28 kD, parvalbumin, α-actinin, and S100/intestinal calcium-binding protein. Eight individual proteins – calcineurin B from *Bos*, troponin C from *Astacus*, calcium vector protein from *Branchiostoma*, caltractin from *Chlamydomonas*, cell-division-cycle 31 gene product from *Saccharomyces*, 10 kD calcium-binding protein from *Tetrahymena*, LPS1 eight-domain protein from *Lytechinus*, and calcium-binding protein from *Streptomyces* – are tentatively identified as unique; that is, each may be the sole representative of another sub-family.

Figure 1 shows relationships among members of the parvalbumin sub-family (Moncrief *et al*., 1990). It is generally accepted that parvalbumin arose from a XY-AB-CD-EF gene intermediate most likely with four canonical EF-hands (Kretsinger, 1987). Roughly 425 million years ago, the parvalbumin family diverged into α and β sublineages (Goodman & Pechére, 1977; Nakayama *et al*., 1992; Moncrief *et al*., 1990). The mammalian genome encodes one isoform from each lineage (Fohr *et al*., 1993). The α and β sublineages differ in isoelectric point (β: pI<5; α: pI>5) and C-terminal helix length (one residue longer in α). There are at least 11 residues characteristically different between α and β forms of parvalbumin. A cystein at position 18 and an aspartic acid at position 61 are characteristic of β-parvalbumins (Kretsinger, 1980). During evolution, before α-parvalbumin and β-parvalbumin (oncomodulin) diverged, the N-terminal XY site of this gene was deleted and the remaining AB site lost its capacity to bind Ca^{2+}. Alternatively, first the AB loop underwent the inactivating mutations and then the XY site was deleted. The tentative reconstruction of the ancestral one-domain 40 amino acid polypeptide was made using the maximum parsimony method (Goodman & Pechére, 1977) and its synthesis was achieved (Maximov et al., 1978). The amino acid sequence of this peptide is as following:

H-Glu-Glu-Thr-Asp-Asp-Glu-Ile-Lys-Glu-Val-Leu-Lys-Ala-Phe-Asp-Lys-Asp-Gly-Gly-Gly-Arg-Ile-Asp-Phe-Glu-Glu-Phe-Val-Lys-Leu-Ile-Leu-Gly-Val-Thr-Gly-Glu-Gly-Ala-Arg-OH

It was found that even such short peptide demonstrates a weak Ca^{2+}-specific binding.

It was found that in the period from 425 to 400 million years ago the parvalbumin evolution has been characterized by extremely high rate of evolution, 89 nucleotide replacements per 100 codons per 100 million years, which was followed by a much slower rate of 8 nucleotide replacements per 100 codons per 100 million years from 400 million years ago to the present (Goodman & Pechére, 1977).

Parvalbumins are acidic proteins: their pI values lay within 3.9 to 6.6. It is of interest that Rehbein *et al*. (2000) suggested to use parvalbumins as marker proteins for native and urea isoelectric focusing. They prepared a lyophilized reference protein material of partially purified parvalbumins of different fish species. The species were selected so as to get parvalbumins with pI values from 3.5 to 5.5.

Parvalbumins are characterized by typical ultraviolet absorption spectra in which fine absorption peaks of phenylalanine residues are clear visible. This is due to the fact that tyrosine is often absent or only present as one residue per protein molecule, whereas tryptophan is encountered as single residue only in parvalbumin III of *Gadidae* species.

Interestingly, the chicken muscle parvalbumin amino acid sequence has ca. 80% sequence identity with α-type parvalbumins from mammalian (rabbit, human and rat) muscle. By contrast, the chicken thymus parvalbumin ("avian thymic hormone") sequence is very similar to reptile (turtle, salamander and frog) muscle β-type parvalbumins. Brewer et al. (1991) hypothesized that the evolutionary appearance of the warm-blooded reptiles was accompanied by recruitment of the β parvalbumin isozyme for promotion of lymphocyte maturation.

Immunochemical reactivity between various parvalbumins from different species was tested using monospecific antisera (Gosselin-Rey, 1974; Piront & Gosselin-Rey, 1974, 1975; Lehky & Stein, 1979; Gerday *et al.*, 1978). It turned out that cross-reactivity appears almost exclusively restricted to closely related species. Moreover, poor or no cross-reaction at all can be observed between the different isotypes present in the same muscle (reviewed by Gerday, 1982). For example, four main parvalbumins isolated from carp white muscle belong to three different families; two different non-cross-reacting families exist in *Gadidae* species, in perch, in pike and lungfish. No common antigenic determinants were found between the two parvalbumins of frog white muscle despite the fact that the two isotypes are obviously located in the same cell (Gillis *et al.*, 1979).

THE OCCURENCE OF PARVALBUMIN

The first study of the distribution of parvalbumin in the animal kingdom (Focant & Pechére, 1965) showed that parvalbumins are present only in the white muscles of fish and amphibian in amounts close to 300 μmol/kg wet weight (Gosselin-Rey & Gerday, 1977; Haiech *et al.*, 1979a) and sometimes in even higher quantities as much as 1200 μmol/kg wet weight as in the swimbladder muscle of toadfish (Hamoir *et al.*, 1980). It took another ten years to get evidence that parvalbumins are present in the muscles of higher vertebrates (Lehky *et al.*, 1974; Pechére, 1974) in amounts of 10-40 μmol/kg wet weight (Blum *et al.*, 1977). Later it became clear that parvalbumins are present in all vertebrates including man.

In fish skeletal muscles concentration of parvalbumin strongly depends upon the muscle type. For example, in carp red muscles, like the lateral line or *supracarinalis*, the amount of parvalbumin is about 10 times less than in white trunk muscle, approaching the amount found in warm blood vertebrates. The muscles differ not only by the overall parvalbumin content but also by the pattern of isotypes. In carp, the component I is especially abundant in red muscles tested (Hamoir *et al.*, 1981) and also in some white muscles like the *extraocular* and *protractor hyoideus lateralis*. The three other components, II, III and IV, found in carp muscles, are also expressed differently in various muscles according to a subtle modulation, which may be related in some way to muscle activity.

Surprisingly, while most of the muscles of fish and amphibian contain several parvalbumin isotypes, the higher vertebrates contain usually only one component. Up to five parvalbumins occur in the muscle of coelacanth (Jauregui-Adell & Pechére, 1978) and lungfish (Gerday, 1980) and three to four in *Cyprinidae* species (Piront & Gosselin-Rey, 1974). This polymorphism can be explained by two mechanisms susceptible to act in synergy (Gergay, 1982). First, it can be explained by polyploidization of the whole genome followed by differentiation. This is the case for many fishes belonging to the *Cyprinidae* family, which have twice as many chromosomes. Second, it can be explained by regional gene duplication giving rise to several loci on the same chromosome.

Interestingly, parvalbumins are absent in smooth muscle and their presence in cardiac muscles is very limited. Although traces of parvalbumin were detected in heart muscle (Gosselin-Rey, 1974), it does not seem that, in general, this organ is characterized by significant amounts of parvalbumin. Three parvalbumin isotypes were isolated from the white muscle of haemoglobin-myoglobin-free fish *Channichthys rhinoceratus* (Laforet *et al.*, 1991). Antibodies against the parvalbumin mixture were raised in rabbits and used for discovery, quantitation and isolation by affinity chromatography of parvalbumin in the cardiac muscle of three fish species: *Channichthys rhinoceratus, Champsocephalus gunnari* and *Notothenia neglecta*. The cardiac muscle of these species contains parvalbumin in concentration close to 1 μmol per kg wet weight. Sometimes the concentration of parvalbumin in cardiac muscle becomes rather high. Le Peuch *et al.* (1978) revealed the presence of high parvalbumin concentration in heart muscle of shrew. The heart of this small animal is characterized by a very fast beat rate evaluated to be 1000 beats/min. It contains a single parvalbumin component in concentration of 19 μmol/kg wet weight, which is comparable to the parvalbumin content found in other mammalian muscles.

In rat and mouse, the type IIB fibers demonstrate the strongest immunoreactivity for parvalbumin with different degrees of intensity. 60-70% of type IIA fibers exhibit moderate staining intensity, while the remaining type IIA and type I fibers lack parvalbumin (Celio & Heizmann, 1982; Füchtbauer *et al.*, 1991; Heizmann, 1984; Müntener *et al.*, 1987).

Schmitt & Pette (1991) showed that type I fibers from rabbit *tibialis anterior* and *vastus lateralis* muscles contain extremely low parvalbumin concentrations (2-5 micrograms/g wet weight). Type IIA fibers displayed slightly higher values with mean values of 17 and 29 micrograms/g wet weight (range 5-65) in *tibialis anterior* and *vastus lateralis*, respectively. Much higher parvalbumin concentrations were found in type IIB fibers with wide ranges from 75-1150 micrograms/g wet weight in *tibialis anterior* and 440-1370 micrograms/g wet weight in *vastus lateralis*. Whereas the IIB fibers of the *tibialis anterior* displayed a continuum, two subgroups were distinguishable according to their parvalbumin contents (means of 590 and 1230 micrograms/g wet weight) in *vastus lateralis*. Possibly, the population with the lower parvalbumin content, which was histochemically defined as type IIB in this study, corresponds to fiber type IID. The finding that parvalbumin is predominantly present in type IIB fibers was also confirmed by the parallel decay of parvalbumin and type IIB fibers during chronic low-frequency stimulation.

The localization of parvalbumin was determined in skeletal muscles with different fiber type compositions from quails (*Coturnix japonica*) and pigeons (*Columba livia*) by sandwich ELISA (Nishida *et al.*, 1995, 1997). Parvalbumin was found in *posterior latissimus dorsi* and *sartorius* in significant amounts, whereas it was undetectable in quail *anterior latissimus dorsi* and pigeon *latissimus dorsi,* which contain exclusively slow-tonic fibers with the lowest lactate dehydrogenase activity or predominant H-type isozyme characteristics. The *pectoralis superficialis* and *pectoralis profundus* muscles from quails and pigeons seem to consist mostly of fast-twitch glycolytic and oxidative-glycolytic fibers because of the highest myoglobin contents and lactate dehydrogenase activities or M-type and H/M-type isozyme patterns. Despite fast-twitch fiber compositions, parvalbumin was absent from these *pectoralis* muscles. The tetanic contraction induced in avian fast-twitch *pectoralis* fibers during flapping flight might be independent of a function of parvalbumin as a relaxing factor. This study confirmed that the avian *pectoralis* muscles, composed mostly of fast-twitch fibers, contain parvalbumin in undetectable amounts in contrast to other avian fast-twitch fibers as well as fish, amphibian or mammalian fast skeletal muscles, in which this protein is specifically present in significant amounts. This fact suggests unique fast-twitch contractile properties of avian *pectoralis* muscles.

Musculatures of two fish species belonging to the suborder *Cyprinoidei* were examined histochemically and immunohistochemically for demonstration of mATP-ase activity and parvalbumin content, respectively (Zawadowska & Supikova, 1992). Immunohistochemical results showed that lateral musculatures of the two species examined possessed parvalbumin II component entirely in the fast-contracting fibres, intermediate and white. This results correlate with mATP-ase activity in the tench musculature and did not correlate in the pond loach with respect to mATP-ase alkali stable red muscle fibres in the fish. It was concluded that fish muscles mATP-ase activity not always correlates with parvalbumin content in contrast to mammalian muscles where such correlation seems to be obvious.

Huriaux *et al.* (1992) proposed that temporal and spatial variations in total parvalbumin concentration and differential expression of parvalbumin isoforms in barbel reflect the functional requirements of the fish axial musculature according to fish size and myomeric location. Significant differences have been found in the relaxation rates of rostral *versus* caudal white muscle fibers of the Atlantic cod *Gadus morhua L.* Using denaturing gel electrophoresis, a series of fresh muscle samples from the dorsal epaxial muscle region was analyzed and several differences were detected (Thys *et al.*, 1998). For example, rostral muscles were found to contain significantly greater amounts of parvalbumin than caudal muscles. Two soluble Ca^{2+}-binding proteins, in addition to parvalbumin, were also detected in the rostral muscle samples yet were absent from the caudal samples.

Parvalbumins were isolated from skeletal muscles of a tropical amphibian, *Leptodactylus insularis*, and three new isotypes were identified (Gerday *et al.*, 1991). The total concentration of parvalbumins in *L. insularis* was the same as the total amounts found in an amphibian from the temperate or variable zone (*Rana temporaria*). Muscles of the thigh and foreleg had the maximum parvalbumin concentration (0.35 mmol.kg wet weight^{-1}). Samples from *pectoralis* and *rectus abdominis* muscles had significantly less parvalbumin concentration (0.29 mmol.kg^{-1}). Three previously unknown parvalbumin isotypes (IV, IIIa, and IIIb) were isolated from the tropical amphibian. They were different from the isotypes (IVa and IVb) predominant in *R. temporaria* skeletal muscle.

The content and concentrations of parvalbumin components change during development of organs. Anuran amphibians, animals that lead a terrestrial life after metamorphosis, start a development of hindlimbs during and after metamorphosis. Metamorphosis is initiated by thyroid hormone and some other hormones, namely, prolactin, growth hormone, and adrenocortical hormones, which antagonize or synergize with the action of thyroid hormone. In order to see whether changes occur in the muscle protein components in the course of postmetamorphic development, Hasebe *et al.* (2003) subjected *Gastrocnemius* muscle extracts from growing froglets to two-dimensional electrophoresis. They revealed changes in level of two parvalbumins: one parvalbumin, α-form pI 5.0, first appears at 45 days after metamorphosis; another one, β-form pI 4.8, is expressed until the former parvalbumin appeared. It was found that as the frogs grow the β-parvalbumin mRNA level decreases whereas that of α-parvalbumin mRNA increases. The appearance of different types of parvalbumin isoforms at different development stages seems to reflect specific physiological needs (for mobility, feeding, etc.) at these stages. The changes in parvalbumin expression levels of α and β isoforms in the bullfrog may be related to the shift of the habitat from aqueous to terrestrial, where locomotion becomes intense, and body weight increases.

Parvalbumin content in muscles changes with age. In the study of Cai *et al.* (2001), the parvalbumin expression profile of *extensor digitorum longus* and *soleus* muscles, representing fast- and slow-twitch skeletal muscles, respectively, was established using high resolution two-dimensional electrophoresis. These experiments revealed that parvalbumin expression in *extensor digitorum longous* muscle is down-regulated during aging. In addition, high-intensity exercise could reverse this age-related change. *Soleus* muscles do not normally express parvalbumin, but high-intensity exercise could ectopically induce its expression in both young and old *soleus* muscles.

Polyacrylamide gel electrophoresis was used to analyse the distribution of parvalbumin isoforms in white muscles of larval, juvenile, and adult *Chrysichthys auratus* (catfish, siluriforms) and to study the kinetics of their synthesis (Chikou *et al.*, 1997). Parvalbumin isoform II was first detected from day 5 post-hatching and was the main "larval" isoform in this species. Parvalbumin III appeared at the beginning of the juvenile stage but always remained the minor isoform, even in adult fish. Young mature specimens (approximately 12 cm long) displayed the highest total parvalbumin content.

As was mentioned above, the parvalbumin family includes α and β sublineages. α-Parvalbumin is most abundant in fast-twitch skeletal myofibrils (Heizmann *et al.*, 1982) and GABAergic neurons (Celio & Heizmann, 1981) (see below) and is also expressed in a number of nonexitatory tissues (reviewed by Heizmann, 1988): endocrine glands (Endo *et al.*, 1985), the brain (de Lecea *et al.*, 1995), bone (Toury et al., 1995), kidney (Kerschbaum *et al.*, 1994; Loffing *et al.*, 2001) and others. In general, the tissue distribution of β-type parvalbumin is more restricted than that of α-type parvalbumin in tetrapods. In β-lineage, chicken parvalbumin 3 is specifically expressed in the thymus (Hapak *et al.*, 1994), and mammalian β-type parvalbumin (oncomodulin) is specifically expressed in the hair cells of the Corti organ (Henzl *et al.*, 1997; Sakaguchi *et al.*, 1998).

The mammalian β-isoform of parvalbumin (oncomodulin) was found first in rat hepatoma (MacManus, 1979) and was subsequently detected in the blastocyst and placental cytotrophoblasts (Brewer & MacManus, 1985, 1987). Oncomodulin was previously believed to be an oncofetal protein, absent from any normal adult mammalian tissue. It was shown to occur in a wide variety of rodent and human tumors (MacManus *et al.*, 1982; MacManus & Whitfield, 1983; Brewer *et al.*, 1984). The name oncomodulin was applied to the protein to reflect its common appearance in neoplasms. Expression of oncomodulin is restricted to early embryonic stages, the placental cytotrophoblasts, and neoplastic tissues.

Early reports (MacManus & Whitfield, 1983) suggested that oncomodulin possesses certain calmodulin-like regulatory properties, nevertheless the function of oncomodulin as a calmodulin-like enzyme modulator is controversial (Blum & Berchtold, 1994). Two types of experiments demonstrate that oncomodulin may act in an analogous fashion to calmodulin in T14 and T10 cancerous cell lines, which both express oncomodulin. First, both oncomodulin transcript and protein levels increase at the G1/S boundary in a manner similar to calmodulin, though not to the same extent, in the chemically transformed rat fibroblast cell line T14 synchronized at mitosis by nocodazole. Second, antisense oligonucleotides specific to the oncomodulin ATG region inhibit growth of T14 in a similar dose-dependent manner as observed with calmodulin-specific antisense probes.

Interestingly, it was found that both α- and β-parvalbumins are contained in the sensory cells of the mammalian auditory organ, the organ of Corti. It was previously revealed that the α-parvalbumin is expressed by the inner hair cells while the β-parvalbumin by the outer hair cells (Eybalin & Ripoll, 1990; Pack & Slepecky, 1995; Sakaguchi *et al.*, 1998). It was revealed also that an acidic calcium binding protein (CBP-15) expressed in the guinea pig organ of Corti is identical to the β-isoform of parvalbumin (oncomodulin) (Henzl *et al.*, 1997). The organ of Corti is apparently the sole site of expression for the β-parvalbumin in adult animals.

It is assumed that Ca^{2+} signaling serves various purposes in different parts of a hair cell. The Ca^{2+} concentration in stereocilia is thought to regulate adaptation and, through rapid transduction-channel reclosure, to provide amplification of mechanical signals. In presynaptic active zones, Ca^{2+} mediates the exocytotic release of afferent neurotransmitter. At efferent synapses, Ca^{2+} activates the K^+ channels that dominate the inhibitory postsynaptic potential. Hair cells of the inner ear contain high concentrations of calcium-binding proteins, including parvalbumin.

Using light and electron microscopic immunostaining, Sakaguchi *et al.* (1998) found that in gerbil, rat, and mouse inner ears oncomodulin is located exclusively in cochlear outer hair cells. They failed to detect immunoreactive oncomodulin in any other adult tissues. Hackney *et al.* (2003) used light microscope immunofluorescence and post-embedding immunogold labeling in the electron microscope for characterization of the distribution of three calcium-buffering proteins in the turtle cochlea. Both calbindin-D_{28k} and β-parvalbumin were found in hair cells, in which they show similar distribution, whereas calretinin is present mainly in hair-cell nuclei and also occurs in supporting cells and nerve fibers.

At the ultrastructural level, oncomodulin immunostaining was found overlying the nucleus, cytoplasm, and the cuticular plate of gerbil outer hair cells (Sakaguchi *et al.*, 1998). Few, if any, immunostaining was present over intracellular organelles and the stereocilia. In turtle cochlea, the hair-cell concentration of calbindin-D_{28k} but not of β-parvalbumin increases from the low- to high-frequency end of the cochlea (Hackney *et al.*, 2003). Cytoplasmic concentration of calbindin-D_{28k} is 0.13-0.63 mM and that of β-parvalbumin is approximately 0.25 mM, but calretinin concentration is an order of magnitude less. Total amount of Ca^{2+}-binding sites on the proteins is at least 1.0 mM in low-frequency hair cells and 3.0 mM in high-frequency cells. mRNA for all three proteins is expressed in turtle hair cells. β-Parvalbumin was found in both inner and outer hair cells of the guinea pig cochlea. Hackney *et al.* (2005) found that in animals with fully developed hearing, inner hair cells have the proteinaceous calcium buffer similar to that of outer hair cells in which the cell body contains β-parvalbumin (oncomodulin) and calbindin-D_{28k} at levels equivalent to 5 mM calcium-binding sites. Both proteins are partially excluded from the hair bundles, which may permit fast unbuffered Ca^{2+} regulation of the mechanotransducer channels. The sum of the calcium buffer concentrations decreased in inner hair cells and increased in outer hair cells as the cells developed their adult properties during cochlear maturation.

The exclusive association of oncomodulin with cochlear outer hair cells in mature tissues is likely to have functional relevance. It was suggested that oncomodulin may be involved in mediation of intracellular responses to cholinergic stimulation, which are known to be Ca^{2+} regulated (Sakaguchi *et al.*, 1998). Hackney *et al.* (2003) suggested that calbindin-D_{28k} and β-parvalbumin may serve as endogenous mobile calcium buffers. Hackney *et al.* (2005) suggested as well that Ca^{2+} has distinct roles in the two types of hair cell, reflecting their different functions in auditory transduction. Ca^{2+} is used in inner hair cells primarily for fast phase-locked synaptic transmission, whereas Ca^{2+} may be involved in regulating the motor capability underlying cochlear amplification of the outer hair cell. The high concentration of calcium buffer in outer hair cells, similar only to skeletal muscle, may protect against deleterious consequences of Ca^{2+} loading after acoustic overstimulation.

Using cDNA subtraction and a gene expression assay based on *in situ* hybridization, Heller *et al.* (2002) detected abundant expression of mRNAs encoding parvalbumin 3 in bullfrog saccular and chicken cochlear hair cells. They cloned cDNAs encoding this protein from the corresponding inner-ear libraries and raised antisera against recombinant bullfrog parvalbumin 3. Immunohistochemical labeling indicated that parvalbumin 3 is a prominent Ca^{2+}-binding protein in the compact, cylindrical hair cells of the bullfrog's sacculus, and occurs as well in the narrow, peanut-shaped hair cells of that organ. Using quantitative Western blot analysis, it was evaluated that the concentration of parvalbumin 3 in saccular hair cells is approximately 3 mM. Soto-Prior *et al.* (1995) cloned and sequenced α-parvalbumin cDNA from the guinea pig cochlea. The deduced amino acid sequence shows greater identity with the rabbit sequence (86.3%) than with other mammalian sequences (< 82%). Using *in situ* hybridization and immunohistochemistry, α-parvalbumin mRNA and the protein were found in primary auditory neurons and inner hair cells, in agreement with the data showing α-parvalbumin mRNA expression in the spiral ganglion and the organ or Corti.

Parvalbumin was isolated and purified from rat kidney (Schneeberger & Heizmann, 1986). Its biochemical and immunological properties are indistinguishable from the muscle counterpart. The protein is localized in part of the distal tubule and proximal collecting duct, similar to calbindin 28kD.

Calcium-binding proteins were found to be involved in the stimulus-secretion coupling mechanisms in secretory glands. They were revealed by means of radioactive ^{45}Ca, after electrophoresis in SDS-PAGE and transference to Zeta probe membranes, in Duvernoy's or venom gland homogenates from three families of South American snakes: *Viperidae* (*Bothrops jararaca* and *Crotalus durissus terrificus*); *Elapidae* (*Micrurus corallinus*), and *Colubridae* (*Phylodrias patagoniensis* and *Oxyrhopus trigeminus*) (Goncalves *et al.*, 1997). A band with an estimated molecular weight of 12 kDa was found in all glands studied and was identified as parvalbumin.

Parvalbumin was found also in thymus. Atoji *et al.* (2000) examined the distribution of parvalbumin in the pigeon thymus by light and electron microscopic immunohistochemistry. Parvalbumin immunoreaction was found in epithelial cells of the cortex, which formed dense mesh-like structures. Parvalbumin-positive epithelial cells were classified into 2 types. The first type comprises elongated cells with spindle-shaped, oval, or triangular nucleus, which have a slightly irregular contour and contain rich heterochromatin peripherally. This type of cell forms the majority of immunoreactive cells. The other cell type consists of polygonal epithelial cells with oval nucleus. Euchromatin occupys a large part of the nucleus. The cytoplasm contains numerous cell organelles compared with the elongated type, in particular, electron-dense vacuoles of various sizes and often bundles of tonofilaments. Both types of epithelial cell are interconnected by desmosomes. It is of interest that avian species express two thymus-specific isoforms of parvalbumin known as avian thymic hormone (ATH) and chicken parvalbumin 3 (CPV3). The CPV3 content of chicken thymus tissue (120 μg/g tissue) is 4 times lower than that of ATH (500 μg/g tissue) (Hapak *et al.*, 1994). The CPV3 sequence exhibits 58% identity with ATH and 52% identity with the chicken muscle parvalbumin isoform. ATH and CPV3 are both confined to the thymic cortex and they are frequently co-expressed by a sub-set of epithelial cells, however, their expression patterns are not completely superimposible. A sub-set of cortical thymocytes exhibits peripheral staining for one or both proteins (Hapak *et al.*, 1996).

The parathyroid glands are of major importance in calcium homeostasis. Small changes in the plasma calcium concentration induce rapid changes in parathyroid hormone secretion to maintain the extracellular Ca^{2+} levels within the physiological range. Extracellular Ca^{2+} concentration is continuously measured by a G-protein-coupled Ca^{2+}-sensing receptor, which influences the expression and secretion of parathyroid hormone. Using immunohistochemistry and Western blot analysis, Pauls *et al.* (2000) detected parvalbumin in normal and in hyperplastic and adenomatous human parathyroid glands. The strongest parvalbumin signal was present in chief cells and water clear cells, whereas only a weak signal was observed in oxyphilic cells. Parvalbumin and parathyroid hormone were found co-localized in the same cell types.

Pohl *et al.* (1995) studied the presence of parvalbumin and its transcripts during the postnatal development of the rat ovary: 13 developmental stages between day 1 and day 83 were examined. Starting from day 18 postpartum, both parvalbumin mRNA and parvalbumin were detected in low amounts, simultaneously with the onset of differentiation of secondary gland cells in the ovarian interfollicular stroma. Parvalbumin and its transcripts were primarily detected in conspicuous patches of interstitial gland tissue and in the differentiated thecal cells around the large follicles, and parvalbumin appeared to be fully expressed 33 days after birth. An involvement of parvalbumin in the steroid metabolism of these cells was suggested.

Chromophobe renal carcinoma is composed of neoplastic cells showing several features similar to those found in the intercalated cells of the collecting ducts. Martignoni *et al.* (2001) studied the immunohistochemical expression of parvalbumin, calbindin-D_{28K}, and calretinin in 140 renal tumors, including 75 conventional (clear cell) carcinomas, 32 chromophobe carcinomas, 17 papillary renal cell carcinomas, and 16 oncocytomas. Parvalbumin was strongly positive in all primary chromophobe carcinomas and in one pancreatic metastasis; it was positive in 11 of 16 oncocytomas and absent in conventional (clear cell) and papillary renal cell carcinomas, either primary or metastatic. Calbindin-D_{28K} and calretinin were negative in all tumors, with the exception of two chromophobe carcinomas, four oncocytomas, and two papillary renal cell carcinomas showing inconspicuous calretinin expression. These results demonstrate that parvalbumin can be used as a suitable marker for distinguishing primary and metastatic chromophobe carcinoma from conventional (clear cell) and papillary renal cell carcinoma.

Possible role of parvalbumin in biomineralization during tooth development was investigated by determining its subcellular localization by immunohistochemical methods (Davideau *et al.*, 1993; Ichikawa *et al.*, 1994). The rat root pulp contains three types of nerve fibers: parvalbumin-immunoreactive smooth fibers, calretinin-immunoreactive smooth fibers and calretinin-immunoreactive varicose fibers. These fibers are directed toward the roof of the pulp chamber and pulp horn without marked ramification. In the sub-odontoblastic layer at the roof of the pulp chamber and pulp horn, parvalbumin-immunoreactive smooth fibers repeatedly ramify and extend varicose terminals into the odontoblastic layer. It was shown that parvalbumin in these tissues is a nuclear and a cytosolic protein, not associated with any particular intracellular organelle. Epithelial and mesenchymal undifferentiated cells show no specific parvalbumin immunolabelling. In differentiated ameloblasts, secretory-pole formation (Tomes' process) is associated with a proximal-distal gradient of parvalbumin, but after the Tomes' process has formed, parvalbumin is evenly distributed throughout the cell. The parvalbumin contents of ruffle-ended and smooth-ended ameloblasts appear to be very different. Differentiated odontoblasts are less heavily labeled than ameloblasts, and the label is restricted to the cell body during the whole dentinogenesis. These data suggest that parvalbumin could contribute to membrane plasticity during differentiation, as it was suggested for dendritic growth in the nervous cells. Moreover, parvalbumin could buffer calcium specifically in the cells producing mineralized enamel and dentine during the later stages of tooth development.

It is well known that enamel cells contain large amounts of calcium, particularly during the developmental phase (maturation) when dental enamel is hypermineralized. Hubbard (1996) found that calcium-binding capacity of enamel cell cytosol increases during development, in parallel with the putative transcellular flux of calcium. At maturation phase, the content of calcium-binding proteins in enamel cells exceeds that in brain and other calcium-oriented tissues, which implies a large calcium burden. A search for likely cytosolic calcium transporters revealed only one high-affinity calcium-binding protein 12 kDa, distinguished from α-parvalbumin, that was up-regulated during maturation, but its low abundance (0.02% of soluble protein) precluded a major calcium transport or cytoprotective role. Two low-affinity calcium-binding proteins up-regulated during maturation (by 1.8-fold and 2.1-fold respectively) were identified as calreticulin and endoplasmin, both residents of the endoplasmic reticulum. Together, calreticulin and endoplasmin constituted an exceptionally high proportion (5%) of soluble protein during maturation, which gives an inferred calcium capacity 67-fold higher than that of the principal cytosolic calcium-binding protein 28-kDa calbindin.

Calcium binding proteins play important roles in formation of bone and cartilage. In the works of Toury *et al.* (1995, 1996) the distribution of α-parvalbumin in the epiphyseal plate cartilage and bone of growing rats was examined by electron microscope immunocytochemistry of undecalcified samples. It was revealed that parvalbumin immunoreactivity increases with maturation of chondrocytes and is maximal in the zone of calcification, which is located in the cytoplasm of chondrocytes, osteoblasts and osteocytes. The immunolabeling is associated with amorphous electron-dense material in the cytoplasm and not bound to membranes. There is moderate parvalbumin immunolabeling over the dense chromatin in the nuclei of chondrocytes and bone cells, but none in the cell processes of mature and hypertrophic chondrocytes, in the matrix vesicles themselves, or in the cell processes of osteoblasts. However, parvalbumin immunoreactivity was found in the cell processes of the osteocytes of compact cortical bone. Thus, immunoreactive parvalbumin is confined to the cell bodies of chondrocytes and osteoblasts, and is unlikely to be directly involved in mineral deposition. According to Toury *et al.* (1995, 1996), the maximal parvalbumin immunoreactivity in the last terminal chondrocytes of the zone of calcification suggests that the protein is involved in buffering intracellular Ca^{2+}, preventing stimulation of degenerative processes by high intracellular calcium. In bone, the osteoblasts and the osteocytes of trabecular and compact cortical bones are immunoreactive for parvalbumin and contain parvalbumin mRNA. Parvalbumin lays in their cytoplasm, but there is no parvalbumin immunostaining in the extracellular uncalcified or mineralized bone matrix. The long processes of osteocytes, in compact bone only, are parvalbumin immunoreactive and osteoclasts contain cytoplasmic parvalbumin immunoreactivity. The pattern of immunoreactive parvalbumin distribution indicates that the protein is not involved in the extracellular mineralization of cartilage and bone matrix. The parvalbumin immunoreactivity in the cell processes of osteocytes of compact cortical bone seems to indicate that parvalbumin may be involved in the regulation of Ca^{2+} fluxes and hence in calcium homeostasis in bone.

EXPRESSION OF PARVALBUMIN

The parvalbumin gene is 15.5 kilobase pairs (kb) long and contains 5 exons of 66 to 585 bp the longest being at the 3' end of the gene (reviewed by Heizmann & Berchtold, 1987). One intron has an exceptional length of 9 kb. Its acceptor site for splicing has a high homology to the corresponding sequence in calmodulin. A deletion of 6 bp in the first Ca^{2+}-binding loop results in the loss of its calcium binding properties.

Two parvalbumin-specific mRNA species, 1100 bp and 700 bp, were revealed in rat *gastrocnemius* (Berchtold & Means, 1985). Both parvalbumin mRNA species are probably translated in *gastrocnemius* muscle since both could be found associated to polysomes in this tissue (Epstein *et al.*, 1986).

Schleef *et al.* (1992) isolated five overlapping genomic parvalbumin clones which overall span 28 kilobase pairs around the Pva locus on mouse Chromosome (Chr) 15. Four introns interrupt the coding sequences at positions corresponding to those in rat and human parvalbumin genes. The transcription start site, 25 bp downstream from the TATA-box, was mapped by oligonucleotide primer extension on poly(A)(+)-RNA. The analysis of 0.4 kb promoter sequence of the mouse parvalbumin gene revealed CCAAT- and TATA-box sequences and a 59 bp GC-rich stretch between positions -59 and -118. Similar motifs were found in the parvalbumin genes of rat and human. A perfect 11-bp repeat upstream to positions -149 and -163 respectively is homologous only to the rat promoter.

Full-length cDNA clones coded for two β-type homologues of parvalbumin genes, pvalb3a and pvalb3b, were isolated from zebrafish (Hsiao *et al.*, 2002). Whole-mount *in situ* hybridization revealed that the spatial and temporal expression of pvalb3a and pvalb3b are distinct and highly development-regulated during early embryogenesis. Unlike their counterparts of parvalbumin 3 in chicken and oncomodulin in mammals, zebrafish pvalb3a transcripts are widely expressed in mucous cells, the olfactory epithelium, anterior pituitary, pharyngeal teeth germ, macrophages, inner ear and lateral line neuromasts, whereas, pvalb3b transcripts are more restrictedly expressed in the yolk syncytial layer, inner ear and pronephric ducts.

At the same time, Friedberg (2005) analysed existing genomic information and concluded that in zebrafish nine genes encode parvalbumin. These genes possess introns that differ in size and show nucleotide variability but they contain the same number of exons, and for each corresponding exon, the number of nucleotides therein are identical in all the paralogs. This rule also applies to the multiple parvalbumin genes of other species e.g. mammals. Each of these genes displays, however, characteristic 5' and 3' UTRs which appear highly conserved between closely related species (so that orthologs among these species can be readily identified) but which show larger numbers of mutations between species that are more distant in evolution. From the constructed tree Friedberg (2005) concluded that the traditional classification of parvalbumins as α or β (based mainly on charge of the protein molecule) is not sustainable. Numbers 1-9 are assigned to the various isoforms. A bifurcation of isoforms into 1 and 4; 2 and 3; 6 and 7; 8 and 9 appears to have occurred simultaneously in more recent time, i.e. perhaps approximately 60 mys ago when primates and rodents branched.

Southern blot analysis of genomic DNA from 25 human-hamster somatic cell hybrids showed that the human gene for oncomodulin resides on chromosome 7 (Ritzler *et al.*, 1992). Analysis of human-mouse hybrids, selectively retaining human chromosome 7 or a portion of it, allowed specific assignment of the gene locus to the p11-p13 region of chromosome 7. By means of gene dosage analysis on Southern blots, it was shown that the gene for human parvalbumin maps distally to the cat eye syndrome marker D22S9 on chromosome 22q. Using somatic cell hybrids containing parts of human chromosome 22, the parvalbumin gene was sublocalized to the region 22q12-q13.1.

Parvalbumin appears late during development and seems to be expressed exclusively in differentiated cells. For this reason parvalbumin has not been found to be expressed in any established cell lines at considerable levels. Therefore, unfortunately, cellular regulation and physiological significance of parvalbumin expression cannot be studied in cell cultures. Alternative method of investigation of effects of parvalbumin in cells and tissues is to ectopically express this protein and study consequences of this manipulation.

For example, Castillo *et al.* (1995) produced transgenic mice expressing rat parvalbumin under the control of the human metallothionein IIA (MTII A), SV-40 early, and neuron-specific enolase (NSE) promoters. As expected, the NSE promoter showed highest activity in brain. However, NSE-driven expression could also be detected to various degrees in all investigated tissues. SV-40-dependent parvalbumin expression showed no tissue preference and varied considerably among different strains. Later these authors generated transgenic mice carrying the full-length complementary DNA (cDNA) of rat parvalbumin under the control of the heavy-metal inducible metallothionein IIA promoter (Castillo et al., 1997). They showed the expression of parvalbumin in endothelial cells lining the liver sinusoids *in situ* and after isolation *in vitro*. Vasoconstriction, thought to be mediated by the Ito cell, was not affected in the transgenic animals, whereas microvascular exchange was markedly decreased in normal mice but virtually not affected in the transgenic animals. This suggests that ectopically expressed parvalbumin is involved in the regulation of Ca^{2+} signals in the sinusoidal endothelial cells.

In the work of Van Do *et al.* (1999), the protein-coding regions of two distinct cDNAs, 14.1 and 24.1, which comprise the entire parvalbumin-encoding regions, were subcloned into an *Escherichia coli* expression vector (pET-19b). Both proteins were expressed and the generated target proteins were localized in both soluble and insoluble fractions of the expression host. The recombinant products in the soluble fraction were purified using the His tag-purification system and analysed on Western blots with anti-salmon parvalbumin polyclonal rabbit sera and sera from patients allergic to fish. Both recombinant products (His10-14.1 and His10-24.1) reacted positively with salmon parvalbumin-specific immunoglobulin G (IgG) from rabbits, and with specific immunoglobulin E (IgE) from the sera of six fish-allergic patients. The allergenicity of His10-14.1 was confirmed using enzyme-linked immunosorbent assay (ELISA). In the next work of Van Do et al. (2003) transcripts of two isotypic parvalbumin genes in Atlantic cod were identified and characterized. Subsequently, subfragments were inserted into the expression vector pET-19b, generating plasmids with coding capacity for complete parvalbumin polypeptides fused to an N-terminal His10 tag. Most of the recombinant products were found in the soluble fraction of the expression host *Escherichia coli.* In order to produce recombinant parvalbumin, the pET-19b expression system was also used to express the recombinant parvalbumin (Van Do *et al.*, 2005). This system allows N-terminal fusion to a cleavable His-tag for rapid affinity purification of the target protein.

Hazama *et al.* (2002) generated transgenic mice that expressed the human interleukin-2 receptor α-subunit-green fluorescent fusion protein (hIL-2R-GFP) using two types of parvalbumin transgene. One contained the hIL-2R-GFP gene downstream of a 16.5-kb 5'-upstream parvalbumin genomic sequence (parvalbumin line). The other comprised the hIL-2R-GFP gene in bacterial artificial chromosome with either a 180-kb (PA line) or 155-kb (PB line) insert encompassing the parvalbumin gene. Independent lines of all transgenic mice showed a faithful hIL-2R-GFP expression in fast-twitch muscle fibers. However, appreciable hIL-2R-GFP expression in the CNS occurred only in the parvalbumin transgenic lines. In one line of parvalbumin transgenic mice, hIL-2R-GFP was properly expressed in parvalbumin-containing neurons in the cerebellum, thalamic reticular nucleus, *globus pallidus* and cerebral cortex, though ectopic expression was observed in a particular subset of cerebellar astrocytes. Another line of parvalbumin transgenic mice showed a selective and mosaic expression of hIL-2R-GFP in parvalbumin-containing Purkinje, basket and stellate cells in the cerebellum. These results indicate that the 16.5-kb parvalbumin genomic sequence is sufficient for fiber-type-selective transcription but additional regulatory sequences comprised in bacterial artificial chromosome DNA are required for proper expression in parvalbumin-containing neurons.

ISOLATION, PURIFICATION AND DECALCIFICATION OF PARVALBUMIN

Traditionally, parvalbumins from skeletal muscles of lower and higher vertebrates are purified by ammonium sulfate fractionation, followed by gel filtration and anion exchange chromatography.

Classical method of extraction and purification of parvalbumins from muscles was provided by Pechére *et al.* (1971). It is a three-step procedure involving $(NH_4)_2SO_4$ fractionation, gel filtration on Sephadex G-75 and chromatography on DEAE-cellulose. The procedure starts with the mincing of white muscles and extraction by the medium containing 0.3 M sucrose, 0.01 M triethanolamine, and 0.003 M EDTA followed by centrifugation at low temperature at 5000 g for 30 min. The supernatant is collected and dialysed for 24 hours against 40 volumes of distilled water in a cold room. Precipitated globulines are removed by centrifugation at low temperature at 5000 g for 30 min. The solution of protein at 10 mg/mL concentration is brought to 70% saturation in $(NH_4)_2SO_4$ at 0°C at neutral pH by addition of 44 g of solid salt and 0.5 mL 1 M NaOH for each 100 mL under gentle stirring. After 20 min the suspension is centrifuged in the cold (5000 g, 30 min) and the supernatant is saturated with $(NH_4)_2SO_4$ at 0°C by addition of 22 g of solid salt per 100 mL of supernatant, under stirring. After 2 hours the precipitate is collected by centrifugation in the cold (16000 g, 30 min) and dialyzed for 1 hour in the cold against a small amount of 0.015 M HCl adjusted to pH 5.7 with piperazine.

The solution is then passed through a column of Sephadex G-75 previously equilibrated in the cold with 0.015 M HCl adjusted to pH 5.7 with piperazine. The fractions corresponding to the second peak, which contains parvalbumins, are pooled and passed through a column of DEAE-cellulose equilibrated in the cold with the same buffer as used during the gel filtration. The elution is carried out with a solution of increasing chloride concentration so that the effluent would linearly reach a total chloride concentration of approx. 0.07 M after 5 dead volumes and approx. 0.14 M after 8 dead volumes. All the parvalbumins are thus desorbed from the resin in the order of their decreasing isoelectric points.

Sometimes heat treatment of the muscle extract is applied, which leaves most Ca^{2+}-binding proteins in solution, followed by the final purification of parvalbumin by two consecutive anion exchange chromatography steps (Blum *et al.*, 1977; Heizmann & Strehler, 1979).

When parvalbumin concentrations are considerably lower than those in the muscles of fish, amphibian or small mammals, as for example in human muscle and in most non-muscle tissues, application of high performance liquid chromatography on reverse phase supports is required for the isolation of parvalbumin (Berchtold *et al.*, 1982b; 1983).

Hutnik *et al.* (1990b) used the following method for isolation of cod III parvalbumin. Their interest to this protein is explained by the fact that cod parvalbumin contains a single tryptophan residue, which is very useful in spectral studies. The protein was isolated from 500 g of frozen fillets of cod by means of acetone precipitation. Acetone was added to a final concentration of 54% (v/v) dropwise at 0-4°C with stirring, followed by additional stirring for 15 min at 0°C. The solution was centrifuged for 75 min at 9500 rpm in a Sorvall RC-5B refrigerated superspeed centrifuge. The supernatant was made 80% in acetone (v/v), and after centrifugation, the pellet was redissolved in 100 mL of 20 mM sodium phosphate buffer, pH 7.3, containing 24 mM NaCl. The solution was gently agitated overnight, centrifuged, and divided into two 50 mL portions. In two separate applications, the 50-mL protein solutions were loaded onto a Sephadex G-75SF column (100 cm x 2.5 cm), eluted at a flow rate of 3 mL/min, and collected in 5-mL fractions. The fractions containing parvalbumins were identified by both their peak position and their calcium binding activity. The fractions were pooled and dialyzed overnight at 4°C against 16 mM piperazine buffer, pH 5.7. After combination of the fractions from gel filtration of the two 50-mL samples, a total volume of 170 mL was applied to the ion-exchange DEAE-52 column (42 cm x 3 cm). The column was eluted with a linear NaCl gradient (500 mL x 500 mL, 0-0.1 M NaCl) and fractions were collected. Fractions containing the parvalbumins were identified on the basis of their absorbance at 280 and 260 nm (differential Phe/Trp content).

At this stage the fractions appeared homogeneous on the sodium dedecyl sulfate – polyacrylamide gel electrophoresis but upon isoelectric focusing in 7.5% polyacrylamide rod gels, 5% ampholytes, pH gradient 3-5, two bands appeared in fractions across the elution profile. Therefore, pooled fractions from the DEAE-52 cellulose ion-exchange column were dialyzed overnight at 4°C against 20 mM Tris, pH 8, containing 1 mM dithiothreitol and applied to a DEAE-Sephacel column (30 cm x 0.9 cm) equilibrated with the same buffer. The protein was then eluted with a linear NaCl gradient (250 mL x 250 mL, 0-0.3 M NaCl). Of the three peaks obtained in the elution profile, peak 2 was found to contain homogeneous parvalbumin III.

Multiple parvalbumin isoforms were detected in the tail (skeletal) muscle of the American alligator (*Alligator mississipiensis*) (Laney *et al.*, 1997). One of these isoforms (APV-1) was highly purified and partially characterized. Protein purification involved mainly gel filtration and anion exchange chromatography, and characterization included gel electrophoresis, amino acid composition analysis, metal ion analysis, MALDI-TOF and ESI mass spectrometry, ultraviolet and fluorescence spectroscopy, and one- and two-dimensional 500 MHz proton NMR spectroscopy. Briefly, the isolation procedure involved the following steps: ammonium sulfate precipitation (the 70-100% ammonium sulfate cut contains the parvalbumins); separation of the alligator parvalbumin fraction from the high molecular weight proteins by gel filtration chromatography (on a 2.5 cm × 90 cm Sephadex G-75 column, using 0.1 M ammonium bicarbonate as an eluant); and a final separation of the alligator parvalbumin isoforms by anion exchange chromatography. DEAE cellulose was used in the anion exchange step, using a buffer prepared by adjusting the pH of a 0.015 M HC1 solution to 5.7 with solid piperazine. Trace myoglobin and the alligator parvalbumin isoforms were eluted from the DEAE column by establishing a linear NaCl gradient (from 0 to 0.20 M in Cl^-).

In the work of Ross *et al.* (1998) parvalbumin isotypes were isolated, on a preparative scale level, by use of size exclusion chromatography (SEC) and anion exchange HPLC. White muscle filets of fish were homogenized under refrigerated conditions with an equal volume of iced deionized water. The solution was centrifuged at 16,000 g for 30 min and the supernatant was filtered off. Then the supernatant was taken to 70% saturation with $(NH_4)_2SO_4$ and the resulting precipitate was separated by centrifugation under prior conditions. The pellet was discarded as the supernatant was taken to 100% $(NH_4)_2SO_4$ saturation. The precipitated pellet was redissolved in 30 mL of 0.05 M NH_4HCO_3 (pH 7.8) buffer and dialyzed at 4°C against 4 L buffer reservoir of the above medium for removal of $(NH_4)_2SO_4$. After that the solution was lyophilized. The crude protein was dissolved in a small volume of 25 mM mono/dibasic phosphate buffer solution (pH 6.8) and chromatographed on a G-75-120 Sephadex column (75 cm x 2.5 cm). Fractions corresponding to low molecular weight components were pooled, lyophilized and redissolved in 25 mM mono/dibasic phosphate buffer solution (pH 6.8) and chromatographed using size exclusion chromatography-anion exchange HPLC on a Pharmacia Superdex 75 HR 10/30 column using the same buffer as the mobile phase. High molecular weight contaminants were removed and a single peak containing parvalbumins and slight contaminants (myoglobin) was collected, lyophilized and reinjected into the same system for refinement of the purification process and single isolated peaks were readily obtained.

Isoelectric focusing in immobilized pH gradients is reported for unequivocal identification of fish species (Esteve-Romero *et al.*, 1996). For clear-cut species identification, two strategies have been used: (i) to perform isoelectric focusing in very narrow (1 pH unit and less) acidic gradients, typically spanning the pH 4-5 range, where fewer proteins are present and the pattern is much clearer; (ii) to focus the analysis on the parvalbumins, since this protein class is highly species-specific and resistant to heat. Thus, not only fresh muscle could be analyzed, but also boiled fish samples. In all cases unambiguous determination of each species could be performed, either by simple visual band inspection or, in the most difficult cases, by densitometric evaluation of the Coomassie-blue stained profiles. The analysis was performed in extracts of single species and also in mixtures of the most closely related species.

Decalcification of parvalbumin can be achieved by means of gel-filtration procedure (Blum *et al.*, 1977). Parvalbumin solution in concentration of 15 mg/mL is incubated for 12 hours at 4°C in 0.1 M EGTA, 0.1 M EDTA at pH 7-8 and then the solution is applied on a plastic column (1.5x40 cm) of Sephadex G-25, prewashed by EGTA and equilibrated by appropriate buffer pH 7-8. Decalcified parvalbumin collected in plastic tubes contains less then 0.03 μg Ca^{2+} per mg of protein.

Decalcification of parvalbumin can be carried out also by precipitation with trichloroacetic acid (Haiech *et al.*, 1979b; Hutnik *et al.*, 1990). Following precipitation and centrifugation of the protein solution, the supernatant is poured off and the precipitated pellet is redissolved in Ca^{2+}-free buffer. Sometimes this method is better compared to the more commonly used method of EGTA treatment.

It should be noted that only plastic or quartz ware prewashed by HCl and deionized water should be used for the work with Ca^{2+}-free parvalbumin. All buffer solutions should be prepared using deionized water. Any contacts of the protein solution with glass should be excluded. Even contacts of the Ca^{2+} free parvalbumin solution with glass pH electrodes can add a significant quantity of contaminating Ca^{2+} (Permyakov & Yarmolenko, 1980).

SECONDARY AND TERTIARY STRUCTURE OF PARVALBUMIN

The three-dimensional structure of parvalbumin is known in details now. The structure of carp parvalbumin pI 4.25 molecule was first determined by Nockolds et al. in 1972 and refined at atomic resolution in 1973-1975 (Kretsinger & Nockolds, 1973; Moews & Kretsinger, 1975).

Parvalbumin is a typical small globular protein. Roughly, a parvalbumin molecule can be represented by an ellipsoid of revolution with major axes of 36 and 30 Å. The external 2.7 Å shell of the molecule is composed of polar and charged groups exposed to solvent. There is a hydrophobic core inside the molecule, which is also an ellipsoid of revolution with the major axes of 18 and 14 Å. The core is composed of seven Phe, five Leu, four Ile, and three Val residues.

Figure 2 shows the ribbon model of the carp parvalbumin pI 4.25 molecule. This protein is characterized by a high content of α-helices and by the presence of β-sheeted structure. 52 of 108 amino acid residues form six α-helices, from A to F.

It was found that the parvalbumin molecule consists of three homologous 30-residue segments, each consisting of central loop flanked by short amphipatic helices: six helices (from A to F) in parvalbumin molecule are connected by three loops, two of which, between the C and D helices and between E and F helices, form calcium-binding sites. C-helix, CD-calcium binding loop, and D-helix (residues 41 to 70) are related to E-helix, EF-calcium binding loop, and F-helix (residues 80 to 108) by the intramolecular two fold axis. Kretsinger (Kretsinger & Nockolds, 1973) proposed to call the structure E-helix – calcium binding loop – F-helix by “EF-hand” (Figure 3): the thumb corresponds to the helix F, the forefinger is the helix E, whereas the clenched middle finger symbolizes the calcium-binding loop located between the two helices (Figure 4). The name ‘EF-hand’ is now widely used in modern literature on calcium binding proteins. The N-terminal domain, encompassing residues 1 through 40, the so-called AB domain, suffered a two-residue deletion that abolished its metal ion binding capacity.

Figure 2. The ribbon model of the carp parvalbumin pI 4.25 molecule (PDB file 5CPVcarpCa).

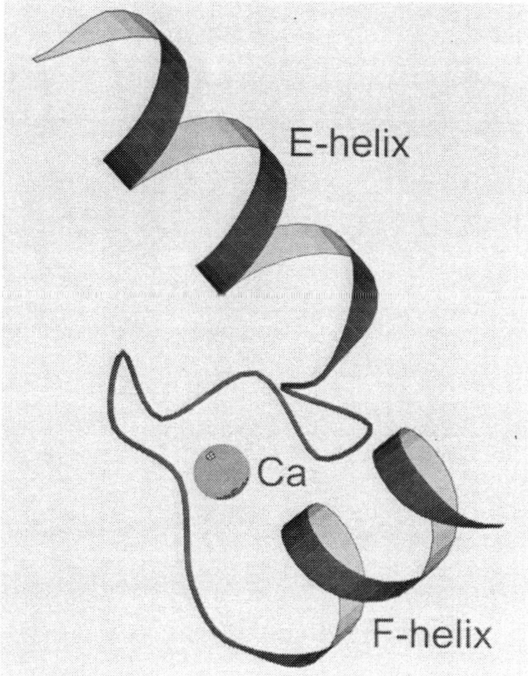

Figure 3. Ca^{2+}-binding domain "EF-hand". Most of the figures were performed by means of the MolScript v2.1 software (Kraulis, 1991).

Figure 4. Professor Robert Kretsinger.

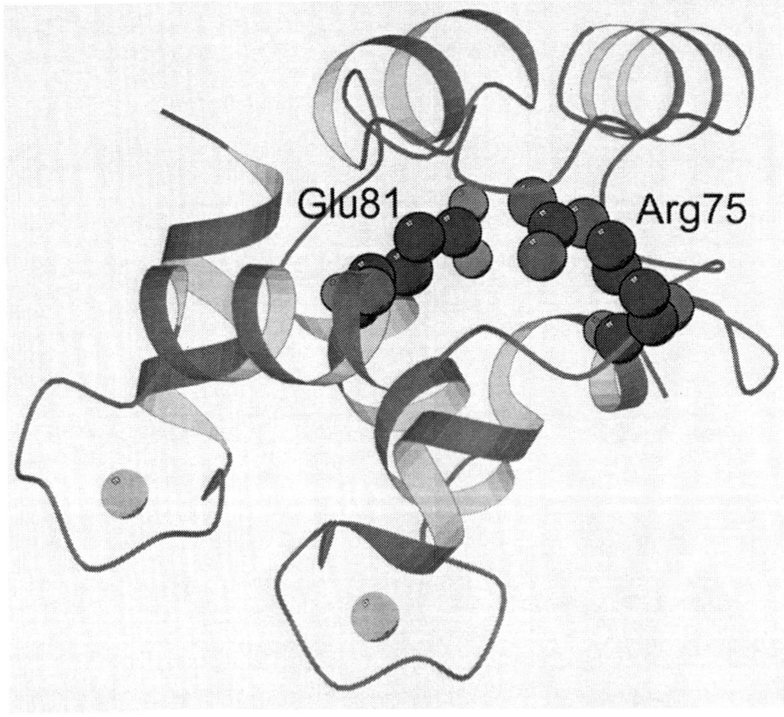

Figure 5. Salt bridge Arg75 – Glu81 in carp parvalbumin pI 4.25 (PDB file 5CPV).

A characteristic feature of the parvalbumin structure is a salt bridge between invariant Arg75 and Glu81 located inside the molecule (Figure 5). This strong dipole is isolated from solvent by the N-terminal segment. It is corroborated by the fact that in the Ca^{2+}-bound state Arg75 does not react with 1,2-cyclohexanedione under conditions of physiological pH and temperature (Coffee & Solano, 1976). In the presence of EDTA and at pH 8, it is readily modified by 1,2-cyclohexanedione. Chemical modification of the single Arg75 in parvalbumin causes a disorganization of a part of the α-helical regions and a pronounced decrease in affinity to calcium (Gosselin-Ray *et al.*, 1973).

Yang *et al.* (2002) analyzed structures of 11 calcium binding sites in different classes of proteins and found that all the natural calcium binding sites of EF-hand (both classical and pseudo-EF-hand motifs) and non-EF-hand proteins can be described using a set of geometric descriptions of the ideal pentagonal bipyramid geometry.

Parvalbumin binds two calcium ions per molecule. Within the CD and EF binding loops, the six oxygen ligands to the bound Ca^{2+} are located at the approximate verteces of an octaedron (Figure 6) or pentagonal bipyramid. Ligands are denoted as +X, +Y, +Z, -X, -Y, and –Z. CD-Ca^{2+} is coordinated by oxygen atoms of six amino acid residues, while EF-Ca^{2+} is coordinated by oxygen atoms of five residues (Table 2). Side chain oxygen atoms furnish the +X, +Y, +Z, and –Z ligands. A main chain carbonyl provides the invariant –Y ligand.

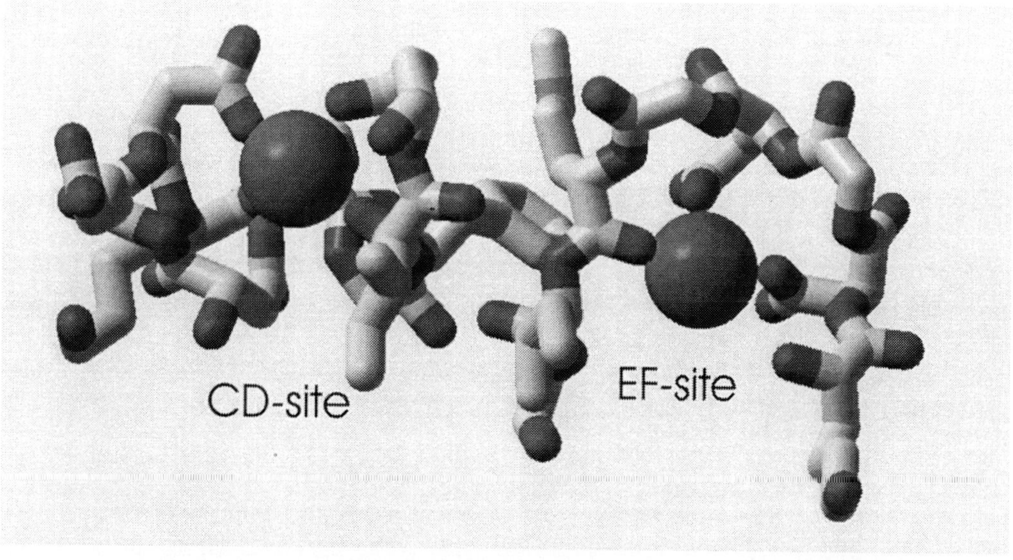

Figure 6. CD and EF calcium binding sites in pike parvalbumin pI 4.1 (PDB file 1PALpike41).

The CD calcium is not coordinated by water, while the –X vertex of the EF octahedron is water oxygen (residue 98 is glycine). The parvalbumin metal ion-binding sites differ at the +Z and –X residues: whereas the CD site employs serine and glutamate (or aspartate), respectively, the EF site employs aspartate and glycine. Molecular dynamic simulations and experimental data show that incorporation of aspartic acid, arginine, cysteine and serine at the 9th position (-X) of the calcium binding loop ("gateway" position) is very disruptive to the structural integrity of the Ca^{2+} coordination site in the whole silver hake parvalbumin (Elkins *et al.*, 2001; Fahie *et al.*, 2002).

Table 3. The CD and EF Ca^{2+}-binding sites in carp parvalbumin pI 4.25.

	CD-site		EF-site	
+X	Asp-51		Asp-90	
+Y	Asp-53		Asp-92	
+Z	Ser-55		Asp-94	
-Y	Phe-57	peptide carbonyl	Lys-96	peptide carbonyl
-X	Glu-59		Gly-98	H_2O
-Z	Glu-62		Glu-101	

Many calcium binding proteins contain the EF-hand domains. A classic EF-hand motif spans 29 residues and consists of a highly conserved calcium binding loop flanked by two helices. A 12-residue loop contains all of the calcium binding ligands. Seven oxygen atoms from the side chains of Asp, Asn and Glu, the main chain and water at sequence positions 1, 3, 5, 7, 9 and 12 of the loop coordinate calcium ion in a pentagonal bipyramidal arrangement. In a typical geometry, the side chain of Asp serves as a ligand on the X-axis in the calcium binding loop (position 1). The –X-axis (position 9) is filled by a bridged water molecule. The –Z axis is shared by two carboxyl oxygen atoms of Glu side chain at position 12 in a bidentate mode to calcium.

An analysis of 567 calcium-binding amino acid sequences, which are similar to the parvalbumin calcium-binding sequence, showed that the position 12 glutamate (62 and 101) is 92% conserved (Falke *et al.*, 1994). The remaining 8% of the EF-hand sequences that do not contain glutamate at position 12 all contain aspartate. It was found as well that an E101D substitution at the EF loop position 12 results in a dramatically less tightly bound monodentate Ca^{2+} coordination by aspartate. Molecular dynamics simulations demonstrate that the aspartate is still capable of attaining a suitable orientation for bidentate coordination but the inherent rigidity of the calcium binding loop prevents bidentate coordination in the parvalbumin E101D mutant (Cates *et al.*, 1999a, 2002). The Ca^{2+}-loaded structure of E101D shows that this mutant cannot provide the seven-fold coordination preferred by Ca^{2+}, presumably because of strain limits imposed by tertiary structure. Analysis of these results supports a model wherein the characteristics of the last coordinating residue and the plasticity of the Ca^{2+}-binding loop delimit the allowable geometries for the coordinating sphere (Cates *et al.*, 1999b). Interestingly, the E101D mutation does not affect the Mg^{2+} coordination geometry of the binding loop, but it does pull the F helix 1.1 Å towards the loop.

According to Rigden & Galperin (2004), the EF-hand represents but one, among many, structural context for the DxDxDG-like Ca^{2+}-binding loops. While the canonical EF-hand domain appears to provide the highest affinity to Ca^{2+}, certain calcium binding proteins display only one of its two characteristic features, either the helix-loop-helix structural motif or the DxDxDG sequence motif.

Ca^{2+} and Mg^{2+} are the most physiologically relevant metal ions that interact with EF-hand proteins. Both Mg^{2+} and Ca^{2+} are small, closed-shell, spherical metal ions, which form more likely ionic than covalent bonds. There are subtle differences between the two divalent cations that the EF-hand binding site can exploit to select between the two. Mg^{2+} favors sixfold, octahedral coordination, whereas Ca^{2+} is most commonly found coordinated by seven or eight ligands. Similarly, Ca^{2+}–ligand bond distances are typically 2.3–2.6 Å, whereas Mg^{2+}– ligand bond lengths average 2.0–2.1 Å in length (reviewed by Martin, 1990). Finally, as a result of its greater surface charge density, Mg^{2+} has a 10^3 times slower desolvation rate than Ca^{2+} (Falke *et al.*, 1994).

In parvalbumin, the highly conserved glutamate at calcium binding loop position 12 changes from a bidentate ligand in the sevenfold ligation of Ca^{2+} to a monodentate ligand in the sixfold coordination of Mg^{2+} (Figure 7) (Cates *et al.*, 2002). It was found that the EF-hand-type binding site is extraordinary in its ability to discriminate between two small cations, Ca^{2+} and Mg^{2+}, which are similar in charge, size, and electronic configuration.

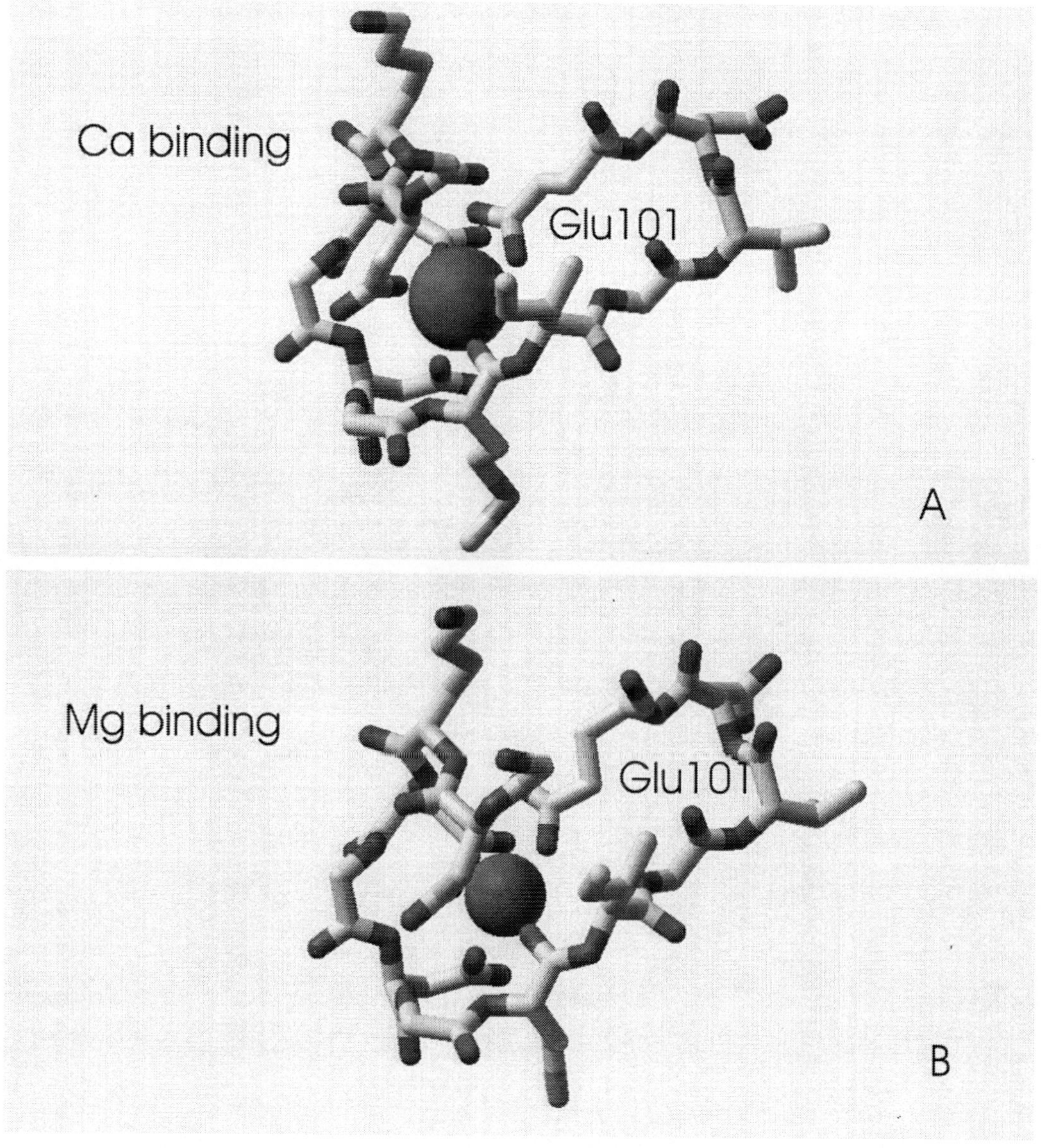

Figure 7. Coordination of Ca^{2+} (A) and Mg^{2+} (B) in pike parvalbumin pI 5.0 (PDB files 3PALpikeCa and 4PALpikeMg).

These binding sites exploit subtle, but crucial, differences in the properties of all the components involved in metal ion binding, including properties dictated by the structural configuration of the binding loop, properties of the amino acid side chain moieties involved in metal ion coordination, and properties of the metal ions themselves.

Energy calculations show that it is easier to remove Ca^{2+} with the donor water from the EF site than to remove Ca^{2+} from the CD site after twisting round the Glu59 (Lockhart & Gray, 1987). The outer region of the CD site is more constricted, having shorter Ca-O distances than the anterior one. The opposite arrangement is found in the EF loop.

A comparison of the ytterbium-substituted structure of carp parvalbumin 4.25 with the native and cadmium-substituted structure shows no significant differences, except around the substituted EF metal-binding region (Kumar *et al.*, 1991). The displacement of calcium by ytterbium at the EF site causes a movement in the polypeptide backbone of Ser91 and Asp92. The movement results in an increase in the number of oxygen ligands bound to ytterbium in the EF site from seven to eight. Experiments with lanthanide ions showed that the CD and EF calcium binding sites in parvalbumin differ both structurally and functionally: while the EF-domain ligands are flexible, expanding or contracting relative to their Ca^{2+}-coordinating positions in response to lanthanide exchange, the ligand cage provided by the CD domain is rigid (Williams *et al.*, 1984).

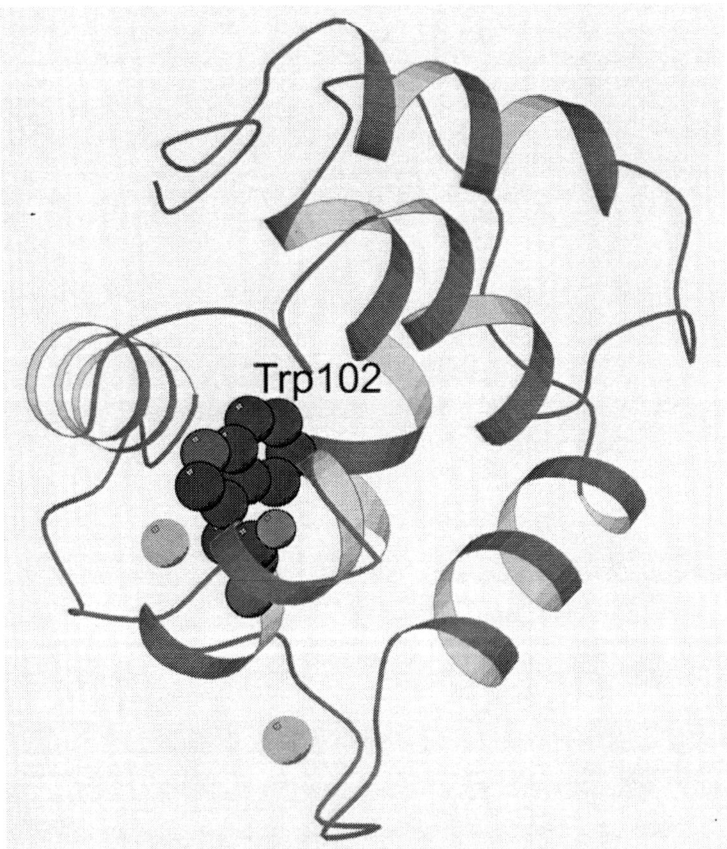

Figure 8. Ribbon model of whiting parvalbumin (PDB structure 1A75whitingPA).

Most parvalbumins do not contain tryptophan residues. The solution structure of a mutant carp parvalbumin bearing a single tryptophan residue at position 102 (F102W) was determined using multi-dimensional NMR (Moncrieffe *et al.*, 2000). The introduced tryptophan residue is located immediately after the last metal ion coordinating residue and is in the same location as the naturally occurring tryptophan residue in whiting parvalbumin (Figure 8). It turned out that the crystal structure of carp parvalbumin and the NMR structure of the mutant F102W are similar. The hydrophobic core is intact in the mutant protein and Trp102 is buried in the hydrophobic core and is close to Phe30, Phe47, Phe70, Phe85 and Phe66.

Moncrieffe *et al.* (2000) used the minimum perturbation mapping technique to explore the possible existence of multiple conformations of the indole moiety of Trp102 of the F102W mutant. The maps for parvalbumin suggest two potential conformations of the indole side-chain. The high energy barrier for rotational isomerization between these conformers implies that interwell rotation would occur on time-scales of milliseconds or greater and suggests a rotamer basis for the heterogeneous tryptophan fluorescence: like for cod and whiting parvalbumins (Permyakov *et al.*, 1985), only two exponential components can fit experimental fluorescence decay curve for the F102W mutant. Just in the terms of the existence of two conformers were interpreted fluorescence decay data for whiting parvalbumin in the work of Permyakov *et al.*, (1985). However, the absence of alternate Trp102 conformers in the NMR data suggests that the heterogeneous fluorescence of Trp102 may arise from mechanisms independent of rotameric states of the Trp side-chain. The backbone and Trp102 side-chain dynamics at 30°C of the F102W mutant was characterized based on an analysis of ^{15}N NMR relaxation data. Tryptophan fluorescence emission and UV difference spectra of the mutant protein indicate that the Trp residue at position 102 is confined to a hydrophobic core and conformationally strongly restricted. Moncrieffe *et al.* (2000) concluded that for the F102W mutant, indole rotational isomerization does not occur on the fuorescence time scale and thus the heterogeneous fuorescence lifetimes of Trp102 apparently arise from quenching mechanisms independent of Trp rotameric states or the existence of two static distributions of protein conformations.

Parvalbumin is characterized by high content of phenylalanine residues (Figure 9). According to the NMR data of Birdsall *et al.* (1979), 10 phenylalanine side chains can be put into classes. There are three phenylalanines whose para proton resonances are expected to experience only small current effects, Phe2, Phe57 and Phe70, and two side chains whose para proton signals are markedly shifted upfield, Phe24 and Phe29.

Microenvironments of the histidines in Ca^{2+}-bound parvalbumins (carp, pI 4.25; pike, pI 5.0; rat, pI 5.5) (Figure 10) were examined with ^{1}H NMR techniques (Williams *et al.*, 1986b). His26, present in all three parvalbumins, is characterized by abnormal pK_a value (from 4.2 for carp to 4.32 for pike and 4.44 for rat) and shows absolutely no photochemically induced dynamic nuclear polarization (photo-CIDNP).

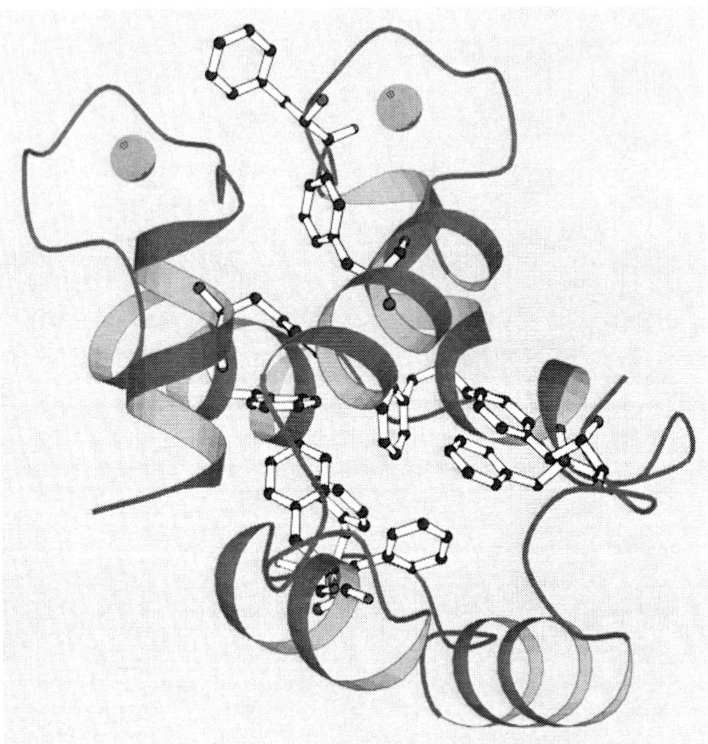

Figure 9. Phenylalanine residues in carp parvalbumin pI 4.25 (PDB file 5CPVcarpCa).

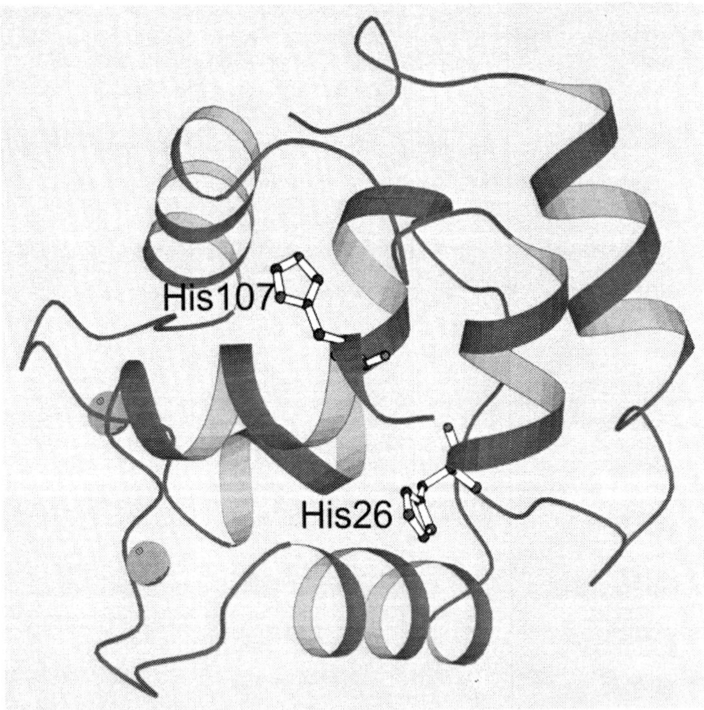

Figure 10. Histidine residues in pike parvalbumin pI 5.0 (PDB file 2PASpike50).

The value of pK_a of His26 in carp parvalbumin pI 4.47 is 5.2 (Birdsall *et al.*, 1979), which is explained by the proximity of His26 to the neighboring Lys27. Although the crystal structure of carp parvalbumin shows that His26 is exposed to solvent, it was concluded that in solution this residue, in its unprotonated state, is a part of the hydrophobic core (Williams *et al.*, 1986b). His48 in rat parvalbumin and His106 in pike parvalbumin show dramatic photo-CIDNP enhancements of their C_2H, C_5H and β-CH_2 ^1H NMR resonances. Since His48 has nearly normal pK_a value 6.14, it was concluded that microenvironment of His48 differs little from random coil exposure. At the same time, the protonation behavior of His106 (pK_a 7.1) in pike parvalbumin suggests that its microenvironment is different from random coil exposure. Interestingly, its C_5H resonance is extremely sensitive to protonation/deprotonation of His25. It was suggested that the protonated form of His106 is involved in a hydrogen bond with a backbone C=O or side chain –OH in the BC-linker region and that the protonated form of His25(26) is expelled from the hydrophobic core.

Parvalbumin contains 0-1 Tyr residues in positions 2 or 48 (Figure 11). The data of photochemically induced dynamic nuclear polarizability obtained on carp parvalbumin pI 3.95 show that its single Tyr2 is either buried or involved in a hydrogen bond *via* its phenolic –OH group (Williams *et al.*, 1986). ^{13}C-NMR data demonstrate that Tyr2 is a clear example of an aromatic group held rigidly within the protein structure when both Ca^{2+}-binding sites are occupied (Nelson *et al.*, 1976).

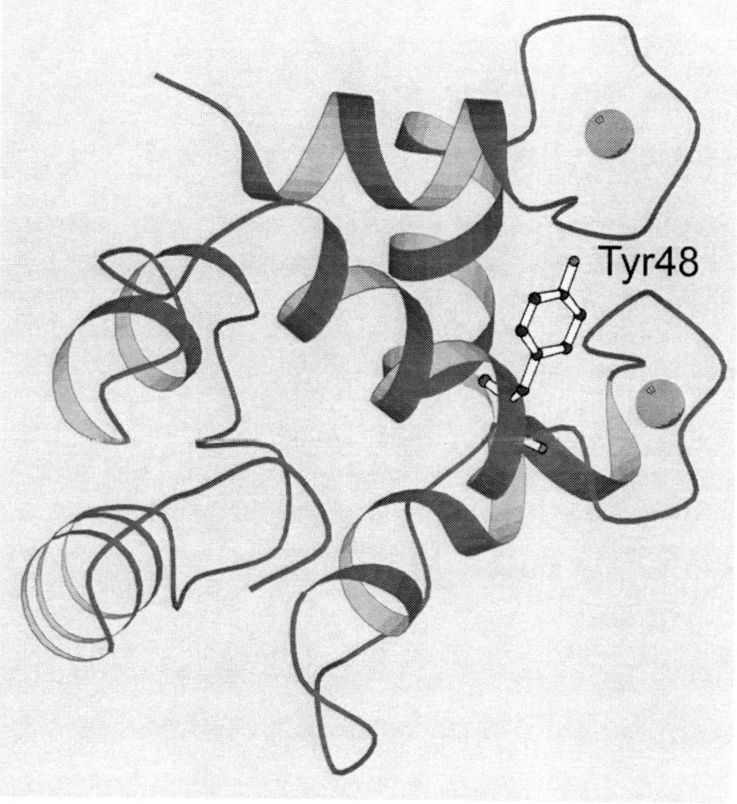

Figure 11. Tyrosine residue in pike parvalbumin pI 4.1 (PDB file 1PALpike41).

The three-dimensional structure of pike parvalbumin pI 4.10 was also established with resolution of 1.93 Å (Declercq *et al.*, 1988). It turned out to be very close to the structure of carp parvalbumin pI 4.25, nevertheless there was an essential difference in Ca^{2+} coordination in these proteins. In carp parvalbumin, Ca^{2+} coordination is octahedral and calcium ion in the CD-site is surrounded by six oxygen atoms at distances less than 3 Å, while calcium ion in the EF-site surrounded by eight oxygen atoms. In pike parvalbumin, Ca^{2+} coordination is octahedral as well but calcium ions are surrounded by seven oxygen atoms each since Glu62 and Glu101 are bidentate ligands in this case. An additional secondary binding site was found in this parvalbumin, which is formed by seven oxygens from Asp53, Glu59, Asp61, and three water molecules and is located near the CD-site.

Later, using a crystal grown under microgravity conditions, cryotechniques (100 K), and synchrotron radiation, Declercq *et al.* (1999) determined the crystal structure of the Ca^{2+}-saturated form of pike (component pI 4.10) at atomic resolution 0.91 Å. Besides a better definition of most of the elements in the protein three-dimensional structure than in previous studies with lower resolution, the high accuracy thus achieved allowed the detection of well-defined alternate conformations, which are observed for 16 residues out of 107 in total. Among them, six residues are located within the hydrophobic core and converge toward two small buried cavities with a total volume of about 60 $Å^3$. They did not find any indications of water molecule present in these cavities. It was suggested that at physiological temperatures there is a dynamic interconversion between these alternate conformations in an energy-barrier dependent manner, a time-dependent remodeling of the void internal space as a part of a slow dynamics regime (millisecond timescales) of the parvalbumin molecule.

X-ray crystal structure of the silver hake major parvalbumin was determined to high resolution as well (Richardson *et al.*, 2000). The main structural features observed in other parvalbumins are also found in this parvalbumin. 114 ordered protein bound water molecules were revealed near the protein surface. 50 water molecules are within hydrogen-bonding distance (<3.2 Å) of the main chain carbonyl oxygens, and 17 water molecules are in contact with backbone amides. 12 water molecules are unique to silver hake parvalbumin and important in stabilizing its structure. The Ca–O distances at the CD site in this structure range from 2.30 to 2.65 Å, while the Ca–O distances at the EF site are within 2.25 to 2.68 Å. The CD calcium in this parvalbumin is coordinated by seven oxygen atoms in a distorted octahedral arrangement, which Kretsinger *et al.* (1989) has described as a pentagonal bipyramid. Glu62 acts as a bidentate ligand, while Asp51 and Asp53 are monodentate ligands. The geometry of the EF site for silver hake major parvalbumin differs from that of the CD site and has been described as a split-vertex octahedron by Kretsinger *et al.* (1989). Similar to the CD site, the coordination involves monodentate ligands (Asp90, Asp92, Asp94) and a bidentate ligand (Glu101). The EF site differs from the CD site in that it has a water molecule in the coordination sphere at position -X.

In addition to the two calcium-binding sites, it has been proposed that an additional secondary cation-binding site exists in parvalbumins. Fluorescence and X-ray crystal studies (Declercq *et al.*, 1991; Roquet *et al.*, 1992) provide evidence for the existence of an additional surface binding site. This site has been shown to bind cations other than calcium and has been described by Declercq *et al.* (1991) to be a satellite of the CD binding site. Refinement data (Richardson *et al.*, 2000) for silver hake major parvalbumin support this work. Three noncontiguous amino acid residues and solvent molecules on the surface of these proteins form the cation-binding pocket. Coordination of calcium in this site involves the oxygen groups of the side chains of residues Asp53, Glu59, and Asp61, as well as two ordered water molecules. This third site seems to be present in both α and β parvalbumins. Two of these residues (Asp53 and Glu59) are always present in α and β parvalbumins, while Glu appears in position 61 of shark parvalbumin as in most α-parvalbumins instead of Asp as in most β-parvalbumins (Richardson *et al.*, 2000).

McPhalen *et al.* (1994) and Bottoms *et al.* (2004) determined the X-ray crystal structure of the rat α-parvalbumin from fast twitch muscle. The overall fold of the polypeptide chain for rat α-parvalbumin is similar to other known parvalbumin structures (root-mean-square deviations in α-carbon atom positions range from 0.60 to 0.87 Å). There are two major Ca^{2+}-binding sites in this parvalbumin, and there is some evidence for a third ion-binding site, adjacent to the CD site, in the rat species. The Ca^{2+} ions in the CD and EF sites are heptacoordinate with pentagonal bipyramidal geometry. These X-ray data allowed determination of structural dynamics in the calcium binding sites. The thermal ellipsoids for Ca^{2+} ions and their oxygen ligands are generally small and nearly spherical in this parvalbumin. Interestingly, the EF site in one of the three molecules located in crystallographic unit displays uncharacteristic flexibility: the ellipsoids for Asp92 are particularly large and non-spherical, and the shape of the Ca^{2+} ellipsoid implies significant vibrational motion perpendicular to the plane defined by the four Y and Z ligands. The level of structural variability among the best-ordered regions of the three independent rat α-parvalbumin molecules in the crystallographic asymmetric unit is two to three times higher than the mean coordinate error (0.10 Å), indicating flexibility in the molecule.

It was found (McPhalen *et al.*, 1994; Bottoms *et al.*, 2004) that sequence differences between α and β-lineage parvalbumins result in a repacking of the hydrophobic core and some shifts in the protein backbone. The shifts are local, however, and entire helices do not shift as rigid units. The high quality of the data for rat parvalbumin permitted 11 residues to be modeled in alternative side-chain conformations, including the two core residues, Ile97 and Leu105. The discrete disorder observed for Ile97 may have functional significance, providing a mechanism for communication between the CD and EF binding loops and between the parvalbumin metal ion-binding domain and the N-terminal AB region.

There were revealed 11 structurally conserved water molecules and 25 conserved protein-water hydrogen bonds in the rat parvalbumin (McPhalen *et al.*, 1994; Bottoms *et al.*, 2004). Most of the conserved hydrogen bonds involve main chain atoms and the carbonyl oxygen, in particular.

The solution structure of human α-parvalbumin was solved by NMR and refined with the help of substitution of the Ca^{2+} ion in the EF site with the paramagnetic Dy^{3+} ion (Baig *et al.*, 2004). A simple ^{1}H-^{15}N heteronuclear single quantum coherence spectrum allowed the NH assignments based on the properties of Dy^{3+}. This allowed these authors to use pseudocontact shifts and residual dipolar couplings for solution structure refinement. They detected small but significant local structural differences with the orthologue protein from rat, whose X-ray structure was available at 2.0 Å resolution. All differences are related to local changes in the amino acid composition. It was found that for residues Leu7, Ser20, Pro73, Asp74, and Ala75 there are striking differences between the two structures and all of the differences are in the loops and not in the helices, as expected. However, some differences are located at the beginning or the end of helices, in such a way that, e.g., one helix is shorter (or longer) in the rat (or in the human) or *vice versa*. There are three distinct regions in human parvalbumin, around Asn8, Ser20, and Pro73, that show a distinct change in the local fold with respect to rat parvalbumin, and in each of these three cases this difference is related to a difference in one amino acid between the two orthologue proteins.

Shark parvalbumin shows the main features of all parvalbumins: the folding of the chain including six α-helices, the salt bridge between Arg75 and Glu81, and the hydrophobic core (Roquet *et al.*, 1992). Compared to the structure of β-parvalbumins from pike and carp, one main difference is observed: the chain is one residue longer and this additional residue, which extends the F helix, is involved through its C-terminal carboxylate group in a network of electrostatic contacts with two basic residues, His31 in the B helix and Lys36 in the BC segment. Furthermore, hydrogen bonds exist between the side-chains of Gln108 (F helix) and Tyr26 (B helix). This results in a "locking" of the tertiary structure through contacts between two sequentially distant regions in the protein and this is likely to contribute to making the stability of an α-parvalbumin higher in comparison to that of a β-parvalbumin.

The authors (Roquet *et al.*, 1992) believe that the lengthening of the C-terminal F helix by one residue appears to be a major feature of α-parvalbumins in general. Besides the lengthening of the C-terminal helix, the classification of the leopard shark parvalbumin in the α-series rests upon the observation of locations of Lys13, Leu32, Glu61 and Val66. In β-pike pI 4.10 parvalbumin, Asp61 is a direct ligand of a third site, the satellite of the CD site. In shark parvalbumin, as in nearly all α-parvalbumins, Glu is located at position 61. Unfortunately, the conformation of the polar head of Glu61 cannot be determined from the X-ray data.

The crystal structure of oncomodulin, a 12 kDa protein isolated from rat tumor, which is homologous to parvalbumin and now called β-parvalbumin, was determined at 1.85 Å resolution (Ahmed *et al.*, 1990; 1993). This protein is found in placenta tissue and in majority of tumor cells, but cannot be detected in most normal adult tissues. Neither the function of oncomodulin during normal development nor its role during carcinogenesis is understood, but it is a member of the parvalbumin family.

The oncomodulin backbone is closely related to that of parvalbumin, but some differences are found with root-mean squire deviations of 1 to 2 Å in residues 2 to 6, 59 to 61 of the CD loop, 87, 90 and 108. Each of the two Ca^{2+} ions that are bound to the CD and EF loops is coordinated by seven oxygen atoms, including one water molecule. The third calcium ion is also seven-coordinated to five oxygen atoms belonging to three different oncomodulin molecules and to two water molecules which form hydrogen bonds to a fourth oncomodulin molecule. The Ca-O distances in all three polyhedra are in the range 2.07 to 2.64 Å. The electron density maps indicate disordered orientations for ten residues on the hydrophilic surface of the molecule.

The extensive molecular aggregation of oncomodulin *via* Ca^{2+} in the crystal structure is a result of larger number of potential calcium-binding oxygen atoms in oncomodulin in comparison with other parvalbumin. Interestingly, the avian parvalbumin called CPV3 readily forms disulfide-linked oligomers (Hapak *et al.*, 1994; Henzl *et al.*, 1995). CPV3 contains two cysteine residues, at positions 18 and 72. The reported three-dimensional parvalbumin structures suggest that the side chain of cysteine-72 should be solvent-accessible. Sedimentation data reveal that CPV3 also undergoes noncovalent self-association. The noncovalent interaction is promoted by either Ca^{2+} or Mg^{2+}, whereas covalent complex formation displays an absolute requirement for the Ca^{2+}-bound protein. Least-squares analysis of sedimentation equilibrium data suggests that 100 μM apo-CPV3 is primarily a mixture of monomeric and dimeric forms. With the addition of Ca^{2+}, the equilibrium becomes exclusively monomer-trimer, with negligible amounts of dimers. A comparable distribution is observed in the presence of Mg^{2+}.

It is intriguing that oncomodulin and parvalbumin are so structurally alike yet so different functionally (Ahmed *et al.*, 1990; 1993). The Cys18 side-chain in the oncomodulin crystal is not in a position that would be easily accessible for molecular dimerization *via* a disulfide bond. The substitution of Glu59, which is preserved in all the determined species of parvalbumin, by Asp59 in oncomodulin seems to break a stabilizing hydrogen bond in the CD loop and make the main-chain in positions 59 to 60 somewhat unstable. This instability in the CD loop, and the strong tendency of oncomodulin for molecular aggregation *via* intermolecular Ca^{2+}, appear to be the two outstanding features that may account for oncomodulin's biological peculiarities.

The aim of the research of Babini *et al.* (2004) was to determine the structure of human β-parvalbumin (109 amino acids, oncomodulin) and to compare it with its paralog and ortholog proteins. The structure was determined in solution using multinuclear and multidimensional NMR methods and refined using substitution of the EF-hand Ca^{2+} ion with a paramagnetic lanthanide. The resulting family of structures had a backbone rmsd of 0.50 Å. Comparison with rat oncomodulin (X-ray, 1.3 Å resolution) as well as with human (NMR, backbone rmsd of 0.49 Å) and rat (X-ray, 2.0 Å resolution) parvalbumins reveals small but reliable local differences, often but not always related to amino acid variability.

The solution structure of human oncomodulin is characterized by the following secondary structure elements (residues in parentheses): α1 (2-4), α2 (7-17), α3 (26-32), α4 (40-50), β1 (57 and 58), α5 (60-63), α6 (64-69), α7 (79-89), β2 (97 and 98), and α8 (99-107). The secondary structure elements found are largely in agreement with the ones of the crystal structure of rat oncomodulin, as expected, with the only exception being helix 5, which starts one amino acid before and is immediately connected to helix 6, giving rise to a single-kink helix, while in the rat structure, the two helices are divided by a two-residue linker. The latter helices are separated by a short connecting linker also in the human and rat parvalbumin structure. All the differences are related to residues located in loops: Leu6, Ser7, Arg75, Glu76, and Leu77.

The internal dynamics of proteins is very important for their biological function as has been shown in protein folding, enzyme catalysis, and protein-ligand interactions. Many functional aspects can be interpreted from static structures, but the extra dimension conferred by internal motions is necessary to understand how proteins function. The characteristic times of motions in proteins at physiological temperatures range over more than 12 orders of magnitude, from below picoseconds to well above seconds. Since proteins are densely packed and their compressibility is low in comparison to liquids and solid polymers conformational variations can only occur in a concerted manner.

The structures of Ca^{2+}-binding proteins, just like structures of any other proteins are subjected to fast fluctuations. Proton NMR data (Cave *et al.*, 1976) show that 8, if not all 10, of the phenylalanine residues of carp parvalbumin pI 4.25 are undergoing rapid rotation or flipping around the C^{β}-C^{1} bond, despite the fact that all are internal residues. Data on structural dynamics obtained by ^{13}C NMR method (Opella *et al.*, 1974; Nelson *et al.*, 1976) show that in parvalbumin α-carbons are fixed rather rigidly while the amino acid side chains possess considerable mobility. Correlation times (τ_c) for rotation around C_β –C_α bonds are about 10 ns for Lys and Val and about 10-20 ns for Asp, Leu, and Phe. Unexpectedly, Phe residues oscillate around C_γ -C_β bonds with τ_c = 4 ns. Rotation of Phe around the C_β-C_α is more restricted sterically by the hydrophobic core of the protein. For comparison, the rotational correlation time for tyrosine residues in calmodulin is 2.3 ns (Lambooy *et al.*, 1982). Besides these fast motions there exist slower rotations of Phe residues by 180° with τ_c = 10-100 ms (Jardetzky, 1978).

Motional properties of the NH bond vectors may be investigated by analysis of the relaxation behavior of the ^{15}N nuclei (Baldellon *et al.*, 1998). Measurement of ^{13}N nuclear relaxation showed the extreme rigidity of the helix-loop-helix EF-hand motifs and the linker segment between them in rat parvalbumin. Continuous variations are observed for the AB loop and for the BC loop which links the AB domain to the more compact CD and EF domains. Both N- and C-termini of the protein are also restricted in their mobility, while the residue at relative position 2 in the Ca^{2+}-binding sites have an enhanced mobility. Part of the specific dynamics in Ca^{2+}-parvalbumin appears to be strongly dependent on the three-dimensional arrangement around the strictly conserved Glu81-Arg75 salt bridge. These key residues are involved in several interactions resulting in a reduced mobility of the DE segment which further locks the CD and EF domains together. Furthermore, the interaction between the AB loop and the salt bridge and hydrophobic contacts involving the A, B, and D helices are known to play an important role in the affinity for Ca^{2+} ions.

Computer simulations of the molecular dynamics of parvalbumin in solution show that the protein changes the structure relative to that of the crystal form, but the helix structures remain largely intact (Ahlström *et al.*, 1987). The average lifetime of hydrogen bonds in helices is about several picoseconds. It could be concluded that parvalbumin is probably the most rigid of the EF-hand proteins. The differences in mobility between parvalbumin and other EF-hand proteins occur mainly at the linker connecting the pair of EF hands and also at the C terminus of the last helix. In Ca^{2+}-parvalbumin, these two regions are characterized by a pronounced rigidity compared to the corresponding more mobile regions in Ca^{2+}-calbindin and Ca^{2+}-calmodulin (Baldellon *et al.*, 1998).

The influence of hydration on the internal dynamics of parvalbumin was investigated in the work of Zanotti *et al.* (1999) by incoherent quasi-elastic neutron scattering (IQNS) and solid-state ^{13}C-NMR spectroscopy using the powdered protein at different hydration levels. Both methods show an increase in protein dynamics upon progressive hydration above a threshold that only corresponds to partial coverage of the protein surface by the water molecules. Selective motions are detected by NMR in the 10-ns time scale for the polar lysyl side chains (externally located), as well as for more internally located side chains (from Ala and Ile). At the same time, IQNS monitors diffusive motions of hydrogen atoms in the protein at time scales up to 20 ps. The combined results suggest that peripheral water-protein interactions influence the protein dynamics in a global manner. There is a progressive induction of mobility at increasing hydration from the periphery toward the protein interior.

STRUCTURES OF RELATED EF-HAND CALCIUM BINDING PROTEINS

Several classes of the EF-hand containing calcium binding proteins are known at present. Three-dimensional structures of many of them were revealed. Let us compare structures of some of these proteins with the structure of parvalbumin.

Calbindin

Calbindin D_{9k} (vitamin D-dependent calcium binding protein) is a small, monomeric protein of the S100 family, predominantly found in the central nervous system (Szebenyi *et al.*, 1981; Dalgarno *et al.*, 1983; Szebenyi & Moffat, 1986). It is also expressed in the epithelial tissue and in kidney (Anderssen *et al.*, 1993). Calbindin D_{9k} seems to play an important role in Ca^{2+} homeostasis of the central nervous system.

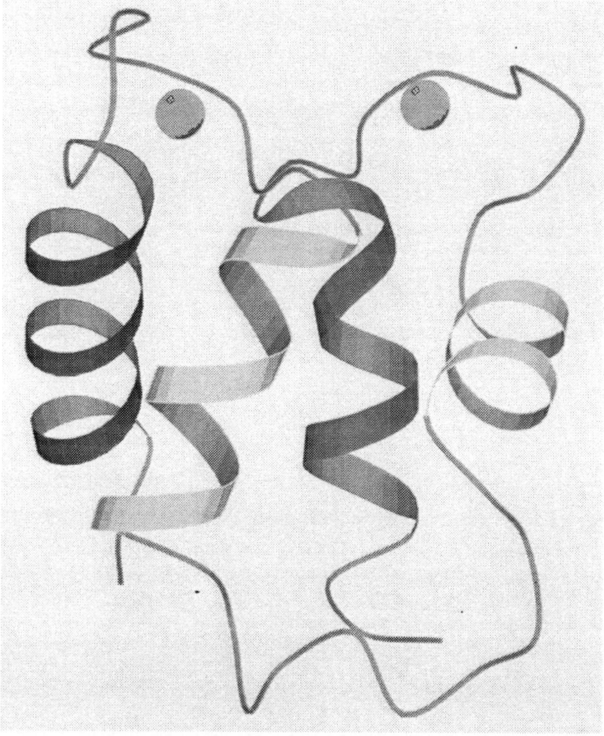

Figure 12. The structure of calbindin D_{9k} (PDB file pdb1b1g).

The molecule of calbindin D_{9k} consists of two calcium binding helix-loop-helix domains connected by a peptide containing one turn of α-helix (Figure 12). Interactions between helices are mostly hydrophobic and the protein possesses some very important hydrogen bonds especially within the calcium binding loops. The C-terminal domain of calbindin D_{9k} is a canonical EF-hand, with ion binding properties similar to those of the parvalbumin EF-hands, but the N-terminal domain is a variant hand (pseudo-EF-hand) whose calcium ligands are mostly peptide carbonyls. Both calcium ions in calbindin are roughly octahedrally coordinated, with protein oxygen atoms at the X, Y, Z, and -Y vertices, a bidentate carboxyl group at -Z, and a water molecule at -X. Laser excited europium luminescence studies of pig intestinal calbindin revealed a strong interaction between the two sites: the fluorescence from site II increases stoichiometrically with the addition not only the first equivalent of europium, but also concomitantly with the fluorescence from site I upon addition of the second equivalent (Hofmann *et al.*, 1988).

On the whole, the structure of calbindin resembles the structure of parvalbumin. Proton NMR data (Drakenberg *et al.*, 1989) show that the structure of this protein in solution corresponds well to its crystal structure. The only difference is a short turn of helix in the loop between helixes II and III in the crystal structure, which is absent in the protein in solution.

The importance of non-covalent interactions in determining the highly specific structure of calbindin D_{9K} was studied making hydrophobic core substitutions in the protein (Julenius *et al.*, 1998; Kragelund *et al.*, 1998). The mutations were found to exert large effects on protein stability, which were evident from positions of midpoins in the urea-induced unfolding varying from 1.8 M for Leu23→Gly up to 6.6 M for Val70→Leu mutations. Good correlation was found between the difference in free energy of unfolding and the change in the surface area of the side chain caused by the mutation. Both increases and decreases in the side chain surface area cause quantitatively equivalent effects on the stability.

S100 Proteins

S100 is a family of Ca^{2+}-modulated proteins of the EF-hand type expressed in vertebrates and implicated in intracellular and extracellular regulatory activities (reviewed by Donato, 2001). S100 proteins (now we know at least 16 members of this protein family), expressed in various cells and tissues, possess EF-hand calcium binding sites. Many functional roles were proposed for S100 proteins and several human disorders such as cancer, neurodegenerative diseases, cardiomyopathies, inflammations, diabetes, and allergies are associated with an altered expression of S100 proteins (reviewed, for example, by Heizmann & Cox, 1998). It is assumed that the function of each of the S100 proteins is associated with their abilities to bind calcium, to dimerize and to interact specifically with target proteins.

Within cells, S100 proteins have been implicated in the regulation of protein phosphorylation, some enzyme activities, the dynamics of cytoskeleton components, transcription factors, Ca^{2+} homeostasis, and cell proliferation and differentiation (reviewed by Donato, 2001). Certain S100 members are released into the extracellular space. Extracellular S100 proteins stimulate neuronal survival and/or differentiation and astrocyte proliferation, cause neuronal death *via* apoptosis, and stimulate or inhibit the activity of inflammatory cells.

The crystal structures of S100B (Matsumura *et al.*, 1998; Smith & Shaw, 1998), holo and apo-S100A6 (Otterbein *et al.*, 2002), S100A7 (Brodersen *et al.*, 1998), S100A8 (Ishikawa *et al.*, 2000), S100A10 (Rety *et al.*, 1999), S100A11 (Rety *et al.*, 2000), S100A12 (Moroz *et al.*, 2001), S100A9 (Itou *et al.*, 2002), S100P (Zhang *et al.*, 2003) and NMR structures of apo-S100A6 (Potts *et al.*, 1995), holo-S100A6 (Sastry *et al.*, 1998) and apo-S100B (Drohat *et al.*, 1996) were described. Despite the numerous putative functions of S100 proteins their three-dimensional strictures are similar.

S100 protein is characterized by the presence of two Ca^{2+} binding motifs of the EF-hand type interconnected by an intermediate region often referred to as the hinge region. The overall fold of the monomers of S100 proteins consists of four helices and three loops. These elements of secondary structure are arranged into two Ca^{2+}-binding motifs known as the S100-hand and the canonical EF-hand connected by a variable linker loop (Figure 13). The C-terminal calcium-binding domain contains the canonical EF-hand sequence and binds Ca^{2+}

with relatively high affinity, whereas the N-terminal S100-hand displays much lower calcium-binding affinity.

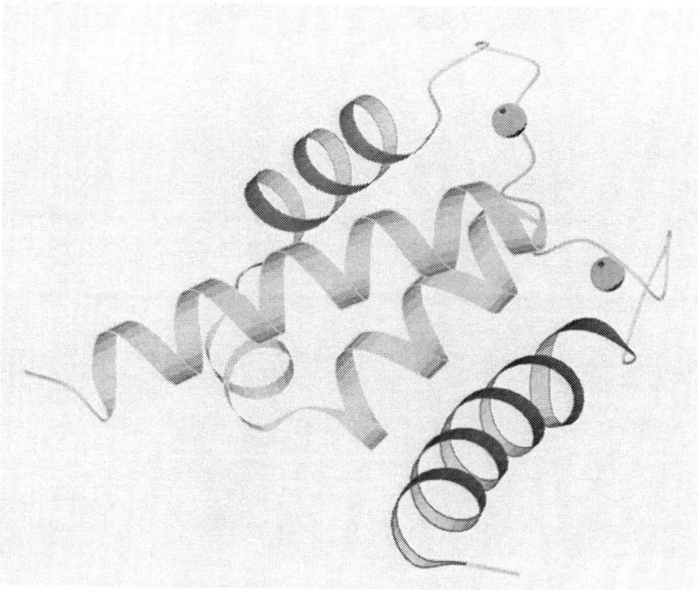

Figure 13. The structure of S100a protein (PDB file pdb1k96).

The N-terminal loop of each S100β subunit has two more amino acid residues than typical EF-hand domains (14 *vs* 12 residues). Furthermore, several of the calcium-liganding residues of this pseudo EF-hand do not conform to those of the EF-hand consensus sequence. It is of interest, that the N-terminal loop in S100A7 and both loops in S100A10 are unable to bind calcium (Brodersen *et al.*, 1998).

The coordination of Ca^{2+} in both sites follows a pentagonal bipiramidal arrangement. In S100A6 (Otterbein *et al.*, 2002), with the exception of the bidentate carboxyl group of Glu33 and water molecule, the coordination of Ca^{2+} in the S100-hand site primarily involves main chain carbonyls. In contrast, Ca^{2+} in the canonical EF-hand site is coordinated by side chains of Asp61, Asn63, Asp65, the bidentate carboxyl group of Glu72, a water molecule, and the main chain carbonyl of Glu67.

Each S100β subunit (91 residues) of Ca^{2+}-bound S100B(ββ) protein contains four helixes (residues 2-20; 29-38; 50-61; 70-83) and one antiparallel β-sheet, which holds the normal and pseudo EF-hands together (Drohat *et al.*, 1998). Helices 1, 1', 4, and 4' associate to form an X-type four-helix bundle at the symmetric dimer interface. The four helices within each S100β subunit form a splayed-type four-helix bundle (four perpendicular helixes) as observed in Ca^{2+}-bound calbindin D_{9k}.

S100P monomer structure contains four helices as well: residues 3-18, 30-40, 53-61, and 71-92 (Zhang *et al.*, 2003). Its two calcium-binding loops contain two short, antiparallel β-strands formed by residues 27-28 and 68-69. Hydrogen bonds between the β-strands provide tight interaction between the loops of the two EF-hands. A hydrophobic core in S100P is formed by three phenylalanine residues Phe15, Phe71 and Phe74 and this structure is conserved in other S100 proteins. In S100P the hydrophobic interaction is extended into helix 1 by residues Ile11, Ile12 and Val14, into helix 2 by Leu36, into loop 1 by Leu28, and into loop 2 by Val69. The hydrophobic interaction is strengthened by other bulky hydrophobic residues. The S100P dimer is formed by hydrophobic interactions between large hydrophobic areas along the N-terminal helix 1 and the C-terminal helix 4 through anti-parallel packing. Hydrogen bonds between hydrophylic residues strengthen the interaction between subunits further.

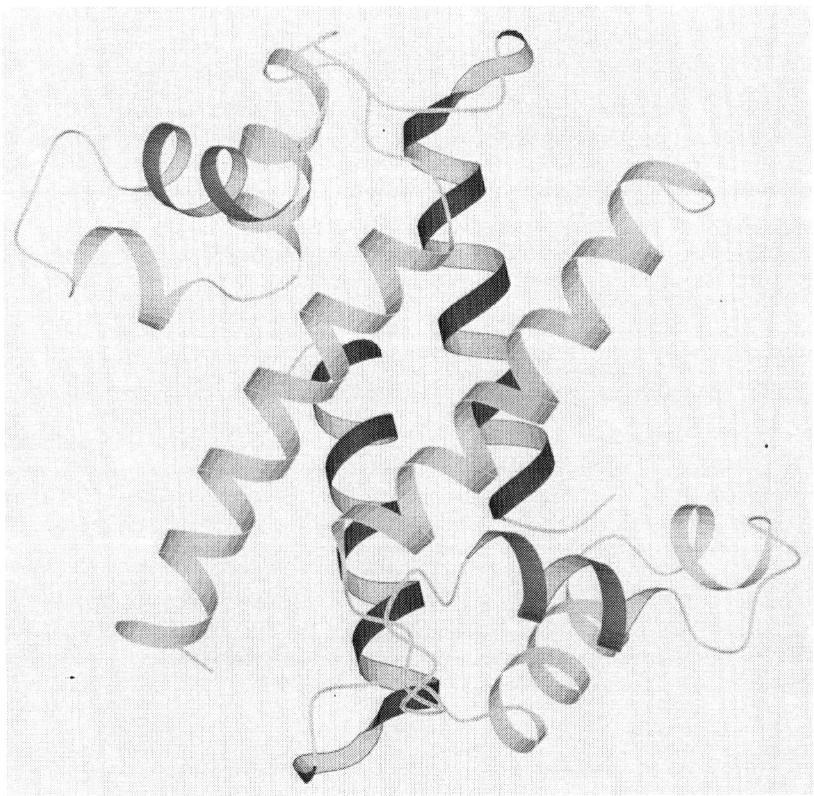

Figure 14. The structure of S100A10 dimer (PDB file 1BT6).

Most of S100 proteins exist as homodimers in which the two monomers are related by a two-fold axis of rotation and are held together by non-covalent bonds (Figure 14) (reviewed by Donato, 2001). This was shown for S100A4, S100A6, S100A7, S100A8, S100A10, S100A11, S100A12 and S100B by NMR spectroscopy and X-ray crystallography. It is reasonable to anticipate that most of S100 proteins form homodimers. One important exception is calbindin D_{9k} which definitely is monomeric. Some S100 proteins form heterodimers, as is the case with the S100A1/S100B, the S100A8/S100A9, the S100B/S100A6, the S100A1/S100A4, and the S100B/S100A11 dimers. Dimerization of S100 proteins seems to be important for their biological function. Upon Ca^{2+} binding, each S100 monomer opens up to accommodate a target protein and the S100 dimer can bind target proteins on opposite sides.

Some S100 proteins bind Zn^{2+} ions with a relatively high affinity and the Ca^{2+} and Zn^{2+} binding sites do not coinside with each other (reviewed by Donato, 2001).

Calmodulin

Calmodulin is present in all eucaryotic cells. This protein activates many enzyme systems and regulates many cellular functions, from the release of neurotransmitters to cell proliferation and DNA repair. Performance of physiological function of calmodulin requires its interactions with many proteins. Most of the interactions occur only in the presence of Ca^{2+} ions, but some of them take place only in the absence of Ca^{2+} ions (Jurado *et al.*, 1999).

Calmodulin is characterized by rather unusual structure. Calmodulin molecule (from rat testis, for example) resembles a dumbbell: two globular lobes connected by a long eight-turn α-helix (Babu *et al.*, 1985; Kretsinger, 1986; Wilson & Brunger, 2000) (Figure 15). The length of the molecule is 65 Å, each globular part has dimensions 25×20×20 Å. Calmodulin molecule contains seven α-helices (from I to VII) comprising residues 7-19, 29-39, 46-55, 65-92, 102-112, 119-128, and 138-148 (α-helical content is 63%). Calmodulin possesses four calcium-binding sites. The calcium binding regions 1 and 4 have typical helix-loop-helix conformation and are similar to the two calcium-binding regions in parvalbumin. Regions 2 and 3 also have the arrangement helix-loop-helix, nevertheless they differ from the calcium binding regions of parvalbumin by the common helix IV. Helix IV is two to three times longer than all the other helices in calmodulin. Calcium binding segments are composed of the residues 20-31, 56-67, 93-104 and 129-140. The molecule is stabilized by multiple interactions between helices and hydrogen bonds between adjacent Ca^{2+}-binding loops.

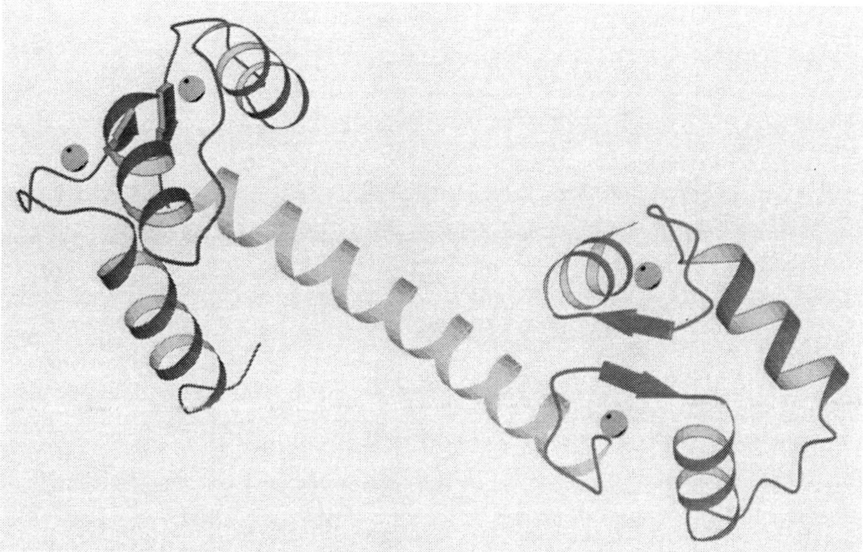

Figure 15. The structure of calmodulin (PDB file 1clm).

Fallon & Quiocho (2003) reported the 1.7 Å X-ray structure of native brain calmodulin that is in a compact ellipsoidal conformation and shows a sharp bend in the linker helix and a more contracted N-terminal domain. They suggested that this conformation may offer advantages for recognition of kinase-type calmodulin targets or small organic molecule drugs.

It is of interest that the forth Ca^{2+}-binding domain of yeast calmodulin is unable to bind Ca^{2+} (Matsuura *et al.*, 1991). Matsuura *et al.* (1993) created chimeric proteins of the yeast and the vertebrate calmodulins: one of their mutants consisted of Ala1 - Ile130 of chicken calmodulin and Asp131 - Lys148 of yeast calmodulin. The mutant showed the yeast-type properties, and its enzyme activation profiles were similar to those of yeast calmodulin. A single substitution of Glu for Gln140 was made in the mutant protein: the resulting mutant bound 4 mol of Ca^{2+} per mol protein and showed the vertebrate-type of enzyme activation. Therefore, alternation of three residues in the Ca^{2+}-binding loop of the yeast-type domain, substitution Ser129 → Asp, insertion of Ile130, and substitution Gln140 → Glu, were enough for the recovery of the Ca^{2+} binding and enzyme activation.

The crystal structure of Ca^{2+}-calmodulin with resolution of 1.0 Å (Wilson & Brunger, 2000) shows that the protein exhibits both anisotropic and anharmonic disorder that spans a range of length-scales, from discrete side chain disorder to domain displacements. The 36 discretely disordered residues are concentrated in the central helix and the hydrophobic pockets, supporting the view that structural plasticity in these regions of the protein is important for target recognition and binding. The manifestations of disorder in Ca^{2+}-calmodulin demonstrate that the protein is in equilibrium between a large set of conformational sub-states even in the crystalline environment at cryogenic temperatures.

Essential individual motions of the two globular domains in calmodulin were revealed (Gryczynski *et al.*, 1988). Computer simulation of molecular dynamics of calmodulin in solution (Wriggers *et al.*, 1998) showed that during the 3-ns simulation, the structure exhibits large conformational changes in the nanosecond time scale. The central α-helix, which was shown to unwind locally upon binding of calmodulin to target proteins, bends and unwinds near residue Arg74, while the two calcium binding domains reorient with respect to each other and α-helices in the N-terminal domain rearrange to make the hydrophobic target peptide binding site more accessible. It was interpreted as a preparative step in the more extensive structural transition observed in the “flexible linker” region 74-82 of the central helix upon complex formation. At the same time, in all molecular dynamics simulations of fully solvated Ca^{2+} saturated calmodulin mutant D129N the four Ca^{2+} ions remained in their binding sites and retained a single water ligand as observed in the crystal structure (Likic *et al.*, 2003). There were found two well defined conformations of the calcium binding loop II, whereas similar effect was not observed for loops I, III, and IV.

Troponin C

Troponin C is a Ca^{2+}-binding subunit of the troponin complex. The binding of Ca^{2+} to troponin C triggers a change in the disposition of the troponin-tropomyosin complex, which may either remove a steric hindrance to the approach of the myosin head to actin, or may allow the release of inorganic phosphate from the actin-myosin head-ADP.P_i complex.

Molecule of troponin C, the closest relative of calmodulin, has a similar dumbbell shape (Figure 16). At present we know the structure of troponin C from various sources (for example, Herzberg & James, 1985; Satyshur *et al.*, 1988). The molecule of turkey skeletal muscle troponin C is 75 Å long with 67% of the amino acid residues in helical conformation. Two globular parts of the molecule are separated by nine-turn helix with no direct intramolecular contacts between the domains.

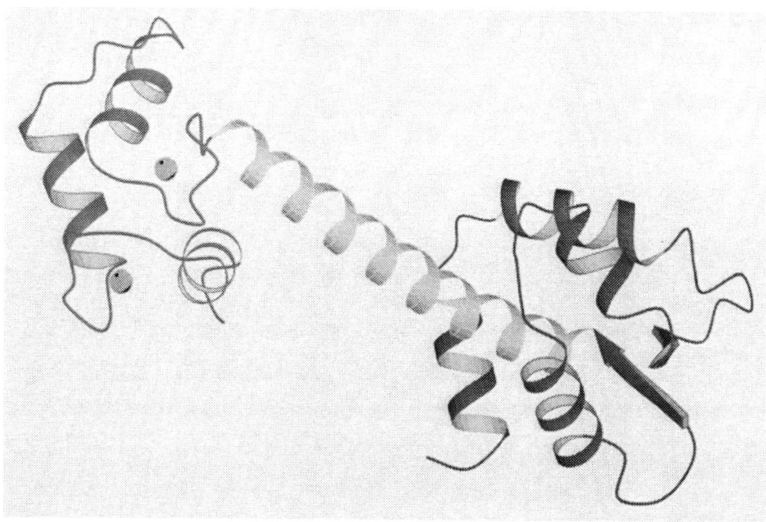

Figure 16. The structure of troponin C (PDB file pdb1top).

Like calmodulin, troponin C has four Ca^{2+}-binding sites: two in each globular part. Although the crystals were obtained from solutions containing excess Ca^{2+} ions sufficient to fill all four Ca^{2+}-binding sites, only the two high affinity sites of the C-terminal domain (III and IV) were occupied by Ca^{2+}. The following residues may be calcium ligands in troponin C: site I, Asp30, Asp32, Gly34 (carbonyl oxygen or H_2O), Asp36, Ser38, Glu41; site II, Asp66, Asp68, Ser70, Thr72 (carbonyl oxygen), Asp74, Glu77; site III, Asp106, Asn108, Asn110, Phe112 (carbonyl oxygen), Asp114, Glu117; site IV, Asp142, Asn144, Asp146, Arg148 (carbonyl oxygen), Asp150, Glu153.

The four helices A-D span the following residues: A, Glu16 - Phe29; B, Lys40 - Met48; C, Lys55 - Asp66; D, Phe75 - Gln85. Loops I and II correspond to the empty low-affinity Ca^{2+}-binding sites. A long helix spanning residues Phe-75 - Asp-106 connects the globular domains. The helix is slightly bent, such that the angle between the D and the E helix axes is about 10°. The C-domain contains four helices: E, Glu96 - Asp106; F, Ile115 - Thr125; G, Glu131 - Asp142; H, Phe151 - Gln159. Loops III and IV are occupied by Ca^{2+}. The helices in the C-domain are arranged in a manner closely similar to the CD EF region of parvalbumin. The arrangement of the helices in the N-domain is different: they tend to be anti parallel, forming a non-ideal up-and-down anti-parallel helix bundle.

Troponin C from skeletal muscles possesses four calcium binding sites while in cardiac troponin C the first calcium binding site is inactive. Putkey *et al.* (1989, 1991) tried to activate the first Ca^{2+}-binding site in cardiac troponin C by deletion of Val28 and conversion of amino acids 29-32 to those found at the similar positions in the active site I of fast skeletal troponin C. At the same time in a series of mutant proteins Ca^{2+}-binding site II was inactivated by mutation of amino acids Asp65, Asp67, and Gly70. All mutated proteins exhibited the predicted calcium-binding characteristics. The single mutation of converting Asp65 to Ala was sufficient to inactivate site II.

The structure of Ca^{2+}-loaded troponin C from cardiac muscles was determined by nuclear magnetic resonance spectroscopy (Sia *et al.*, 1997). The overall solution structure of cardiac troponin C, like the solution structure of skeletal troponin C, resembles a dumbbell in shape, which is not a surprise. It is of interest however that the regulatory N-domain of Ca^{2+}-loaded cardiac troponin C is significantly more compact than the N-domain of Ca^{2+}-loaded skeletal troponin C. In particular, the B-helix of defunct site I exists in the "closed" conformation, exibiting an A-B interhelical angle of 142°.

Several works were carried out to study effects of central helix on the structural and functional properties of troponin C. Five mutations including deletion in the D/E linker region designed to change the length of the central helix and the orientation of the Ca^{2+}-binding domains relative to each other resulted in proteins, which were sensitive to Ca^{2+} and did not extensively alter the thermal transition of troponin C (Dobrovolski *et al.*, 1991). All mutants demonstrated Ca^{2+}-dependent interactions with troponin I and troponin T. All similar D/E linker deletion mutants of chicken troponin C, except one (ΔKG) equally restored force development and Ca^{2+} regulation to troponin C-depleted skinned muscle fibers (Sheng *et al.*, 1991). The 2-residue deletion in the central helix (ΔKG), which would be expected to produce a 160° rotation in the α-helix, significantly affected troponin C activity therefore it was suggested that the change in orientation of the two Ca^{2+}-binding domains appears to be a major parameter affecting troponin C activity. All troponin C mutants with insertions in the central helix, resulted in elongation of molecule compared to wild type troponin C as determined by Stokes' radius, were defective in the activation of the regulated actomyosin ATPase activity in the presence of Ca^{2+} when compared to the wild type protein (Ramakrishnan & Hitchcock-DeGregori, 1995). The native length and structure of the central helix are evidently optimal for normal regulatory function and connectivity alone is insufficient for troponin C function.

The long helix connecting two globular lobes and fully exposed to the solvent is the most unusual feature of the calmodulin and troponin C structures. The central part of the helix is stabilized by salt bridges (Sundaralingam *et al.*, 1985). It is assumed that the exposed helix, a rare feature in protein structures, tottering on the edge of conformational stability, has exactly the required sensitivity to respond to energy changes produced by the calcium binding (Schutt, 1985). The bend in the long helix in troponin C occurs at Gly92. Calmodulin has no Gly in this position.

Later on, the bend in the central helix of the complexed troponin C was found experimentally. The crystal structure of troponin C in complex with the N-terminal fragment of troponin I (residues 1 to 47) was determined (Vassilyev *et al.*, 1998). It turned out that the long central connecting α-helix observed in the structure of uncomplexed troponin C is unwound at the center (residues Ala87, Lys88, Gly89, Lys90, and Ser91) and bent by 90° in the complex. As a result, troponin C molecule in the complex has a compact globular conformation allowing direct interactions between the N- and C-terminal lobes.

According to the data obtained by 3D and 4D heteronuclear NMR spectroscopic techniques, the structures of N-terminal and C-terminal domains of chicken troponin C in solution are highly converged, however, the orientation of one domain with respect to the other is not well-defined, and thus each domain appears to be structurally independent (Slupsky & Sykes, 1995). It is of interest that mutations in the N and D-helices of the N-terminal domain of troponin C affect the C-terminal domain: mutations of Arg11 (N-helix) and Glu76 (D-helix), which form a salt bridge, to Cys reduce stability of the N-terminal domain in the absence of divalent cations, increase the calcium affinity and reduce the cooperativity of the Ca^{2+},Mg^{2+}-sites in the C-terminal domain; moreover they alter the Ca^{2+}-induced conformational change in the C-terminal domain. It shows that the N-helix and C-terminal domain are physically close to each other in solution (Smith *et al.*, 1999). Measurements of the temperature stability of the single tryptophan mutants of chicken skeletal troponin C reveals as well that events occurring in the N-terminal domain affect stability of the C-terminal domain and vice versa, which strongly suggests that there are interactions between the N- and C-terminal domains of throponin C (Moncrieffe *et al.*, 1999).

Heidorn & Trewhella (1988) and Hubbard *et al.* (1988) compared the structures of calmodulin and troponin C in crystal and in solution. Their results obtained by small angle X-ray scattering show that despite of the similarity in dimension and shape of the globular domains in troponin C and calmodulin in solution and in crystal, the distance between them in solution is several Angstroms longer than that in crystal. The approach of the globular domains to each other requires some structural changes in the long connecting helix.

It is of interest to analyze structural and functional correlations between troponin C and calmodulin (Gulati *et al.*, 1993). Compared with troponin C, calmodulin lacks the N-terminal α-helical arm in the N-helix, and its central helix is shorter due to the absence of 88 to 90 residues. Deleting both these regions in troponin C concomitantly lead to the appearance of calmodulin-like regulation as tested by smooth muscle contractivity and by the activation of phosphodiesterase. Surprisingly, the Ca^{2+}-binding capacity of the mutant and the effect on maximally calcium-activated tension of skinned rabbit *psoas* muscle fibers were both conserved. When the troponin C-characteristic region ^{85}EDAKG^{90} was replaced by calmodulin-specific DTD residues, it generated even more effective calmodulin mimic whether or not the N-helix was also retained. In continuation of this work the authors (Gulati *et al.*, 1995) mutated Arg11 in rabbit skeletal troponin C to Ala because the interactions of Arg11 with distal residues in the N-terminal domain seem to link the N-terminal helix to the rest of the structure. The mutant exhibited calmodulin-like function in its ability to activate phosphodiesterase (50% of calmodulin). If, in addition, the KGK triplet (residues 91-93) was also deleted, phosphodiesterase activation increased up to about 80%. Both constructs retain their troponin C function to nearly 100%. Computer simulations of the molecular dynamics of 4Ca-calmodulin, 4Ca-troponin C and 4Ca-R11A mutant showed two types of structural changes, which are considered to have a functional role in calmodulin: (i) a compaction to a more globular form and (ii) a reorientation of the calcium-binding domains around the central connecting helix. Both the wild type and the R11A mutant were compacted, but differently, so that reorientation of the N- and C-terminal domains was found only in the simulated structure of the mutant.

An analysis of the tertiary structures of parvalbumin, troponin C and calbindin carried out by Herzberg & James (1985) showed that the two calcium binding loops occupied by Ca^{2+} in troponin C have conformation similar to that of the two homologous loops in parvalbumin and the loop III-IV in calbindin. This specific arrangement guarantees the proper location of the calcium ligands. It consists of two reversive turns on each side of the loop and four Asx turns (cyclic hydrogen-bonded structure including oxygen of the side chain of residue *n* and amide nitrogen of the main chain of residue *n+2*). The fourth coordination position in both loops of troponin C is occupied by a water molecule, which is located within the distance of a hydrogen bond from aspartic acid and it results in indirect interaction between the cation and negatively charged carboxylate. The same framework is characteristic for the loops in parvalbumin and calbindin in spite of the variability in nature of side chains in equivalent positions. The locations of Ca^{2+} and coordinating water molecule are conserved in all these cases.

Recoverin

One more protein with four potential EF-hand calcium-binding sites is recoverin (Dizhur *et al.*, 1991; Stryer, 1991). Recoverin prolongs the lifetime of photoexcited rhodopsin by inhibiting rhodopsin kinase only at high Ca^{2+} levels. Hence, recoverin makes the desensitization of rhodopsin sensitive to Ca^{2+}, and the shortened lifetime of photoexcited rhodopsin at low Ca^{2+} levels may promote visual recovery and contribute to the adaptation to background light. Recoverin (23 kDa) is present in both rod and cone photoreceptors and was named recoverin since it promotes recovery of the dark state.

The amino acid sequence of recoverin exhibits four potential calcium binding sites characteristic of the EF-hand superfamily. An interesting feature of this protein is that it is N-myristoylated. Upon binding of Ca^{2+}, recoverin undergoes a conformational change that allows it to associate with membranes in a manner that requires N-myristoyl translocation. NMR data indicate that the myristoyl group is in contact with residues in the hydrophobic claster in Ca^{2+}-free recoverin and that it is exposed to solution in the Ca^{2+}-bound conformation (Hughes *et al.*, 1995; Ames *et al.*, 1995).

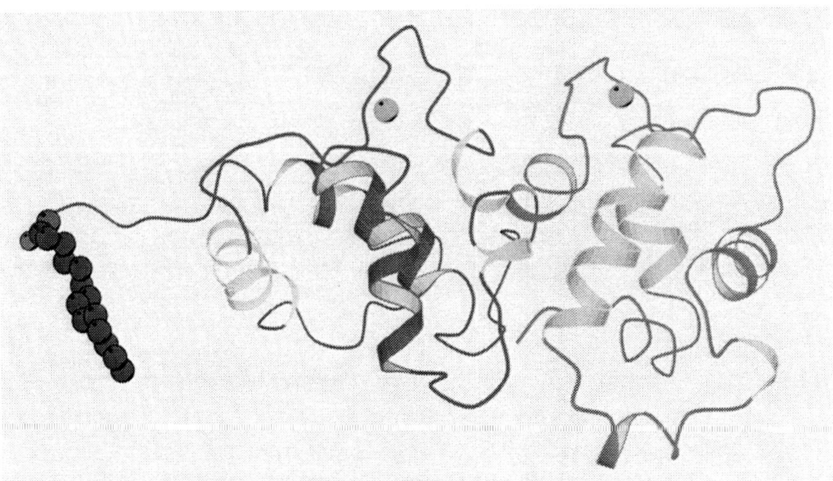

Figure 17. The structure of recoverin (PDB file 1jsa).

The three-dimensional structures of the myristoylated and unmyristoylated forms of recoverin in the Ca^{2+}-free and Ca^{2+}-bound states have been determined by X-ray crystallography (Flaherty *et al.*, 1993) and NMR spectroscopy (Ames *et al.*, 1994, 1997; Tanaka *et al.*, 1995, 1998). The four EF hands of the protein are arranged in a compact globule that contrasts with the dumbbell shape of calmodulin and troponin C (Figure 17). Only EF hands 2 and 3 can bind calcium ions, while the other two EF hands have structural features that prevent calcium binding. The EF-hand 1 is distorted from a favorable Ca^{2+}-binding geometry by Cys39 and Pro40, and the EF-hand 4 contains a salt bridge between Lys-161 and Glu-171. The myristoyl group is in a slightly bent conformation. Hydrophobic groups of Leu28, Trp31, and Tyr32 form a claster that interacts with the front end of the myristoyl (C1-C8), whereas residues Phe49, Phe56, Tyr86, Val87, and Leu90 interact with the tail end (C9-C14) (Tanaka *et al.*, 1998).

Chapter 3

"PHYSIOLOGY" OF PARVALBUMIN

CALCIUM BINDING MECHANISM

It is well known now that parvalbumin molecule has two strong Ca^{2+}, Mg^{2+} binding sites. The most general scheme of the binding of metal cation Me^{2+} to a protein P with two binding sites looks like this:

$$\begin{array}{ccccc} & \nearrow & MeP & \searrow & \\ K_{11} & & & & K_{22} \\ P & & & & MePMe \\ K_{12} & & & & K_{21} \\ & \searrow & PMe & \nearrow & \end{array} \qquad (1)$$

where K_{11}, K_{12}, K_{22} and K_{21} are equilibrium binding constants for the two binding sites:

$$K_{11}=[MeP]/[P][Me] \qquad (2)$$

$$K_{22}=[MePMe]/[MeP][Me] \qquad (3)$$

$$K_{12}=[PMe]/[P][Me] \qquad (4)$$

$$K_{21}=[MePMe]/[PMe][Me] \qquad (5)$$

Concentrations of the reagents are related with each other by the equations of material balance:

$$[P]+[MeP]+[PMe]+[MePMe]=P_t \qquad (6)$$

$$[Me]+ [MeP]+[PMe]+2[MePMe]=Me_t, \qquad (7)$$

where P_t and Me_t are respectively protein and metal cation total concentrations.

The equilibrium binding constants are related with each other by an equation:

$$K_{11}K_{22}=K_{21}K_{12}\, . \qquad (8)$$

If $K_{11}=K_{21}=K_1$, then $K_{12}=K_{22}=K_2$. It means that the binding sites are independent and the binding equilibrium in each site occurs independently from the binding equilibrium in the other site.

If $K_{11}>>K_{12}$, then $K_{21}>>K_{22}$. In this case the general scheme reduces to the scheme of sequential binding:

$$P \overset{K_{11}}{\rightleftharpoons} MeP \overset{K_{22}}{\rightleftharpoons} MePMe \qquad (9)$$

In this scheme the second site cannot be filled before the first one.

If K_{11} is comparable with K_{12}, then K_{21} is comparable with K_{22}. In this case we usually measure apparent binding constants:

$$K_{1app}=([MeP]+[PMe])/[P][Me]=K_{11}+K_{22} \qquad (10)$$

$$K_{2app}=[MePMe]/\ ([MeP]+[PMe])[Me]=K_{11}K_{22}/(K_{11}+K_{12}) \qquad (11)$$

Most intact and recombinant pavalbumins bind two calcium ions per molecule with high affinity, while human parvalbumin mutants containing inactivating substitutions in the calcium binding loops (E101V, D90A, E62V, D51A, D90A/E101V, E62V/E101V, D51A/D90A, D51A/E62V, D51A/E62V, D90A/E101V) bind from 0 to 1 Ca^{2+} per molecule, which was determined by electrospray ionization mass spectrometry (Troxler *et al.*, 1999). Only alligator parvalbumin isoform pI 5.0 has atypically high Ca^{2+} content (3 Ca^{2+} per molecule) (Laney *et al.*, 1997).

According to NMR data (Nelson *et al.*, 1976; Opella *et al.*, 1974), the removal of Ca^{2+} from the EF-site of carp parvalbumin pI 4.25 shifts the carbonyl resonance of Lys96 (Lys96 takes part in the coordination of Ca^{2+} in the EF-site). At the same time, it does not change the resonance of Phe57 (Phe57 takes part in the coordination of Ca^{2+} in the CD-site). Such behavior of the resonances of Lys96 and Phe57 is considered as an evidence of selective removal of Ca^{2+} from the EF-site. Similar results were obtained in the NMR experiments of Birdsall *et al.* (1979): resonances of His26 (located near the EF-site) and resonances of para-H of Phe24 and Phe29 are sensitive only to the second step of the substitution of Ca^{2+} by paramagnetic cations.

The results of studies of the interaction of parvalbumin with lanthanide ions, such as praseodyum and ytterbium, are also in favor of sequential filling of the binding sites (Lee & Sykes, 1980a-c, 1981, 1983; Lee *et al.*, 1979; Breen *et al.*, 1985a). When the Yb^{3+}:protein molar ratio is in the region from 0 to 1, the proton NMR spectrum of carp parvalbumin pI 4.25 contains one set of the lanthanide-shifted resonances. An increase in the Yb^{3+}:protein ratio from 1 to 2 causes disappearance of some of the resonances corresponding to the nuclei located near the both sites and appearance of some new resonances corresponding to the nuclei located near the second site. At the same time, some of the resonances, which seem to correspond to the nuclei located near the first site, remain unchanged.

It was found (Donato & Martin, 1974) that one of the two Ca^{2+} ions in carp parvalbumin pI 4.25 may be removed by dialysis against EGTA without significant change of the far UV circular dichroism spectrum, which suggests a sequential scheme of the binding of calcium ions.

Intrinsic tryptophan fluorescence data for whiting and cod parvalbumins reveal the existence of an intermediate corresponding to the one-calcium state in the course of spectrofluorimetric Ca^{2+}-titration experiment (Permyakov *et al.*, 1980).

Reid & Hodges (1980) suggested that the cation binding sites in parvalbumin and other EF-hand calcium binding proteins have no preformed cavities such as those found in the macrocyclic organic chelating agents and antibiotic ion carriers and that this cavity is formed during the stepwise substitution of the solvent molecules on the cation. They assumed that calcium and magnesium binding to the helix-loop-helix domain is controlled by amino acid residue side chains in the loop region of this arrangement. This control is achieved through amino acid – cation interaction during ion pair formation, cation dehydration and substitution of water by the dentates. It is through control of the cation hydration shell that cooperativity exists between such sites. The different calcium and magnesium binding constants of apparently similar helix-loop-helix regions are explained by differences in the residues involved in the hydration control mechanism and differences in the side chain requirements of optimal calcium or magnesium chelation.

Thus, the results of many authors obtained by various methods suggest the scheme of sequential binding of Ca^{2+} to parvalbumin:

$$P \leftrightarrow PCa \leftrightarrow CaPCa \qquad (12)$$

An analysis of the shape of tryptophan fluorescence spectra of whiting parvalbumin shows that in spite of the fact that this parvalbumin contains only one tryptophan residue per molecule, the widths of the tryptophan fluorescence spectra of its P, PCa, and CaPCa states are larger than the widths of the spectra of free tryptophan and its derivatives with corresponding maxima positions (Permyakov *et al.*, 1980, 1982). This suggests that though this protein contains only one tryptophan residue per molecule, its fluorescence spectrum consists of at least two components. Fluorescence decay curves of whiting parvalbumin in all states studied are approximated not by a single exponential but by a sum of two exponentials with different lifetimes (Permyakov *et al.*, 1985). Measurements of the decay curves at various wavelengths allow decomposition of the steady-state fluorescence spectrum into components with different lifetimes. The steady-state fluorescence spectrum of parvalbumin with two bound calcium ions consists of two emission components with different maximum positions. The spectrum of the protein with one bound calcium ion contains almost the same components, but the measurements in this case are complicated by the presence in the total emission spectrum of contributions of not only the protein with one bound calcium, but of its another states as well. The spectrum of apo-parvalbumin consists of two red-shifted components.

It is reasonable to identify the observed spectral components with emission from two parvalbumin conformers P and P^* with different environments of the single tryptophan residue. If we take into account an equilibrium between the conformers, the scheme (12) will be transformed into the following scheme:

$$\begin{array}{ccccc} P & \leftrightarrow & PCa & \leftrightarrow & CaPCa \\ \uparrow\downarrow & & \uparrow\downarrow & & \uparrow\downarrow \\ P^* & \leftrightarrow & P^*Ca & \leftrightarrow & CaP^*Ca \end{array} \quad (13)$$

Correctness of such interpretation is confirmed by the fact that iodide I^- ions quench the fluorescence only of the conformer with the shorter lifetime, making the decay curve single-exponential. Substitution of H_2O by D_2O disturbs the equilibrium between the conformers, shifting it toward the conformer with the shorter wavelength position of the fluorescence spectrum.

Later Eftink & Wasylewski (1989) tried to fit the fluorescence decay curve for cod parvalbumin also by two exponents and by a set of exponents with a distribution of lifetimes. The model with the lifetime distribution gives better fit to the decay curve and to the data on acrylamide quenching for Ca^{2+}-loaded parvalbumin. Hutnik *et al.* (1990a,b) found as well that the fluorescence decay of the calcium-loaded native cod parvalbumin III is best described by two decay time components. Three lifetime components are necessary to describe the fluorescence decay of the metal-free protein. It was interpreted as a reflection of the existence of the tryptophan residue in at least two conformational states in the presence of Ca^{2+} and in at least three conformational states in its absence. Ferreira (1989) also found that fluorescence decay of the single tryptophan in whiting parvalbumin is nonexponential in both Ca^{2+}-free and Ca^{2+}-loaded state. He described the fluorescence decay by Lorentzian lifetime distributions centered around two components: a major long-lived component at 2-5 ns and a small sub-nanosecond component. An increase in temperature from 8 to 45°C results in a decrease in both the center and width of the major lifetime distribution component and an increase in these parameters for the small lifetime distribution component. Direct anisotropy decay measurements of local tryptophan rotations yielded an activation energy of 2.3 kcal/mol for Ca^{2+}-free parvalbumin and indicated a correlation between rotational rates and lifetime distribution parameters. The binding of calcium ions results in a decrease in the width of the major lifetime distribution component and a decrease in tryptophan rotational mobility within the protein.

Surprisingly, mutant rat F102W parvalbumin demonstrates monoexponential fluorescence intensity and anisotropy decay kinetics of its single tryptophan residue (Feinstein *et al.*, 2003).

Protein isoforms were found also for mutant oncomodulin with a single tryptophan residue as well. Moreover, it was found that in solution calbindin D_{9K} exists as an equilibrium mixture of isoforms with *trans* (75%) and *cys* (25%) isomers of the peptide bond at Pro43 (Kordel *et al.*, 1989).

Many researchers studied the binding of Ca^{2+} to parvalbumin by equilibrium and flow dialysis methods using the Scatchard plots. Cox *et al.* (1977), Moeshler & Shaer (1979), Robertson & Kerric (1978), Lehky *et al.* (1974), Blum *et al.* (1977), and Heizmann & Strehler (1979) did not find any deviations from linearity in the Scatchard plots for parvalbumin. Therefore, these authors believe the binding of Ca^{2+} to parvalbumin is a non-cooperative process. It should be noted however that Scatchard plots often may not give unambiguous information concerning cooperativity of the ion binding. In particular, in the case of the sequential cooperative binding of two ions to a protein molecule, the Scatchard plots may be linear over a wide range of free ion concentrations and binding constants values. For this reason the linearity of the Scatchard plots can not be used as an unambiguous proof of the binding non-cooperativity.

CATION INDUCED CHANGES IN STRUCTURE

Experimental data obtained by various methods show that Ca^{2+}-induced changes in parvalbumins spread over the whole protein globule. Computer simulations showed that a new structure is obtained within approximately 60 ps after Ca^{2+} removal. As a whole, the parvalbumin globule in the cation-free form becomes less compact; accessibility of many of its groups to the solvent increases (reviewed by Permyakov, 1985, 1993).

At the same time, NMR data (Williams *et al.*, 1986a) show that the preferred solution conformation of rat apo-parvalbumin at 25°C is nearly identical with those of its metal-bound forms. The tertiary structure of this form seems entirely intact, several ^{1}H NMR resonances from Phe24, His26, Phe29, Phe47, His48, and Val106 serving as specific monitors of the integrity of the hydrophobic core. The secondary structural elements, six helical segments, are present in the apo-protein. The apparent content of α-helices in apo-parvalbumin is about 42% at 25°C (54% for the Ca^{2+}-loaded parvalbumin). However, back titration of Ca^{2+}-parvalbumin with EGTA at pH 8.1 leads to numerous NMR spectral changes (Birdsall *et al.*, 1979). Removal of both calcium ions results in increased conformational fluctuations within the molecule, with equilibrium between the decalcified and native protein forms taking place within the scale of the NMR experiment. NMR data indicate a slow conformational transition involving a reorganization of the fold of the protein towards a more loosely packed structure.

A circular dichroism spectrum reflects asymmetry of the optical center, and circular dichroism method has been exhaustively used to monitor changes in the secondary and tertiary structure of proteins and polypeptides. In the far-UV region, circular dichroism monitors transitions assigned to peptide chromophores sensitive to secondary structural changes, and in the near-UV region, circular dichroism reflects changes occurring in the aromatic chromophore environment. The ellipticity at 222 nm is commonly used as an α-helical determinant.

The far UV circular dichroism spectra show that removal of calcium from parvalbumin results in a decrease in the α-helical content of the protein. Most part of the α-helical content decrease occurs during the removal of Ca^{2+} from the CD-site (Eberspach *et al.*, 1988; Laberge *et al.*, 1997). At least 50% of the Ca^{2+}-bound α-helical structure is retained in the Ca^{2+}-free protein. According to the data of Donato & Martin (1974), the removal of calcium from carp parvalbumin pI 4.25 changes its helical content from 47% down to 39%.

Fourier transform infrared (FTIR) spectroscopy is now increasingly used as a probe of secondary structure especially in the amide stretch region. The amide I' vibrational region (1600-1700 cm^{-1}) is mostly monitored because protein absorption is usually intense within its range. It corresponds to C=O stretches weakly coupled to C-N stretches and N-H bending modes. In a deuterated solvent, the amide II' region (1500-1600 cm^{-1}) displays absorption assigned to C-N stretches strongly coupled to N-H bending. The amide III' region features

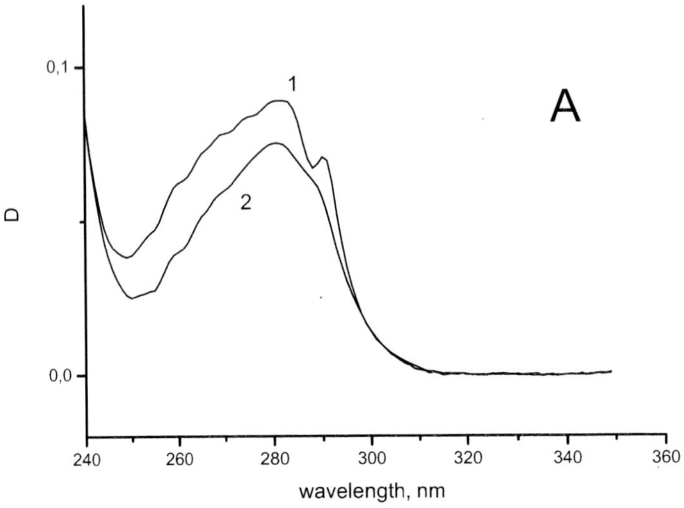

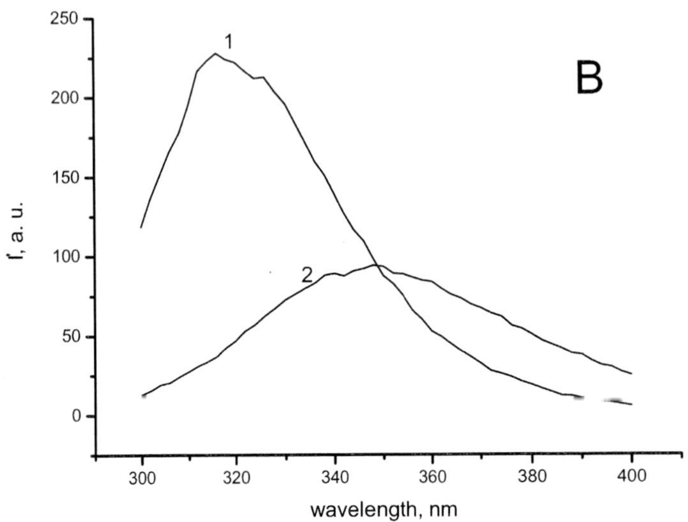

Figure 18. Absorption (A) and fluorescence (B) spectra (excitation at 280 nm) of cod parvalbumin in the Ca^{2+}-loaded (curves 1) and Ca^{2+}-free (curves 2) states.

The amide III' region features weaker absorption in the 1200-1350 cm^{-1} range, assigned to N-H in-plane bending modes coupled to C-N stretches and C-H/N-H deformation modes.

Calcium removal causes serious changes in FTIR spectrum of parvalbumin (Laberge *et al.*, 1997). The band in the amide I' region is drastically changed upon Ca^{2+} removal from parvalbumin, as is the band in the amide III' region. The positions of the bands in the amide regions depend strongly on the local environment; for instance, an anomalously high amide I frequency is usually attributed to a distorted α-helix. Infrared amide I' and III' regions for parvalbumin are very much affected by Ca^{2+} removal (disappearance of the shoulder at 1555 cm^{-1} assigned to the COO^{-} groups in bidentate ligation with calcium, and disappearance of two bands at 1334 and 1316 cm^{-1} associated with helical structure), which reflects significant perturbation of the secondary structure.

Ca^{2+}-loaded parvalbumin shows a remarkable stability against trypsin digestion (Cox *et al.*, 1979). Even at trypsin to parvalbumin ratios as high as 1:40, Ca^{2+}-loaded perch parvalbumin II is unaffected over a 24 hours incubation period. Removal of calcium leads to a marked susceptibility to proteolysis.

For some amino acid residues the calcium-induced changes in environment can be very pronounced. Figure 18 shows absorption and fluorescence spectra of the single tryptophan residue of cod parvalbumin in the presence and in the absence of Ca^{2+}. The huge blue shift of fluorescence spectrum and appearance of fine vibrational structure in the absorption spectrum upon Ca^{2+} binding reflect a decrease in accessibility of the tryptophan residue to water molecules and a creation of a rigid nonpolar environment around it. Gradual increase of Ca^{2+} concentration causes a pronounced shift of fluorescence spectrum of the single tryptophan residue in cod or whiting parvalbumins toward shorter wavelengths by more than 20 nm and an increase in fluorescence quantum yield (Figure 19, Permyakov *et al.*, 1980) which can be used for evaluation of calcium binding constants.

For Ca^{2+}-loaded cod parvalbumin the decay of the triplet state of tryptophan is exponential. In contrast, the triplet decay of the calcium-free protein is nonexponential over the time range of microseconds to milliseconds, a result that may indicate that the metal-free protein is molten-globule-like (Sudhakar *et at.*, 1993, 1995b). It was concluded that the photochemistry of tryptophan must take into account the existence of two excited triplet species and that there are quenching moieties within the protein matrix that decrease the phosphorescence yield in a dynamic manner for the Ca^{2+}-depleted parvalbumin. In contrast, for Ca^{2+}-loaded parvalbumin, the tryptophan environment is rigid on the time scale of milliseconds. It is worth noting that the tryptophan residue in these parvalbumins is located near the calcium binding sites (see Figure 8).

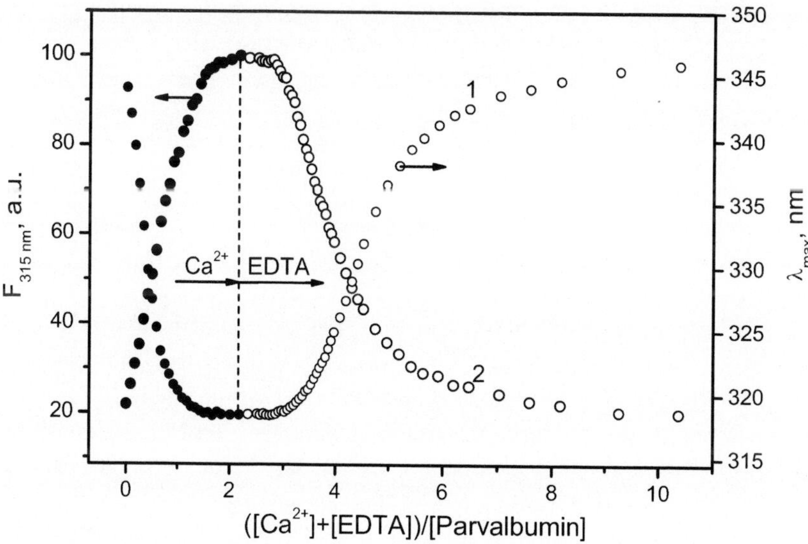

Figure 19. Spectrofluorimetric Ca^{2+}- and EDTA-titration of cod parvalbumin. Curve 1 – fluorescence maximum position; curve 2 – fluorescence intensity at 315 nm.

The UV photolysis of the aromatic amino acid, tryptophan (Trp), in cod parvalbumin, type III, was studied using electron paramagnetic resonance (EPR) spectroscopy in the temperature range 4-80 K (Angiolillo & Vanderkooi, 1996). For the Ca^{2+}-bound protein, irradiation with UV light (250-400 nm) resulted in the generation of atomic hydrogen with a hyperfine splitting of 50.9 mT, whereas in the Ca^{2+}-free form, where the Trp is exposed to solvent, the trapped atomic hydrogen was not found. In the same spectra, the radical signal in the g = 2.00 region could be detected. The line shape of the Ca^{2+}-bound form is similar to the EPR line shape obtained for Trp in micellar systems. In contrast, the EPR line shape for the Ca^{2+}-free form is essentially featureless up to 80 K.

Since native oncomodulin is devoid of tryptophan, site-specific mutagenesis was performed to create a mutant protein in which tryptophan was placed in the identical position (residue 102) as the single tryptophan residue in cod parvalbumin III (Hutnik *et al.*, 1990a). The results showed that in the region probed by tryptophan-102, cod parvalbumin III experiences significantly more pronounced changes in conformation upon decalcification compared to the oncomodulin mutant, F102W. Addition of 1 equivalent of Ca^{2+} produces greater than 90% of the total fluorescence response in F102W, while in cod parvalbumin III, only 74% of the total is observed. Cod parvalbumin III displays a negligible response upon Mg^{2+} addition. In contrast, F102W did respond to Mg^{2+}, but the response is considerably less when compared to Ca^{2+} addition. These results demonstrate that both cod parvalbumin III and oncomodulin undergo Ca^{2+}-specific conformational changes. However, oncomodulin is distinct from cod parvalbumin III in terms of the electronic environment of the hydrophobic core, the magnitude of the Ca^{2+}-induced conformational changes, and the number of calcium ions required to modulate the major conformational changes.

The purpose of the study of Pauls *et al.* (1993) was to construct α-parvalbumin with a unique Trp at position 102 in order to compare α- and β-parvalbumins including oncomodulin with respect to cation-dependent conformational changes. Tryptophan fluorescence emission and UV difference spectra of the mutant protein showed that the Trp residue at position 102 in it is confined to a hydrophobic core and conformationally strongly restricted: the fluorescence spectra of the metal-free and metal-bound forms of F102W showed an emission maximum at 313 nm for the Trp fluorescence and very high fluorescence quantum yield indicating that the Trp is buried in the hydrophobic core of the protein and there are probably no water molecules in its vicinity. There are very few natural proteins with such a blue-shifted Trp fluorescence, the best known being that of azurin whose emission maximum is at 308 nm (Burstein et al., 1977). It is of interest that upon Ca^{2+} or Mg^{2+} binding the structural organization of the region around the Trp in the mutant protein is hardly affected, but significant changes in its electrostatic environment take place (for example, a short distance displacement of Lys44 toward or away from the Trp). A comparison of the environment within a 6-Å radius of the Trp102 in F102W mutant showed that the nearest neighbor residues to Trp102 are very hydrophobic with the exception of the single Lys44. At the same time, in cod parvalbumin the 11 nearest neighbor residues to Trp102 are considerably less hydrophobic. The conformational change upon Ca^{2+} and Mg^{2+} binding in cod parvalbumin, as monitored by UV difference spectrophotometry, increases linearly from 0 to 2 cations bound, indicating that the binding of both ions contributes equally to the structural organization.

The removal of Ca^{2+} from parvalbumin changes the environment of its single tyrosine residue (if it contains Tyr), which is reflected in a decrease of tyrosine fluorescence quantum yield as was shown for carp parvalbumin pI 3.95 (Permyakov *et al.*, 1983c) and pike parvalbumin pI 4.2 (Permyakov *et al.*, 1983a; Eberspach *et al.*, 1988). It is of interest that this change occurs at nearly constant fluorescence decay time, which is indicative of mainly static quenching of the tyrosine fluorescence, probably by the non-coordinating carboxylate groups.

The accessibility of cystein residues in parvalbumin to thiol reagents strongly depends upon the binding of Ca^{2+}. Whereas the single cysteine group of Ca^{2+}-loaded perch parvalbumin II reacts sluggishly with the thiol reagent Nbs_2, the reaction is completed within 30 min when calcium is removed (Cox *et al.*, 1979). The binding of both Ca^{2+} and Mg^{2+} protects the cystein group.

Di- and trivalent members of the lanthanide series substitute for Ca^{2+} in the parvalbumin binding sites because they have similar binding characteristics and ligand specificities. The luminescent isomorphous Ca^{2+} analogue, Tb^{3+}, can be bound in the 12-amino acid metal binding sites of proteins of the EF hand family, including parvalbumin, and its luminescence can be enhanced by excitation energy transfer from a nearby aromatic amino acid. Tb^{3+} can be used as a sensitive luminescent probe of the structure and function of these proteins. Excitation in the ultraviolet spectral region of the aromatic amino acids phenylalanine, tyrosine, and tryptophan can lead to the radiationless transfer of excitation energy to a nearby Tb^{3+}, followed by luminescence of Tb^{3+}. The Tb^{3+} luminescence arises from transitions between the $^5D_4$ to $^7F_6$ states (490 nm) and the $^5D_4$ to $^7F_5$ states (545 nm). Factors affecting the quantum yield of Tb^{3+} luminescence in calcium binding protein chelates include spectral overlap of donor emission with Tb^{3+} absorption, donor-Tb^{3+} distance, and the efficiency of the chelator to exclude water from the coordination sphere of Tb^{3+}.

The effect of changing the molecular environment around Tb^{3+} on its luminescence was studied using native cod parvalbumin III and site directed mutants of both oncomodulin and calmodulin (Hogue *et al.*, 1992). Tryptophan in the binding loop position 7 best enhances Tb^{3+} luminescence in the oncomodulin mutant Y57W. Excitation spectra of Y57F, F102W, Y65W oncomodulin mutants and cod parvalbumin III revealed that the principal Tb^{3+} luminescence donor residues are phenylalanine or tyrosine located in position 7 of a loop, despite the presence of other nearby donors, including tryptophan. The spectra also revealed conformational differences between the Ca^{2+}- and Tb^{3+}-bound forms. An alternate binding loop, based on Tb^{3+} binding to model peptides, was inserted into the CD loop of oncomodulin by cassette mutagenesis. The order of filling of the binding sites by Tb^{3+} in this protein reversed, with the mutated loop filling by Tb^{3+} first. This indicates a much higher affinity for the consensus-based mutant loop.

The Eu^{3+} $^7F_0 \rightarrow ^5D_0$ excitation spectra of parvalbumin and oncomodulin are strongly pH-dependent (Henzl *et al.*, 1985). For example, the sharp doublet observed near 579.5 nm for fully calcium saturated pike parvalbumin at pH 6.0 is replaced at higher pH values by a broad spectrum having a maximum at 578.4 nm. The fact that the signals from both the CD and EF sites diminish concomitantly, with an apparent $pK_o = 8.2$, suggests that both sites exhibit this pH-dependent behavior. Nevertheless, results obtained with the site-specific variant of oncomodulin D59E necessitated substantial revision of these ideas (Trevino *et al.*, 1990). It appears that the pH-dependent behavior is actually confined to the CD site. Moreover, these authors observed no corresponding change in the number of O-H oscillators coordinated to the bound Eu^{3+} ions in the pH range over which they observed the spectroscopic changes. It is likely that such behavior results from deprotonation of one or more carboxyl groups clustered at the COOH-terminal end of the CD domain.

Excitation spectroscopy of the $^7F_0 \rightarrow ^5D_0$ transition of Eu^{3+} and diffusion-enhanced energy transfer were used to study metal-binding characteristics of parvalbumin from codfish (Cronce & Horrocks, 1992). Energy is transferred from Eu^{3+} ions occupying the CD- and EF-binding sites to the freely-diffusing Co(III) coordination complex energy acceptors: $[Co(NH_3)_6]^{3+}$, $[Co(NH_3)_5H_2O]^{3+}$, $[CoF(NH_3)_5]^{2+}$, $[CoCl(NH_3)_5]^{2+}$, $[Co(NO_2)_3(NH_3)_3]$, and $[Co(ox)_3]^{3-}$. In the absence of these energy acceptors, the excited-state lifetimes of Eu^{3+} bound to the CD and EF sites are indistinguishable, even in D_2O; however, in the presence of the positively charged energy acceptor complexes, the Eu^{3+} probes in the cod parvalbumin have different excited-state lifetimes due to a greater energy-transfer from Eu^{3+} in the CD site than from this ion in the EF site. The observation of distinct lifetimes for Eu^{3+} in the two sites allows the study of the relative binding site affinities and selectivity, using other members of the lanthanide ion series. It was found that in the course of a titration of the metal-free protein, Eu^{3+} fills the two sites simultaneously. Eu^{3+} is competitively displaced by other Ln^{3+} ions, with the CD site showing a preference for the larger Ln^{3+} ions while the EF site shows little, if any, competitive selectivity across the Ln^{3+} ion series.

Ca^{2+}/Mg^{2+} exchange in the binding sites of parvalbumin results in conformation changes monitored by various methods. As initially was established by X-ray crystallography, a remarkable conformational rearrangement occurs upon Ca^{2+}/Mg^{2+} exchange in the C-terminal EF-hand site of parvalbumin, (Figure 7; Declercq *et al.*, 1991). Such a conformational rearrangement is characterized by the following features: (i) the co-ordination number decreases from seven oxygen atoms in the Ca^{2+}-loaded form to six oxygen atoms in the Mg^{2+}-loaded form (ii) Glu101, at the relative position 12 in the EF-hand loop sequence (gateway 'Glu12'), acts as a bidentate ligand in the Ca^{2+}-loaded form and as a monodentate ligand in the Mg^{2+}-loaded form. As part of the conformational rearrangement, the χ_1 dihedral angle undergoes a *gauche*(+) (χ_1=-75°) to *gauche*(-) (χ_1=+60°) rotamer transition upon substitution of Ca^{2+} by Mg^{2+}, whereas the χ_2 angle remains practically unchanged and the χ_3 angles in both forms adopt a nearly mirror image relationship. The locus of the cation remains practically invariant within the protein tertiary structure, independent of its occupancy by Ca^{2+} or Mg^{2+}. The simple rotation of the Glu101 side-chain around its C^α-C^β bond, allows a change in the co-ordination number of the central cation, the Glu101 residue contributing as a bidentate ligand for Ca^{2+} and as a monodentate ligand for Mg^{2+}. In the CD site of parvalbumin the homologous residue Glu62 displays a behavior similar to that of Glu101 upon Ca^{2+}/Mg^{2+} exchange, as far as the χ_1 angle is concerned, based on NMR evidence (Blancuzzi *et al.*, 1993) and, as suggested by an infrared study, as far as the bidentate ↔ monodentate switch is concerned (Nara *et al.*, 1994). Both glutamyl residues Glu62 and Glu101 occupy homologous positions in the CD and EF sites of parvalbumin, corresponding with the relative position 12 in the canonical EF-hand loop. The residue Glu12 is highly conserved in all EF-hand loops, Glu being found substituted only by Asp, and in only 8% of all known sequences.

The X-ray crystallography data are in a good agreement with the NMR results. Two forms of parvalbumin, i.e., the fully Ca^{2+}-loaded form $PaCa_2$ and the fully Mg^{2+}-loaded form $PaMg_2$, were investigated by 2D ^1H NMR in solution (Blancuzzi *et al.*, 1993). A detailed analysis of the resonances, which belong to residues involved in direct coordination of Ca^{2+} and Mg^{2+}, shows that the sixth ligand, a highly conserved Glu residue at the relative position 12 in both cation-binding sites CD and EF, undergoes a conformational rearrangement through a 120° rotation of its side chain about the C_α-C_β bond with $PaMg_2$ adopting the less energetically favored g-conformation. Similarly, chemical shift effects, which selectively involve NH and C_α H resonances (as well as side-chain resonances) in both CD and EF sites, demonstrate a symmetrical behavior of both cation-binding sites upon Ca^{2+}/Mg^{2+} exchange.

Nara *et al.* (1994) examined metal-ligand interactions in the Ca^{2+}-binding sites of pike parvalbumin (pI = 4.10) by Fourier-transform infrared spectroscopy. The spectral region of the COO^- antisymmetric stretch provides useful information on the types of coordination of the COO^- groups to the metal ions in the Mg^{2+}-, Mn^{2+}-, and Ca^{2+}-bound forms. In the spectrum of the Ca^{2+}-bound form, two bands are observed at 1,582 and 1,553 cm^{-1}, whereas, in the spectra of the Mg^{2+}- and Mn^{2+}-bound forms, bands are observed only in the region around 1,582 cm^{-1} and no band is found in the region around 1,553 cm^{-1}. It was shown that the 1,553-cm^{-1} band of the Ca^{2+}-bound form reflects the bidentate coordination of the COO^- groups of both Glu62 in the CD site and Glu101 in the EF site to the Ca^{2+} ions. The absence of such a band in the spectrum of the Mn^{2+}-bound form is consistent with the X-ray structure of this form where both of the two COO^- groups are uni-dentate. These unidentate COO^- groups of Glu62 and Glu101 in the Mn^{2+}-bound form seem to give rise to a band at 1,577-1,574 cm^{-1}. The infrared spectrum of the Mg^{2+}-bound form is also consistent with the 'pseudo-bridging' coordination of the COO^- group of Glu101 reported in the X-ray structure of a form where the Mg^{2+} ion occupies only the EF site, and the same spectrum is further indicative of the 'pseudo-bridging' coordination of the COO^- group of Glu62.

In order to understand the molecular mechanisms underlying such a conformational rearrangement, Allouche *et al.* (1999) undertook a theoretical study using the free energy perturbation (FEP) method, starting from high-resolution crystal structures of the same parvalbumin (pike 4.10 isoform) differing by the substitution of their two cationic sites CD and EF, i.e. the 1pal structure with CD(Ca^{2+}) and EF(Ca^{2+}), the 4pal structure with CD(Ca^{2+}) and EF(Mg^{2+}). When Mg^{2+} is 'alchemically' transformed into Ca^{2+} within the EF-4 site of 4pal, the calculation correctly predicts the conformational rearrangement of Glu12. When Ca^{2+} is transformed into Mg^{2+} within the CD site of 4pal, the FEP calculation predicts the topology of the fully Mg-loaded form. As expected, Glu62 (at the relative position 12 in CD loop) is predicted to be a monodentate residue within a regular octahedral arrangement of six oxygen atoms around Mg^{2+}.

The amide proton exchange rates were measured for the pike parvalbumin loaded either with calcium ($PaCa_2$) or with magnesium ($PaMg_2$) by using 2-D total correlation spectroscopy experiments (Baldellon *et al.*, 1992). The differences in the exchange rates observed between these two species were unexpected when compared with relatively small conformational changes induced in parvalbumin by the Ca/Mg exchange. For the calcium-loaded protein ($PaCa_2$), a significant difference was observed for the amide proton exchange rates of residues located in the N-terminal domain AB in contrast to the slower exchange rates for the residues in the CD and EF domains. Such a difference does not exist for $PaMg_2$, where faster exchange rates are observed over all the sequence. Since amide proton exchange rates are related to the solvent's accessibility in proteins, these authors interpreted their results in terms of difference of the equilibria between 'closed-states' and 'opened-states' for individual amide protons of the protein when calcium is replaced by magnesium. The CD and EF domains, and to a lesser extent the AB domain, would be more rigid when the protein is loaded with calcium ions. For the magnesium-loaded parvalbumin ($PaMg_2$) the faster exchange rates could be explained by a more flexible structure than in the case of the $PaCa_2$.

Equilibrium Metal Binding Constants

An equilibrium interaction of a protein P with a metal cation Me (or any other low molecular or high molecular weight substance) such as

$$P + Me \leftrightarrow PMe \qquad (14)$$

is characterized by association (binding) constant (measured in M^{-1}):

$$K_a = [P] \times [Me]/[PMe], \qquad (15)$$

where [P], [Me], and [PMe] are equilibrium concentrations. It is possible to use dissociation constant $K_d = 1/K_a$ measured in M.

The most common method of determination of equilibrium binding parameters is the method of equilibrium or flow dialysis followed by further analysis of the data obtained on the basis of the Scatchard plots. In this method equilibrium concentration of free metal cations is measured in a solution, which does not contain protein and is separated from it by a porous membrane permeable to the cations and not permeable to the protein. It should be noted that usually this procedure takes a lot of time and protein since free cation concentration is measured by such methods as atomic absorption spectrophotometry or by means of radioactive isotopes. Moreover, the use of membranes can give artifacts due to interactions of the protein with membranes. In the case of Ca^{2+} such method allows measurement of only rather low binding constants since all preparations and chemicals, including deionised water, contain rather large quantities of contaminating calcium which can be removed only with great difficulties. For this reason the Ca^{2+}-EGTA or Ca^{2+}-EDTA buffer systems are commonly used for the determination of Ca^{2+} binding constants. This requires, however, use of high concentrations of the chelators, which sometimes is not good because of their possible direct interaction with some proteins.

These and another drawbacks of the existing experimental methods for determination of equilibrium cation binding parameters of proteins forced Permyakov et al. (Permyakov *et al.*, 1980; Permyakov, 1993b) to create a new more simple and more reliable methods. One of such methods is a method based on the use of the metal ion-induced changes of intrinsic protein fluorescence and the use of the so called chelator "back-titration". In principle, instead of intrinsic fluorescence, one could use emission of some fluorescent labels, circular dichroism, chemical shifts or widths of resonances in NMR spectrum or some other physical parameters.

The interaction of proteins with metal cations usually causes essential reorganization of their structure, which can change the environments of side chains of some aromatic amino acid residues (tryptophan, tyrosine, and phenylalanine). As a rule, this induces changes in parameters of their fluorescence, which can be used for measurement of bound cation concentration. The main condition for such measurements is the existence of an easily measurable fluorescent response to the binding of a substance to a protein. In principle, it is possible to have a situation when a protein binds a substance, but it does not change its fluorescence, i.e. the change in conformation induced by the binding does not touch the environment of protein chromophores. Therefore, the absence of a fluorescent or any other local response to the addition of a substance to the protein solution cannot be used as an evidence of the absence of the binding of this substance to the protein.

Figure 19 shows the results of spectrofluorimetric Ca^{2+}- titration of Ca^{2+}-free cod parvalbumin. The binding of Ca^{2+} induces a pronounced blue shift of its tryptophan fluorescence spectrum and an increase in fluorescence quantum yield. Quantitative evaluations of the binding parameters require the use of fluorescence parameters linearly related to the fraction of conversion of the protein. Such parameter is, for example, fluorescence quantum yield, q. It is clearly seen from Figure 19 that the dependence $q([Ca^{2+}]/P_0)$ is a straight line with a break at $[Ca^{2+}]/P_t = 2$ (P_t is total protein concentration). This means that fluorescence is sensitive to the binding of two Ca^{2+} ions per molecule and that the experiment was carried out in the conditions when $P_t >> 1/K_a$. The limited sensitivity of the spectrofluorimeter does not allow decrease of the protein concentration so that it would be comparable with $1/K_a$. Therefore we cannot determine the Ca^{2+}-binding constant from the direct Ca^{2+}-titration of the protein. Nevertheless, the measurement of the binding constant is yet possible in this case, but in order to do it we should use a sufficiently strong chelator of divalent cations with known cation-binding properties. For Ca^{2+} it may be ethylene glycol-bis(β-aminoethyl ether) N,N,N',N'-tetraacetic acid (EGTA) or ethylenediaminetetraacetic acid (EDTA).

Figure 19 shows the results of the spectrofluorimetric EGTA-titration of Ca^{2+}-loaded cod parvalbumin. The increase in EGTA concentration induces spectral changes opposite to those caused by the Ca^{2+} binding. It is very important for this experiment that the effective Ca^{2+} binding constant of the chelator should be comparable with the Ca^{2+}-binding constant of the protein. Only in this case we can evaluate the binding constant with sufficient accuracy from the obtained competition data. If the chelator binds Ca^{2+} much more tightly then the protein, the curve of spectrofluorimetric EGTA-titration of the protein will not differ from a straight line, and in this case evaluation of the binding constant would be impossible. In the opposite extreme case, in which the chelator binds Ca^{2+} much worse than the protein, the curve of the spectrofluorometric EGTA-titration will be very gently sloped and it will take too much of the chelator to remove Ca^{2+} from the protein. Sometimes it is not good since some proteins are able to bind EGTA and EDTA though with low affinity, which will prevent carrying out the experiment at high chelator concentrations.

The effective Ca^{2+}-binding constant of EGTA and EDTA can be changed by pH (Schwarzenbach & Flaschka, 1965). If properties of a protein permit it, by changing pH one can reach the effective Ca^{2+}-binding constant of chelator, which is comparable with the Ca^{2+}-binding constant of the protein.

Table 4. Equilibrium parameters of Ca^{2+}, Mg^{2+}, Na^{+}, and K^{+} binding to parvalbumins. pH 8, 20°C. (Permyakov *et al.*, 1980, 1981, 1983).

	Ca^{2+}		Mg^{2+}		Na^{+}	K^{+}
Parvalbumin	$-\log K_1$	$-\log K_2$	$-\log K_1$	$-\log K_2$	K(mM)	K(mM)
Pike *E.lucius L.* pI 5.0 pI 4.2	8.6 8.5	8.8 5.3	5.6 5.4	5.1 3.6	26 28	111 53
Whiting *Gadus merlangus L.*	7.7	6.9	3.6	4.3	40	111
Gadus morhua narcocephalus Til.	9.0	9.7	5.8	4.8	20	91
L. herzensteini Jordan et Snyder	9.2	9.9	4.9	5.1		
Limonada Aspera	9.9	10.2	6.0	5.4		
P.quadrotubercu-latus (Pallas)	7.8	8.5	4.5	4.9		
L.obscura (Herz)	9.1	5.4	5.1	3.9		
Ophiocephalus argus Warachowskii	9.3	10.0	5.4	3.7		
Peccothus glehni Dybowski	8.3	8.2	3.7	3.3		
Eleginus gracialis	9.3	9.3	3.9	2.9		
Carp *C.carp L. pI 4.25* *pI 4.47* *pI 3.95*	8.8 8.5 7.4	7.3 6.7 7.1				
Theragra chalcograma (Pall)	9.7	9.4	4.8	4.0		

The simplest scheme of the competition of calcium chelator *E* and parvalbumin *P* for Ca^{2+} looks like this:

$$\begin{aligned} P + Ca^{2+} &\leftrightarrow PCa(II) && K_{a1} \\ PCa(II) + Ca^{2+} &\leftrightarrow Ca(II)PCa(II) && K_{a2} \\ E + Ca^{2+} &\leftrightarrow ECa(II) && K_E \end{aligned} \qquad (16)$$

where E is EGTA or EDTA. One can evaluate K_{a1} and K_{a2} with sufficient accuracy by fitting the theoretical curve computed according to the scheme (16) to the experimental points by means of variation of K_{a1} and K_{a2}.

Table 5. Equilibrium parameters of Ca^{2+}, Mg^{2+}, Na^{+}, and K^{+} binding to parvalbumins obtained by various authors and by various methods.

Parvalbumin	Ca^{2+}		Mg^{2+}		K_{Na} mM	K_K mM	Method, conditions	Reference
	$logK_1$	$logK_2$	$logK_1$	$logK_1$				
Pike parvalbumin pI 4.2	8.2	5.8					Intrinsic fluorescence; 25 mM Pipers, pH 6.2; 21°C	Eberspach et al. 1988
Hake parvalbumin pI 4.36	7.0	7.0					^{45}Ca-Chelex 100; 60 mM KCl; 2 mM $MgCl_2$, 30 mM imidazole pH 6.7	Benzonana et al., 1972
Hake parvalbumin pI 4.36	8.4	7.8	4.5	4.5			Equilibrium dialysis, UV-spectroscopy, flow dialysis; 150 mM KCl, 25 mM HEPES, pH 7.55	Haiech et al., 1979a
Carp parvalbumin pI 4.25	8.4	8.4					Equilibrium dialysis; 100 mM KCl; pH 7.5	Potter et al., 1977
Carp parvalbumin pI 4.25	6.9	6.9					Equilibrium dialysis; 100 mM KCl, 3 mM $MgCl_2$; pH 7.5	Potter et al., 1977
Carp parvalbumin pI 4.25	9.4	9.4					Equilibrium dialysis; 80 mM KCl; pH 7.4	Moeschler et al., 1980
Carp parvalbumin pI 4.25	7.4	7.4					Equilibrium dialysis; 80 mM KCl, 1 mM $MgCl_2$; pH 7.4	Moeschler et al., 1980
Carp parvalbumin pI 4.25	7.6	7.6					Fluorescence; 50 mM cacodilate, pH 7.0; 30 mM KCl; 20°C	Iio & Hoshihara, 1984
Bullfrog parvalbumin pI 4.78	7.0	7.0	3.0	3.0			Fluorescent indicator tetramethylmurexide ; 20°C; 70 mM KCl; 20 mM MOPS, pH 6.8	Ogawa & Tanokura, 1986a
Bullfrog parvalbumin pI 4.78	6.8	6.8	2.9	2.9			Fluorescent indicator tetramethylmurexide ; 20°C; 70 mM KCl; 20 mM MOPS, pH 6.8	Ogawa & Tanokura, 1986a

Table 5. Continued

Parvalbumin	Ca^{2+}		Mg^{2+}		K_{Na} mM	K_K mM	Method, conditions	Reference
	$logK_1$	$logK_2$	$logK_1$	$logK_1$				
Frog parvalbumin pI 4.88	6.4	6.4					^{45}Ca-Chelex 100; 60 mM KCl; 2 mM $MgCl_2$, pH 6.7	Benzonana et al., 1972
Frog parvalbumin pI 4.88	8.1	8.1	4.7	4.7			Equilibrium dialysis, UV-spectroscopy, flow dialysis; 150 mM KCl, 25 mM HEPES, pH 7.55	Haiech et al., 1979a
Frog parvalbumin pI 4.50	6.7	6.7					^{45}Ca-Chelex 100; 60 mM KCl; 2 mM $MgCl_2$, pH 6.7	Benzonana et al., 1972
Frog parvalbumin pI 4.50	8.7	8.7	4.6	4.6			Equilibrium dialysis, UV-spectroscopy, flow dialysis; 150 mM KCl, 25 mM HEPES, pH 7.55	Haiech et al., 1979a
Rabbit parvalbumin pI 5.55	8.2	8.2	4.8	4.8			Equilibrium dialysis, UV-spectroscopy, flow dialysis; 150 mM KCl, 25 mM HEPES, pH 7.55	Haiech et al., 1979a
Rat parvalbumin pI 4.44	8.2	8.0	4.7	4.9			NMR; 10 mM piperazine, pH 7.7; 150 mM KCl	Williams et al., 1986a
WT rat parvalbumin	7.4		4.5				Flow dialysis; 50 mM Tris-HCl, pH 7.5; 150 mM KCl; 25°C	Pauls et al., 1993
F102W rat parvalbumin	7.4		4.6				Flow dialysis; 50 mM Tris-HCl, pH 7.5; 150 mM KCl; 25°C	Pauls et al., 1993
Rat β parvalbumin	7.4	6.2	4.0	2.2			Isothermal titration calorimetry; 25 mM Hepes, pH 7.4; 150 mM NaCl; 25°C	Henzl et al., 2003
WT rat α parvalbumin	8.1	8.1	4.0	4.0			Isothermal titration calorimetry; 25 mM Hepes, pH 7.4; 150 mM NaCl; 5°C	Henzl et al., 2003

Table 5. Continued

Parvalbumin	Ca^{2+}		Mg^{2+}		K_{Na} mM	K_K mM	Method, conditions	Reference
	$logK_1$	$logK_2$	$logK_1$	$logK_1$				
Rat α S55D/E59D parvalbumin	9.2	8.5	5.4	4.3			Isothermal titration calorimetry; 25 mM Hepes, pH 7.4; 150 mM NaCl; 5°C	Henzl et al., 2003
Rat α parvalbumin	8.4	7.8	4.3	3.6			Isothermal titration calorimetry; 25 mM Hepes, pH 7.4; 150 mM NaCl; 5°C	Henzl et al., 2004
Rat α parvalbumin	9.5	8.8	5.3	4.6	1.6		Isothermal titration calorimetry; 25 mM Hepes, pH 7.4; 150 mM KCl; 5°C	Henzl et al., 2004
Rat α parvalbumin					8.3		Isothermal titration calorimetry; 1 mM imidazole-EDTA, pH 7.4; 25°C	Henzl et al., 2000
Rat β parvalbumin	7.4	6.2	4.0	2.2			Isothermal titration calorimetry; 25 mM Hepes, pH 7.4; 150 mM NaCl; 5°C	Henzl et al., 2004
Rat β parvalbumin					16.7	18.2	Isothermal titration calorimetry; 1 mM imidazole-EDTA, pH 7.4; 25°C	Henzl et al., 2000
Rat β parvalbumin	7.5	6.6	4.1	2.5			Isothermal titration calorimetry; 25 mM Hepes, pH 7.4; 150 mM KCl; 5°C	Henzl et al., 2004
WT oncomodulin	6.1	7.4	2.5	3.8			Flow dialysis; 25 mM Hepes, pH 7.4; 150 mM NaCl; 25°C	Hapak et al., 1989
WT oncomodulin	6.1	7.4					Flow dialysis; 25 mM Hepes, pH 7.4; 150 mM NaCl; 25°C	Trevino et al., 1991

Table 5. Continued

Parvalbumin	Ca^{2+}		Mg^{2+}		K_{Na} mM	K_K mM	Method, conditions	Reference
	$logK_1$	$logK_2$	$logK_1$	$logK_1$				
WT oncomodulin	6.1	7.3					Flow dialysis; 25 mM Hepes, pH 7.4; 150 mM NaCl; 25°C	Henzl et al., 1998
S55D oncomodulin	7.2	7.6					Flow dialysis; 25 mM Hepes, pH 7.4; 150 mM NaCl; 25°C	Henzl et al., 1998
D59G oncomodulin	4.7	7.1					Flow dialysis; 25 mM Hepes, pH 7.4; 150 mM NaCl; 25°C	Henzl et al., 1998
D94S oncomodulin	6.3	4.4					Flow dialysis; 25 mM Hepes, pH 7.4; 150 mM NaCl; 25°C	Henzl et al., 1998
G98D oncomodulin	6.1	8.4					Flow dialysis; 25 mM Hepes, pH 7.4; 150 mM NaCl; 25°C	Henzl et al., 1998
S55D/ D59G oncomodulin	6.1	7.3					Flow dialysis; 25 mM Hepes, pH 7.4; 150 mM NaCl; 25°C	Henzl et al., 1998
D94S/G98D oncomodulin	6.6	5.5					Flow dialysis; 25 mM Hepes, pH 7.4; 150 mM NaCl; 25°C	Henzl et al., 1998
S55D/ D59G/ D94S oncomodulin	6.6	4.9					Flow dialysis; 25 mM Hepes, pH 7.4; 150 mM NaCl; 25°C	Henzl et al., 1998
S55D/ D59G/ D94S/ G98D oncomodulin	6.8	5.9					Flow dialysis; 25 mM Hepes, pH 7.4; 150 mM NaCl; 25°C	Henzl et al., 1998
N52K oncomodulin	5.9	7.2					Flow dialysis; 25 mM Hepes, pH 7.4; 150 mM NaCl; 25°C	Palmisano et al., 1990

Table 5. Continued

Parvalbumin	Ca^{2+}		Mg^{2+}		K_{Na} mM	K_K mM	Method, conditions	Reference
	$logK_1$	$logK_2$	$logK_1$	$logK_1$				
Q54K oncomodulin	6.1	7.2					Flow dialysis; 25 mM Hepes, pH 7.4; 150 mM NaCl; 25°C	Palmisano et al., 1990
Y57F oncomodulin	6.1	7.4					Flow dialysis; 25 mM Hepes, pH 7.4; 150 mM NaCl; 25°C	Palmisano et al., 1990
Y57F oncomodulin	6.2	7.4					Flow dialysis; 25 mM Hepes, pH 7.4; 150 mM NaCl; 25°C	Trevino et al., 1991
G60E oncomodulin	6.2	7.3					Flow dialysis; 25 mM Hepes, pH 7.4; 150 mM NaCl; 25°C	Trevino et al., 1991
G60E oncomodulin	6.3	7.3	2.9	3.5			Flow dialysis; 25 mM Hepes, pH 7.4; 150 mM NaCl; 25°C	Palmisano et al., 1990
D59E/G60E oncomodulin	6.4	7.5	3.1	3.6			Flow dialysis; 25 mM Hepes, pH 7.4; 150 mM NaCl; 25°C	Palmisano et al., 1990
L58I oncomodulin	5.7	7.2					Flow dialysis; 25 mM Hepes, pH 7.4; 150 mM NaCl; 25°C	Trevino et al., 1991
D59E oncomodulin	6.2	7.4					Flow dialysis; 25 mM Hepes, pH 7.4; 150 mM NaCl; 25°C	Trevino et al., 1991
D59E oncomodulin	6.3	7.4	3.0	3.7			Flow dialysis; 25 mM Hepes, pH 7.4; 150 mM NaCl; 25°C	Hapak et al., 1989
D45K oncomodulin	6.3	7.3					Flow dialysis; 25 mM Hepes, pH 7.4; 150 mM NaCl; 25°C	Trevino et al., 1991

Table 5. Continued

Parvalbumin	Ca^{2+}		Mg^{2+}		K_{Na} mM	K_K mM	Method, conditions	Ref.
	$logK_1$	$logK_2$	$logK_1$	$logK_1$				
K69G oncomodulin	6.2	7.4					Flow dialysis; 25 mM Hepes, pH 7.4; 150 mM NaCl; 25°C	Trevino et al., 1991
D59E/G60E oncomodulin	6.4	7.5					Flow dialysis; 25 mM Hepes, pH 7.4; 150 mM NaCl; 25°C	Trevino et al., 1991
Y57F/K69G oncomodulin	5.9	7.3					Flow dialysis; 25 mM Hepes, pH 7.4; 150 mM NaCl; 25°C	Trevino et al., 1991
Y57F/ D59E/G60E oncomodulin	6.2	7.4					Flow dialysis; 25 mM Hepes, pH 7.4; 150 mM NaCl; 25°C	Trevino et al., 1991
D59E/G60E/ K69G oncomodulin	6.4	7.4					Flow dialysis; 25 mM Hepes, pH 7.4; 150 mM NaCl; 25°C	Trevino et al., 1991
Y57F/ D59E/G60E/ K69G oncomodulin	6.6	7.4					Flow dialysis; 25 mM Hepes, pH 7.4; 150 mM NaCl; 25°C	Trevino et al., 1991
WT Oncomodulin	7.3	6.2	3.6				Flow dialysis; 20 mM Tris-HCl, pH 7.4; 150 mM NaCl; 1 mM DTT; 25°C	Cox et al., 1990
C. carpio parvalbumin	8.2						Fluorescence of fluo-3; 20 mM Hepes, pH 7.2; 150-250 mM KCl; 25°C	Erickson et al., 2005
G. gibberifrons parvalbumin	7.7						Fluorescence of fluo-3; 20 mM Hepes, pH 7.2; 150-250 mM KCl; 25°C	Erickson et al., 2005
avian thymic hormone, β-parvalbumin	8.4	8.0	4.3	4.1			Isothermal titration calorimetry	Henzl & Agah, 2006
CPV3-C72S, β-parvalbumin	7.7	7.4	4.7	4.3			Isothermal titration calorimetry	Henzl & Agah, 2006

Table 4 contains values of effective binding constants of Ca^{2+}, Mg^{2+}, Na^{+}, and K^{+} for various parvalbumins measured by means of the intrinsic fluorescence method (Permyakov *et al.*, 1980, 1981, 1983; Permyakov, 1985). Table 4 contains the effective binding constants of

Ca^{2+}, Mg^{2+}, Na^{+}, and K^{+} for various parvalbumins measured by means of various methods and by various authors.

The values of the Ca^{2+}-association constants (measured by fluorescence method in the absence of any competing metal ions) in the Table 3 lie mostly within the region from 10^8 to 10^{10} M^{-1}. Other methods give similar values of Ca^{2+} association constants (Table 5). Tables 4 and 5 show that the values of Mg^{2+}-binding constants for parvalbumins are mostly within the range from 10^4 to 10^5 M^{-1}. It is of importance that all the values in the Table 3 were obtained by the same authors by the same method since the association constants measured by different authors for the same protein even by the same method and in approximately the same conditions may differ by an order of magnitude or more (see Table 5). If the measurements are carried out by different methods the values of binding constants may differ even more.

It should be noted that different parvalbumins might essentially differ in their Ca^{2+}-binding affinity. Even different parvalbumins from the same animal may have very different binding constants. This is clearly seen in the case of carp parvalbumins pI 4.25, 4.47, and 3.95 and pike parvalbumins pI 5.0 and 4.2. These results suggest that in order to construct various schemes of regulation of Ca^{2+} flows in the muscle cell sarcoplasm one should know the exact quantities of different parvalbumins in the cell and their binding constants. The idea of a "mean" parvalbumin may be rather incorrect.

It is well seen from the Tables 4 and 5 that Ca^{2+}-affinities of the CD and EF sites in some parvalbumins are the same (mostly for α-parvalbumins). In other parvalbumins (β-parvalbumins), Ca^{2+} affinity of one of the sites is lowered, as is the case, for example, of pike parvalbumin pI 4.2 (see also Eberspach *et al.*, 1988) and toad parvalbumin (Ogawa & Tanokura, 1986a,b). Measurements of distances between the strong and the weak sites and the single Tyr48 in pike parvalbumin pI 4.2 showed that the CD site is the strong and the EF site is the weak metal binding site in this parvalbumin (Eberspach et al., 1988).

Surprisingly, rat parvalbumins α and β show widely different metal ion binding properties. While the CD and EF sites of rat α-parvalbumin are Ca^{2+}/Mg^{2+} sites with similar affinities (Pauls *et al.,* 1993; Eberhard & Erne, 1994), the CD and EF sites of rat β-parvalbumin (oncomodulin) are distinctly non-equivalent (Hapak *et al*., 1989): whereas the EF site is a Ca^{2+}/Mg^{2+} site, the CD site is a Ca^{2+}-specific one, which is rather unusual for parvalbumins. According to the ^{15}N NMR relaxation data (Henzel *et al*., 2002), whereas rat α parvalbumin tends to display the higher order parameter in helical regions, rat β parvalbumin tends to display the higher value of order parameter in the loop regions: one region is located in the AB domain, in the vicinity of Pro21 and Pro26; another one occurs in the CD-binding loop between residues 57-62. Interestingly, the Ca^{2+}-free form of rat β parvalbumin displays an average order parameter greater than that of the Ca^{2+}-bound form.

According to the data of Eberhard & Erne (1994), two equivalent Ca^{2+}/Mg^{2+} binding sites in rat α-parvalbumin are characterized by dissociation constants for Ca^{2+} and Mg^{2+} in Hepes buffer containing 150 mM K^{+} at 35°C and pH 7.2 11.0±1.8 nM and 41±8 μM, respectively. Dissociation constants values of Ca^{2+} binding were found to be about fourfold larger in the presence of Na^{+} as compared with K^{+}, indicating that Na^{+} distinctly influences Ca^{2+} binding to rat parvalbumin. Both Ca^{2+} and Mg^{2+} bindings to parvalbumin are exothermic.

Birds express three parvalbumins, one α isoform and two β isoforms. The latter are known as avian thymic hormone (ATH) and avian parvalbumin 3. Henzl & Agah (2006) carried out a detailed Ca^{2+}- and Mg^{2+}-binding studies on recombinant ATH and the C72S variant of CPV3, using global analysis of isothermal titration calorimetry data. In Hepes-buffered saline, ATH binds Ca^{2+} with apparent microscopic binding constants of 2.4±0.2x10^{8} and 1.0±0.1x10^{8} M^{-1}. The corresponding values for CPV3-C72S are substantially lower, 4.5±0.5x10^{7} and 2.4±0.2x10^{7} M^{-1}, a 1.9-kcal/mol difference in binding free energy. Interestingly, despite its decreased Ca^{2+} affinity, CPV3-C72S exhibits increased affinity for Mg^{2+}, relative to ATH: whereas the latter has Mg^{2+}-binding constants of 2.2±0.2x10^{4} and 1.2±0.1x10^{4} M^{-1}, CPV3-C72S yields values of 5.0±0.8x10^{4} and 2.1±0.3x10^{4} M^{-1}.

Thus, the β-parvalbumin lineage displays a spectrum of Ca^{2+}-binding affinity, with ATH and the mammalian β isoform at the high- and low-affinity extremes and CPV3 in the middle (Henzl & Agah, 2006).

It should be noted that in most cases the so called Ca^{2+}-specific sites in calcium binding proteins are usually characterized by lowered Ca^{2+} affinities compared to the Ca^{2+}/Mg^{2+}-sites. Perhaps they still can bind Mg^{2+} ions but with very low affinities and some experimental methods are not sensitive enough to monitor this binding. Just for this reason such sites seem to be considered as Ca^{2+}-specific ones.

The genetic engineering methods have been widely used for studies of the binding of metal cations to parvalbumin. Most point mutations were made in the calcium binding loops of parvalbumins. Since most parvalbumins contain neither tryptophan no tyrosine, some investigators genetically replaced phenylalanine residues by tryptophan residues to use them as fluorescent labels to study metal binding properties.

For example, in the work of Pauls *et al.* (1993) rat parvalbumin was expressed in *Escherichia coli* and mutated by replacing a Phe at position 102 with a unique Trp (F102W mutant) in order to introduce a distinct fluorescent label into the protein. Mass spectroscopy and NMR data indicate that the recombinant wild-type and F102W mutant proteins have the expected molecular weight and retain the native structure. Both proteins contain two non-cooperative Ca^{2+}/Mg^{2+}-binding sites with intrinsic affinity constants for Ca^{2+} and Mg^{2+} of 2.4x10^{7} M^{-1} and of 2.9x10^{4} M^{-1}, respectively, for wild type parvalbumin, and similar constants of 2.7x10^{7} M^{-1} and of 4.4x10^{4} M^{-1}, respectively, for the F102W mutant. In the next study of Pauls et al. (1994) three new parvalbumin mutant proteins, derived from F102W and containing alterations at positions essential for Ca^{2+} binding in either one of the two Ca^{2+}-binding sites (PV-CD and PV-EF) or in both sites (PV-CD/-EF), were expressed and purified. Intrinsic tryptophan fluorescence revealed that in the three mutant parvalbumins the residue Trp102 is deeply buried in the hydrophobic core. It was established that PV-CD/-EF binds neither Ca^{2+} nor Mg^{2+}, while both PV-CD and PV-EF bind one Ca^{2+} with affinity constants of 1.1x10^{7} and 3.2x10^{6} M^{-1}, respectively. Both mutants bind one Mg^{2+} with constant 8x10^{4} for PV-CD and 3x10^{3} M^{-1} for PV-EF. These data indicate that inactivation of the EF site, much more than of the CD site, impairs divalent cation binding. The binding of Ca^{2+} and Mg^{2+} is mutually exclusive, indicative of a Ca^{2+}/Mg^{2+} mixed site. However, as for F102W, the magnesium binding constant values obtained from the competition equation are approximately 40-fold lower than the affinities measured by direct binding.

Simultaneous replacement of Asp-94 with serine and Gly-98 with glutamate in rat α-parvalbumin creates a CD-site ligand array in the context of the EF-site binding loop but this engineered site has markedly reduced Ca^{2+} affinity (Tanner et al., 2005). Seeking an explanation for this phenomenon, the authors obtained the crystal structure of the α D94S/G98E variant. Surprisingly, the Ca^{2+} coordination within the engineered EF site of the 94/98E variant is nearly identical to that within the CD site, suggesting that the attenuated affinity of the EF site in 94/98E is not a consequence of suboptimal coordination geometry. It was found that the reduced affinity for divalent ions is evidently not the result of heightened monovalent ion competition. The thermodynamic analysis indicates that the less favorable Gibbs free energy of binding reflects a substantial enthalpic penalty. Significantly, the crystal structure reveals a steric clash between Phe57 and the C_γ atom of Glu98. The consequent displacement of Phe57 also produces a close contact with Ser55. Thus, steric interference may be the source of the enthalpic penalty.

As was shown above, despite striking sequence homology (53% identity) between parvalbumin and oncomodulin (β-parvalbumin), the ion-binding sites in the letter display far lower affinity for Ca^{2+}. Their CD sites, in particular, differ in affinity by a factor of about 400. Trevino *et al.* (1991) attempted to identify the structural basis for this difference by systematically substituting the parvalbumin residue for oncomodulin residue at points of nonidentity. Replacement of Asp45 in the C-helix of oncomodulin by Lys (D45K) slightly increases the association constant for Ca^{2+} at the CD site from $1.2x10^6$ to $1.9x10^6$ M^{-1}. Replacement of Lys69 in the D helix by glycine (K69G) similarly only slightly increases the association constant up to $1.7x10^6$ M^{-1}. Combined mutations in positions 57, 59, 60, and 69 of oncomodulin increase the association constant up to $4x10^6$ M^{-1}, reflecting a 3-fold increase in affinity relative to the wild type protein. As a part of an effort to understand the structural basis for the differences in calcium and magnesium binding properties in parvalbumin and oncomodulin, Palmisano et al. (1990) prepared site-specific variants of oncomodulin in which the amino acid residues at positions 52, 54, 57, 59, and 60 were replaced with the residues present at the corresponding positions in rat parvalbumin. The proteins with the single-site substitutions at residues 52, 54, and 57 were indistinguishable from the wild-type protein. By contrast, the substitutions at residues 59 and 60 perturb both the Eu^{3+} luminescence parameters and the Ca^{2+} and Mg^{2+} affinities, and these differences are amplified when both replacements are simultaneously incorporated into the protein. Consistent with the increased parvalbumin-like character, calcium association constant of the double mutant is slightly increased from $1.3x10^6$ M^{-1} (for the wild-type protein) to $2.4x10^6$ M^{-1}, and magnesium association constant is increased from 286 to $1.3x10^3$ M^{-1}. Nevertheless, the affinity of the CD binding domain in D59E/G60E for Ca^{2+} remains almost 2 orders of magnitude lower than of the corresponding site in rat parvalbumin, strongly suggesting that residues besides those present in the binding loop are involved in dictating the metal ion-binding properties of the oncomodulin CD site. It should be noted however that such small changes in association constants found in both cited works are practically within experimental errors of most experimental methods used for determination of the metal binding constants.

A serious difference between parvalbumin and oncomodulin is the difference in selectivity of metal binding to their CD sites, a Ca^{2+}/Mg^{2+}-site in parvalbumin and Ca^{2+}-specific in oncomodulin. Pauls *et al.* (1996b) substituted selected amino acid residues in the CD site of parvalbumin by those present at identical positions in oncomodulin. For these studies they also used the mutant rat parvalbumin with inserted single tryptophant residue as a conformational-sensitive fluorescent probe. One of the mutants had one new substitution Ile66 by Phe ([F66, W102]PV), and the other one had four new substitutions, namely Val46→Ile, Leu50→Ile, Ile58→Leu, and Ile66→Phe ([I46, I50, L58, F66, W102]PV), which did not change significantly the hydrophobic core. Both mutant proteins have two metal-binding sites of identical affinities with intrinsic calcium affinity constants of $2.9x10^7$ M^{-1} for [F66, W102]PV and $1.7x10^7$ M^{-1} for [I46, I50, L58, F66, W102]PV and intrinsic magnesium affinity constants of $3.1x10^4$ M^{-1} for [F66, W102]PV and $1.9x10^4$ M^{-1} for [I46, I50, L58, F66, W102]PV. Thus, the multi-residue substitution, but not the two-residue one, results in a slight decrease of affinity compared to [W102]PV ($2.7x10^7$ M^{-1} for Ca^{2+}, $4.4x10^4$ M^{-1} for Mg^{2+}). The effects are practically within the limits of experimental accuracy. It was shown that Mg^{2+} effect on the two Ca^{2+}-binding sites of [F66, W102]PV is the same and quantitatively very similar to that described for [W102]PV. At the same time, in [I46, I50, L58, F66, W102]PV, Mg^{2+} antagonizes the binding of the second Ca^{2+} (likely at the EF site) much more than that of the first Ca^{2+} (likely the CD site). According to the competition scheme, the two sites display Mg^{2+} binding constant values of 390 M^{-1} and $3.9x10^3$ M^{-1}, respectively. Thus, it turns out that the single Ile66→Phe mutation does not change the cation binding parameters, while multiple mutations in the hydrophobic core still do not change the affinity for Ca^{2+} and Mg^{2+}, but strongly affect the Mg^{2+} antagonism and probably the selectivity of the CD site making it oncomodulin-like.

To reveal to what degree the Ca^{2+}/Mg^{2+} specificity and affinity of the EF-hand motif in a protein is intrinsically determined by its sequence, the complete CD sites of rat parvalbumin and oncomodulin were exchanged, yielding two chimeras, [S41-Q71]parvalbumin and [D41-S71]oncomodulin (Durussel *et al.*, 1996). Direct Ca^{2+} and Mg^{2+} binding monitored by flow dialysis and gel filtration revealed that [S41-Q71]parvalbumin binds only one Mg^{2+} with an intrinsic affinity of $3.0x10^4M^{-1}$ and two $Ca^{2+}$with an identical binding constant of $4.4x10^6M^{-1}$, whereas [D41-S71]oncomodulin binds two Mg^{2+} with a mean binding constant of $2x10^4$ M^{-1} and two Ca^{2+} with a binding constant of $1.3x10^7$ M^{-1}. The calcium binding constant of the CD site of [S41-Q71]parvalbumin was 2.5-fold higher than that of the CD site in [W102]oncomodulin, but 5-6-fold lower than that of the CD site in [W102]parvalbumin. In [D41-S71] oncomodulin, calcium binding constant of the CD site was two-fold lower than in [W102] parvalbumin, but eight-fold higher than in [W102]oncomodulin. These results indicate that the sequence of the CD site determines its Ca^{2+}/Mg^{2+}-specificity, whereas its affinity for Ca^{2+} is influenced by the protein into which the CD site is inserted. It is of interest that the inserted CD site in turn influences the affinity of the EF site to which it is inserted in the host protein and the paired sites display an equalized affinity for Ca^{2+}. Mg^{2+} decreases the affinity of the chimeras for Ca^{2+}, but not according to a simple competition model.

Rhyner *et al.* (1996) studied human recombinant α-parvalbumin and nine mutant proteins, containing inactivating substitutions at positions essential for Ca^{2+}binding in the CD binding site (E62V, D51A, D51A, D62V), the EF site (E101V, D90A, D90A, E101V) or in both (E62V,E101V, D51A,D90A, D51A,E62V,D90A,E101V). Binding of either Ca^{2+} or Mg^{2+} to wild type parvalbumin or to mutants with an inactivated EF site lead to a 1.8-fold decrease in phenylalanine fluorescence intensity (the protein contains neither tryptophan no tyrosine), whereas the mutants with an inactivated CD site show only a very slight decrease in phenylalanine fluorescence intensity upon binding of Ca^{2+} or Mg^{2+}. Flow dialysis data revealed that wild type parvalbumin binds 2 Ca^{2+} with equal binding constants, of $2.3x10^7M^{-1}$ and that Mg^{2+} competes with Ca^{2+} with a binding constant of $4.9x10^3$ M^{-1}. The three mutants with an inactivated CD site bind one Ca^{2+} with binding constant, of (2.0 - 2.3)$x10^7$ M^{-1} and magnesium binding constant of (3.4 - 4.6)$x10^3$ M^{-1}, i.e. very similar to those of wild type parvalbumin. The mutants with an inactivated EF site bind one Ca^{2+} with lowered binding constant values of $7.9x10^6$, $4.5x10^6$ and $3.6x10^6$ M^{-1} for D91A, E102V and E101V,D91A, respectively. The magnesium binding constant values of these mutants are about 4-times lower than in the wild type protein. The three mutants with both sites inactivated bind neither Ca^{2+} nor Mg^{2+}.

As in other parvalbumins, the liganding residues in the CD and EF sites of oncomodulin differ at the +Z and -X coordination positions: serine and aspartate, respectively, in the CD site; aspartate and glycine in the EF site. Kauffman *et al.* (1995) prepared a series of oncomodulin variants in which the +Z and/or -X residue(s) from one site were replaced by the corresponding residue(s) from the other. Simultaneous replacement of Ser55 by aspartate and Asp59 by glycine gives the CD site with a coordination sphere superficially equivalent to that of the EF site. As was previously observed for the S55D mutation, the Eu^{3+} $^7F_0 \rightarrow {}^5D_0$ spectrum of the 55/59 mutant is pH independent. At the same time, replacement of Asp94 by serine at the +Z position of the EF site of 55/59 imparts pH dependent behavior to the EF site. The identical mutation in the wild-type background likewise imparts pH dependence to the EF site, giving a protein in which both sites display broad signals near 578.2 nm at pH 8. These results show that the presence of a hydroxyl group at the +Z position is sufficient to confer pH dependence on the $^7F_0 \rightarrow {}^5D_0$ spectrum of the parvalbumin EF-hand domain.

To obtain an insight into the structural factors responsible for the reduction in binding affinity of the EF-site in β-parvalbumin (oncomodulin), Henzl *et al.* (1998) examined oncomodulin mutants in which the CD and EF site ligand arrays were exchanged. They have found that binding affinity may be dictated either by ligand identity or by the binding site environment. For example, the Ca^{2+} affinity of the quasi-EF site resulting from the combined S55D and D59G mutations is substantially lower than that of the authentic EF site. This finding shows that other local environmental factors (e.g. binding loop flexibility, electrostatic potentials) within the CD binding site supersede the influence of ligand identity. At the same time, the CD site ligand array does not become a high-affinity site when imported into the EF site, as in the D94S/G98D mutant. Instead, it retains its Ca^{2+}-specific nature, implying that this constellation of ligands is less sensitive to placement within the protein molecule. The D59G and D94S single mutations substantially lower binding affinity, consistent with removal of a liganding carboxylate. By contrast, the S55D and G98D mutations substantially increase binding affinity. Significantly, the Ca^{2+} affinity of the oncomodulin CD site is increased by mutations that weaken binding at the EF site, indicating a negatively cooperative interaction between the two sites. The introduction of a fifth carboxylate into the ligand array of the CD site (combined S55D and E59D mutations) or the EF site (G98D mutation) of rat α-parvalbumin substantially increases divalent ion affinity (Henzl et al., 2004b). This behavior, while in conflict with that seen in model peptide systems, agrees with existing data for rat β-parvalbumin. Whereas the D59E mutation has minimal influence on β-parvalbumin CD site affinity, E59D has a major impact on the α-parvalbumin CD site, lowering the apparent association constant by a factor of 14. When the α-parvalbumin CD array is imported into the EF site, it acquires a low-affinity phenotype. However, when the EF ligand array is introduced into the α-parvalbumin CD binding loop, it retains a high-affinity signature. This result, contrary to that observed in β-parvalbumin, suggests that the influence of the parvalbumin CD site environment supersedes the intrinsic behavior of the ligand array.

To determine whether these mutations produce a variation on the archetypal EF-hand coordination scheme, Lee *et al.* (2004) obtained high-resolution X-ray crystallographic data for the S55D/E59D mutant of rat α-parvalbumin. As anticipated, the aspartyl carboxylate replaces the serine hydroxyl at the +Z coordination position. The Asp59 carboxylate abandons the role it plays as an outer sphere ligand in wild-type rat β-parvalbumin, rotating away from the Ca^{2+} and, instead, forming a hydrogen bond with the amide of Glu62. Superficially, the coordination sphere in the CD site of rat α-parvalbumin S55D/E59D resembles that in the EF site, however, the orientation of the Asp59 side chain is predicted to stabilize the D-helix, which may contribute to the heightened divalent ion affinity. In 0.15 M KCl and 25 mM Hepes-KOH (pH 7.4) at 5°C, the macroscopic Ca^{2+} binding constants for the mutant protein are $1.8x10^{10}$ and $2.0x10^{9}$ M^{-1}. The corresponding Mg^{2+} binding constants are $2.7x10^{6}$ and $1.2x10^{5}$ M^{-1}.

In the work of Henzl *et al.* (2003b) samples of rat β-parvalbumin were titrated with Ca^{2+}, with Mg^{2+}, with Ca^{2+} at several fixed levels of Mg^{2+}, with Ca^{2+} in the presence of EDTA, and with Ca^{2+} in the presence of EGTA in isothermal titration calorimeter. Simultaneous global least-squares analysis of titration data was carried out to evaluate calcium and magnesium binding constants. Calcium binding constants determined by this method for rat β-parvalbumin are $2.4x10^{7}$ M^{-1} and $1.6x10^{6}$ M^{-1}, while magnesium binding constants for this parvalbumin are $9.8x10^{3}$ M^{-1} and $1.8x10^{2}$ M^{-1} at 25°C. For rat α-parvalbumin calcium binding constants equal to $1.2x10^{8}$ M^{-1} and $1.2x10^{8}$ M^{-1}, while magnesium binding constants are $9.5x10^{3}$ M^{-1} and $9.5x10^{3}$ M^{-1} at 5°C. According to their data, mutations S55D/E59D increase binding constants up to $1.65x10^{9}$ M^{-1} and $3.0x10^{8}$ M^{-1} for calcium ions and up to $2.5x10^{5}$ M^{-1} and $2.1x10^{4}$ M^{-1} at 5°. The mutations increased overall standard free energy for Ca^{2+} binding by 2.1 kcal/mol, 1.6 kcal/mol of which is attributable to heightened CD site affinity.

It was found also (Henzl *et al.*, 2002) that although the pentacarboxylate mutations, S55D and G98D, both increase the Ca^{2+}-binding affinity of rat β parvalbumin, the mutations have rather different effects on backbone dynamics. G98D increases the average order parameter; S55D decreases it. Whereas the favorable free energy of calcium binding to S55D reflects a more favorable enthalpy of binding, that of G98D is almost entirely entropic.

It is well known that both α-parvalbumin and β-parvalbumin (oncomodulin) contain three EF-hand motifs, but the first site (AB) cannot bind Ca^{2+}. Cox *et al.* (1999) tried to recreate the putative ancestral proteins rat [D19-28E]parvalbumin and rat [D19-28E]oncomodulin by replacing the 10-residue-long nonfunctional loop in the AB site by a 12-residue canonical loop. To create a fluorescent conformational probe they used the F102W replacement. Surprisingly, in none of the proteins did the mutation reactivate the AB site. The AB-remodeled parvalbumins bind two Ca^{2+} ions with strong positive cooperativity (Hill coefficient n_H= 2) and moderate affinity ($[Ca^{2+}]_{0.5}$ = 2 μM), compared with $[Ca^{2+}]_{0.5}$ = 37 nM and n_H = 1 for the wild-type protein. Increasing Mg^{2+} concentrations changed n_H from 2 to 0.65, but without modification of the $[Ca^{2+}]_{0.5}$-value. Circular dichroism spectroscopy revealed that the Ca^{2+} and Mg^{2+} forms of the remodeled parvalbumins lost one-third of their α-helix content compared to the Ca^{2+} form of wild-type parvalbumin, while the microenvironment of single Trp residues in the hydrophobic cores remains the same. The metal-free remodeled parvalbumins possess unfolded conformations. The AB-remodeled oncomodulins also bind two Ca^{2+} with $[Ca^{2+}]_{0.5}$=43 μM and n_H=1.45. Mg^{2+} does not affect Ca^{2+} binding. Again, the Ca^{2+} forms display two-thirds of the α-helical content in the wild-type, while their core is still strongly hydrophobic as monitored by Trp and Tyr fluorescence. The metal-free oncomodulins are partially unfolded and seem do not possess a hydrophobic core. The authors suggested that the predicted evolution of the AB site from a canonical to an abortive EF-hand may have been dictated by the need for stronger interaction with Mg^{2+} and Ca^{2+}, and a high conformational stability of the metal-free forms.

The values of Na^+ and K^+ dissociation constants for parvalbumins in the Tables 3 and 4 lie within the region from 12 to 20 mM and Na^+ binds more tightly in comparison with K^+. At the same time, according to the ^{23}Na NMR data (Grandjean *et al.*, 1977; Parello *et al.*, 1979), pike parvalbumin binds K^+ more tightly than Na^+: dissociation constants 11 and 8 mM, respectively. Isothermal titration calorimetry data for β isoform of rat parvalbumin give dissociation constants 17 and 18 mM for Na^+ and K^+, respectively (Henzl *et al.*, 2000).

It is of interest that rat β- and α-parvalbumins have distinct monovalent cation-binding properties (Henzl *et al.*, 2000, 2004b). Differential scanning calorimetry data suggest that, at physiological pH and ionic strength, the β isoform binds roughly 2 equiv of Na^+ or a single equivalent of K^+ with moderate affinity. Under comparable conditions, the α isoform apparently binds just 1 equiv of Na^+ and essentially no K^+. Isothermal titration calorimetry experiments suggest that the bound monovalent ions occupy the EF-hand motifs. In 0.15 M K^+, at pH 7.4, the stability of the apo-β-parvalbumin exceeds that of the α isoform by approximately 2.6 kcal/mol at 37 °C and by approximately 3.0 kcal/mol at 25 °C. Ca^{2+} abolishes these binding events, suggesting that the monovalent ions occupy the EF-hand motifs. Monovalent cations seriously affects α-parvalbumins (Henzl *et al.*, 2004b). In Hepes-buffered NaCl, at 5 °C, the macroscopic Ca^{2+}-association constants are $2.6x10^8$ and $6.3x10^7$ M^{-1} and the Mg^{2+} constants, $1.8x10^4$ and $4.3x10^3$ M^{-1}. In Hepes-buffered KCl, the Ca^{2+} values increase up to $3.3x10^9$ and $5.0x10^8$ M^{-1} and the Mg^{2+} values to $2.0x10^5$ and $3.3x10^3$ M^{-1}. From these data it was calculated that Na^+ association constant for α-parvalbumins is $6.3x10^2$ M^{-1} and was shown that divalent ion-binding is positively cooperative. NMR data suggest that the lone Na^+ ion occupies the CD loop. Monovalent cations have a smaller impact on β-parvalbumins. In Na^+, the Ca^{2+} binding constants for the EF and CD sites are $2.3x10^7$ and $1.5x10^6$ M^{-1}, respectively; the Mg^{2+} constants are $9.1x10^3$ and $1.7x10^2$ M^{-1}. In K^+, these values shift to $3.1x10^7$ and $3.8x10^6$ M^{-1} and the latter to $1.4x10^4$ and $2.9x10^2$ M^{-1}.

Taking into consideration high concentrations of the monovalent cations in a cell one can assume that in spite of their low affinity to parvalbumin their association can essentially influence the binding of Ca^{2+} and Mg^{2+}. Intrinsic fluorescence studies showed that in the presence of 14 mM Na^+ and 155 mM K^+ (modeling of the physiological conditions) whiting parvalbumin binds Ca^{2+} and Mg^{2+} worse than in the absence of these cations (Permyakov *et al.*, 1983b). The physiological concentrations of Na^+ and K^+ decrease apparent Ca^{2+} and Mg^{2+} association constants for the weaker, probably EF, site by an order of magnitude, though the association constant of the other site slightly decreases as well. The decrease of the apparent Ca^{2+} and Mg^{2+} association constant for the CD-site is explained by common competition between divalent and monovalent cations for this site. The decrease of the apparent constants for the EF-site is more pronounced than would follow from the competition scheme. Perhaps this is explained by the binding of Na^+ and K^+ to some secondary sites or by effects of ionic strength, which become significant at "physiological" concentrations of the monovalent cations.

The structural changes in the major isotype of parvalbumin from the toad (*Bufo bufo japonicus*) skeletal muscle caused by Ca^{2+} and Mg^{2+} binding were analyzed by microcalorimetric titration method (Tanokura, 1990). Parvalbumin was titrated with Ca^{2+} in both the absence and presence of Mg^{2+} and with Mg^{2+} in the absence of Ca^{2+}, at pH 7.0, and at 5°, 15°, and 25°C. It was found that the reactions of parvalbumin with Ca^{2+} are exothermic at every temperature in both the absence and presence of Mg^{2+}, but those with Mg^{2+} are always endothermic except for the binding to site 1 at 25°C. No major conformational changes were noted in this work between Ca^{2+}- and Mg^{2+}-bound forms of toad parvalbumin. It was revealed that the metal-free form is much less stable than any form with Ca^{2+} (or Mg^{2+}) bound at one site at least. On Mg^{2+}-Ca^{2+} exchange, the vibrational as well as hydrophobic entropy is only slightly increased in a parallel manner. In contrast, on Ca^{2+} (or Mg^{2+}) binding, the hydrophobic entropy increases but the vibrational entropy decreases; the former indicates the sequestering of nonpolar groups from the surface to the interior of a molecule, and the latter suggests that the overall structures are tightened on Ca^{2+} (or Mg^{2+}) binding but loosened on Mg^{2+}-Ca^{2+} exchange.

Enthalpy changes upon calcium binding to parvalbumin were measured in the works of several authors. In the work of Cox *et al.* (1990) Ca^{2+} binding to the wild type recombinant oncomodulin was studied by equilibrium flow dialysis in the absence and presence of 1, 2, and 10 mM Mg^{2+}. Direct Mg^{2+}-binding experiments were carried out by the Hummel-Dryer gel filtration technique. It was found that in the absence of Mg^{2+} oncomodulin binds two Ca^{2+} with binding constant 2.2×10^7 and 1.7×10^6 M^{-1}, respectively. In the absence of Ca^{2+} the protein binds only one Mg^{2+} with binding constant 4.0×10^3 M^{-1}. Ca^{2+} binding to the low affinity site is only slightly affected by Mg^{2+}, so that in the presence of 2-3 mM Mg^{2+} the two sites have apparently equal affinity for Ca^{2+}. Microcalorimetry measurements showed that, in spite of the different affinities of the two Ca^{2+}-binding sites, ΔH_0 for the binding of each Ca^{2+} is identical and exothermic for -18.9 kJ/site. It follows that the entropy gain upon binding of Ca^{2+} is +77.1 J K^{-1} $site^{-1}$ for the high affinity Ca^{2+}-Mg^{2+} site and +56.0 J K^{-1} $site^{-1}$ for the low affinity Ca^{2+}-specific site. Mg^{2+} binding is endothermic for +13 kJ/site with an entropy change of +111 J K^{-1} $site^{-1}$.

Batch microcalorimetry measurements on carp parvalbumin pI 4.25 give values -37.2 and -12.1 kJ/mol site for ΔH of Ca^{2+} and Mg^{2+} binding, respectively (Moeschler *et al.*, 1980). It should be noted that this method does not resolve metal binding to separate binding sites.

In the works of Erickson *et al.* (2005) and Erickson & Moerland (2006) calcium association constants were measured as a function of temperature within the region from 0 to 25°C for parvalbumins from two Antarctic (*Gobionotothen gibberifrons* and *Chaenocephalus aceratus*) and two high temperate zone fish species (*Cyprinus carpio* and *Micropterus salmoides*). The same was done for parvalbumin from Arctic cod (*Boreogadus saida*). The water temperature range for Florida bass and carp locations is 7 to 30°C, with a mean yearly temperature of about 21°C. Conversely, *G. gibberifrons* and *C. aceratus* (representatives of families *Nothotheniidae* and *Channichthyidae*, respectively) are restricted to the Southern Ocean, where water temperatures range only between -1.9 and +1.5 °C. Measurements by fluorescence method show that calcium association constant values for parvalbumins from the Antarctic species are significantly higher at all assay temperatures and are less sensitive to temperature relative to carp and bass parvalbumins. Calcium association constants for parvalbumin from B. saida were fundamentally similar to those for parvalbumins from Antarctic species, but significantly different from temperate zone species. It is of importance that estimates of calcium association constant are fundamentally similar for parvalbumins from the Antarctic and high temperate zone species when measured at their native physiological temperature. Variation in pH and ionic strength within a physiologically relevant range had only modest effects on calcium dissociation constants. Thermodynamics of calcium binding to parvalbumin from *G. gibberifrons* and *C. carpio* was measured by isothermal microcalorimetry. When measured at 15°C, the Gibbs free energy change (ΔG) was significantly greater for calcium binding to parvalbumin from *G. gibberifrons* than from carp (-43.4±1.5 kJ mol^{-1} and -46.6±3.0 kJ mol^{-1}, respectively), and the relative contribution of entropy to ΔG for calcium binding to parvalbumin from the Antarctic species was about twice that of carp (ΔS=16.0±0.8 J $°C^{-1}$ mol^{-1} for *G. gibberifrons*; ΔS=7.5±0.8 J ° C^{-1} mol^{-1} for *C. carpio*). This study has demonstrated that calcium affinity, binding enthalpy, and binding entropy are conserved among parvalbumins from species native to different thermal environments. Conservation of calcium affinity may preserve the absolute speed of the contraction/relaxation cycle of fast muscles.

Table 6. Binding of calcium to fragments of pike parvalbumin pI 5.0.

Fragment	K_1 (M^{-1})	K_2 (M^{-1})
1-108	3.3×10^8	5×10^8
38-108	10^6	10^7
75-108	217	
Ala74-108	500	
Arg74-108	1.6×10^3	

It is of importance to know the significance of different parts of a protein molecule in the Ca^{2+} binding process. Such information can be obtained from the studies of calcium binding properties of various fragments of the protein. Table 6 contains data on calcium binding for pike parvalbumin pI 5.0 (Medvedkin *et al.*, 1987).

The fragment 75-108, containing two helices and the calcium binding loop in the native protein, the main structural element of many calcium binding proteins, the "EF-hand", has a rather low affinity to Ca^{2+} ions: the equilibrium association constant is 217 M^{-1} according to intrinsic fluorescence data and 833 M^{-1} according to dansyl fluorescence data. It should be noted that this fragment does not contain Arg74 (or 75), which forms the salt bridge with Glu80 (or 81) in the native protein and plays a very important role in maintaining the native structure (Figure 4). An attachment of the missing Arg to the fragment increases the fragment's affinity to Ca^{2+} by an order of magnitude (Table 6; Medvedkin *et al.*, 1987). The attachment of Ala instead of Arg does not produce such effect. This means that the interaction between Arg74 and Glu80 occurs even in such a short peptide and this interaction makes the conformation more similar to that of the native protein. This is reflected also in an increase in ellipticity at 222 nm, which characterizes the helical content of the peptide. Moreover, the Ca^{2+}-induced increase in ellipticity at 220 nm is more pronounced for the peptide with the arginine residue. The binding of calcium to the 75-108 fragment of pike parvalbumin pI 4.2 induces very small changes in ellipticity at 220 nm. This is also the case for the EF fragment 76-108 of silver hake parvalbumin (Revett *et al.*, 1997). NMR and circular dichroism methods indicate significant secondary structure promotion in the EF fragment in the presence of the higher charge-density trivalent cations. Sedimentation equilibrium analysis shows that the EF fragment of silver hake parvalbumin exists in a monomer-dimer equilibrium when complexed with La^{3+}.

In the work of Franchini & Reid (1999) the -X glutamate in a 33-residue model peptide comprising the CD site of carp parvalbumin 4.25 (ParvCD) was replaced with aspartate (ParvCD-XD) and the effect on calcium-dependent dimerization and calcium affinity assessed. The peptide ParvCD has a 10^5-fold lower calcium affinity than the same site in the native protein. Both the ParvCD and ParvCD-XD model peptides did not bind magnesium. Replacement of the -X glutamate with an aspartate resulted in a two-fold increase in the calcium affinity of both the monomer and dimer forms and a twofold increase in the calcium dependent dimerization of the peptide.

Fragment 1-74, containing the CD-calcium-binding site, has also rather low affinity for Ca^{2+}, but it is still higher than that of the fragment 75-108 (Table 6; Medvedkin *et al.*, 1987). The presence in it of the N-terminal region seems to support the intact-like conformation. Similar fragment of carp parvalbumin pI 4.47 binds Ca^{2+} with association constant 10^4 M^{-1} (Derancourt *et al.* 1978).

The fragment 38-108 of pike parvalbumin pI 5.0, which differs from the intact protein by the absence of the helices A and B and the loop between them that does not take part directly in the binding of calcium, binds two Ca^{2+} ions with association constants 10^6 and 10^7 M^{-1} (Table 6; Medvedkin *et al.*, 1987). These values are almost two orders of magnitude lower than those for the native intact protein. The data of circular dichroism method demonstrate low helical content of the apo-state of the fragment. The binding of Ca^{2+} results in a pronounced increase in helical content. In contrast to this, the binding of Ca^{2+} to the intact protein causes much less pronounced effect. An attachment of Trp or Met to the N-terminus of the 38-108 peptide decreases its affinity to Ca^{2+}. It is of interest that the intact protein contains Met in this position. X-ray data show that the 38-108 fragmet keeps the native parvalbumin fold, but is more compact, having a well-structured linker (Thepaut *et al.*, 2001). The fragment has no stable apo-form, has lower affinity for Ca^{2+} than full-length parvalbumin, and does not bind Mg^{2+}. Structural differences in the hydrophobic core are particularly responsible for lowering the calcium-binding affinity of the truncated protein. It means that the whole protein structure, even those parts, which do not take part in the Ca^{2+} coordination directly, is needed for normal Ca^{2+} binding process.

It is surprising that fragments of parvalbumin preserve some affinity to each other. Permyakov *et al.* (1991) and Henzl *et al.* (2004a) observed the formation of noncovalent complex between domain AB and two-site domain CD*EF (association constant 10^5 M^{-1} for pike parvalbumin, $7.7x10^7$ M^{-1} for rat α-parvalbumin and $9.1x10^5$ M^{-1} for rat β-parvalbumin in saturating Ca^{2+}), accompanied by an increase of the affinity for Ca^{2+}. This result suggests that the AB domain can modulate the Ca^{2+} affinities of the CD and EF sites. Similar phenomenon is characteristic also for N- and C-terminal tryptic fragments of *Nereis* sarcoplasmic calcium-binding protein (Durussel *et al.* 1993). It is of interest also that a synthetic 33-residue EF-hand peptide with the sequence of carp parvalbumin CD site dimerize upon the binding of Ca^{2+} with the dimerization association constant of about $1.8x10^4$ M^{-1} (Franchini & Reid, 1999). Calcium association constants for the monomer and dimer forms of the peptide are $2.3x10^3$ and $2.1x10^4$ M^{-1}, respectively.

Attempts to reconstitute the fragments 1-75 and 76-108 of carp parvalbumin, under variety of conditions, into a functional complex which binds calcium were unsuccessful (Coffee & Solano, 1976).

Henzl et al. (2003a) examined the metal ion-binding CD-EF domains from rat α and β parvalbumin. It turned out that the CD-EF fragments differ markedly in their tendency to self-associate. Whereas Ca^{2+}-free α CD-EF is monomeric, the Ca^{2+}-free β peptide dimerizes weakly (association constant 2400±200 M^{-1}). In the presence of 1.0 mM Ca^{2+}, the apparent dimerization constant for β CD-EF (191 000±29 000 M^{-1}) is more than 50 times that of α CD-EF (3400±200 M^{-1}). According to the titration calorimetry data α CD-EF binds two Ca^{2+} with positive cooperativity with binding constants of 3.7(0.1)x10^3 M^{-1} and 8.6(0.2)x10^4 M^{-1}. β CD-EF also binds two Ca^{2+} cooperatively but with lower affinity. Equilibrium dialysis yields Adair constants of 4.2(0.1)x10^3 and 6.1(0.2)x10^3 M^{-1}. Significantly, the difference in Ca^{2+} affinity is substantially smaller than that observed for the full-length proteins suggesting that the AB domain can modulate divalent ion affinity.

The experiments on the parvalbumin fragments demonstrate that the whole protein structure is needed for the Ca^{2+} binding process but some amino acid residues play especially important role. This concerns, in particular, such residues as Arg-74 (75) and some others: removal of two C-terminal residues (Ala108 and Lys107) in carp parvalbumin pI 4.25 only slightly influences its affinity to Ca^{2+}, but the removal of the next one or two residues (Val106 and Leu105) seriously decreases its affinity to metal cations (Corson *et al.*, 1986). The F-helix in carp parvalbumin appears bound to the protein fold through a periodic series of hydrophobic contacts between its side chains (Val99, Phe102, Leu105, and Val106) and those of residues from helix B, the BC-linker region and helix C. Corson *et al.* (1986) emphasized the importance of two structural features of parvalbumin for maintenance of its native structure: ionic interaction between the C-terminal carboxyl and the NH_3^+ group of Lys27 and hydrophobic interaction of the internal part of the F-helix with residues of the hydrophobic core. The removal of the Lys107-Ala108 peptide produces a surrogate C-terminus, which is still well positioned to reestablish the ion pair bond with Lys27. Hydrolysis of the Val106-Lys107 peptide bond also produces a charged C-terminus, however, its orientation is not good enough to reestablish the ion pair bond with Lys27.

Apart from Ca^{2+}, Mg^{2+}, Na^+, and K^+, parvalbumin is able to bind some other metal cations. For example, Cave *et al.* (1979a,c) found a secondary Mn^{2+} binding site in carp parvalbumin pI 4.25 with equilibrium association constant 1.7x10^3 M^{-1}. The secondary site is located near the EF primary site and binds also Ca^{2+}, Mg^{2+}, and Cd^{2+} but with lower affinities. It was found that Cd^{2+} and Sr^{2+} ions compete with Ca^{2+} for the CD and EF sites in parvalbumin (Table 7; Cave *et al.*, 1979a).

It is well known that lanthanide ions easely replace Ca^{2+} in the CD and EF binding sites. Table 7 contains data on the binding constants of various lanthanides for parvalbumins. Sometimes spectral and nuclear spin properties of lanthanides are used for studies of metal binding sites in calcium binding proteins.

Calcium-saturated parvalbumin binds one equivalent of copper with K_a (Cu) ~5x10^5 M^{-1}; while apo-parvalbumin has K_a (Cu) ~1.7x10^5 M^{-1}, i.e. its affinity for copper is slightly lower than is that of the calcium loaded protein (Permyakov *et al.*, 1988c). The binding of copper to parvalbumin decreases the accessibility of its thiol groups to 5,5'-dithio bis(2-nitrobenzoic acid) and increases its affinity to the fluorescent probe bis-ANS. The binding constant of Zn^{2+} evaluated from the competition of copper and zinc for the same site is 5x10^4 M^{-1}

(Permyakov *et al.*, 1988c). Evidently, Zn^{2+} and Cu^{2+} binding sites in parvalbumin do not coincide with the Ca^{2+} binding sites.

Table 7. Parameters of interactions of parvalbumins with various metal cations.

Parvalbumin	Cation	$\log K_1$	$\log K_2$	Method, conditions	Reference
Pike pI 4.2	Tb^{3+}	11.3	7.2	Intrinsic fluorescence; 25 mM Pipers, pH 6.2; 21°C	Eberspach et al. 1988
Carp pI 4.25	Yb^{3+}	10.9	8.8	Optical stopped-flow, NMR; 15 mM PIPES, pH 6.8; 150 mM KCl; 23°C	Corson et al., 1983
Carp pI 4.25	Yb^{3+}	9.3	9.4	NMR; 15 mM PIPES, pH 6.6; 150 mM KCl; 10 mM DTT; 20°C	Lee & Sykes, 1981
Carp pI 4.25	Ce^{3+}	10.5	10.3	NMR; 10 mM PIPES, pH 5.9; 100 mM KCl; 55°C	Williams et al., 1984
Carp pI 4.25	Gd^{3+}	11.3	11.3	NMR; 50 mM cacodylate, pH 6.0	Cave et al., 1979a
Carp pI 4.25	Cd^{2+}	9.7	9.7	NMR; 50 mM cacodylate, pH 6.0	Cave et al., 1979a
Carp pI 4.25	Sr^{2+}	8.0	8.0	NMR; 50 mM cacodylate, pH 6.0	Cave et al., 1979a

KINETICS OF THE Ca^{2+} AND Mg^{2+} BINDING.

Equilibrium binding constants reflect relative affinities of a protein for various cations and their knowledge is very important when we consider equilibrium situation in solution. Apart from the equilibrium metal binding constants (K_{ass}), it is of importance to know kinetic association (k_{on}, $M^{-1}s^{-1}$) and dissociation (k_{off}, s^{-1}) rate constants related to the equilibrium constant by the equation:

$$K_{ass} = k_{on}/k_{off} \tag{17}$$

Their knowledge is extremely important when we want to study kinetics of metal ion changes in a cell.

Permyakov *et al.* (1987b) used fluorescence stopped flow method to measure kinetics of Ca^{2+} and Mg^{2+} dissociation from cod parvalbumin. In these experiments metal-loaded protein was mixed with EDTA. All the kinetic curves for this parvalbumin measured in the temperature range from 10 to 30°C were found to be best fitted by a sum of two exponential terms with time constants τ_1 and τ_2. Figure 20 represents τ_1 and τ_2 as functions of temperature. The two exponential terms in the experimental kinetic curves for parvalbumin were identified as arising from the dissociation of calcium and magnesium ions from the two binding sites of the protein. It was found that the values of Ca^{2+} association rate constants for parvalbumin in the temperature range from 10 to 30°C are within the range from 10^7 to 10^9 $M^{-1}s^{-1}$, i.e. they are close to the diffusion-controlled limit. Mg^{2+} association rate constants for parvalbumin are several orders of magnitude lower. At the same time, the dissociation rate constants for Ca^{2+} and Mg^{2+} are rather close to each other. So, the differences in values of equilibrium Ca^{2+} and Mg^{2+} association constants are due to the differences in association rate constants.

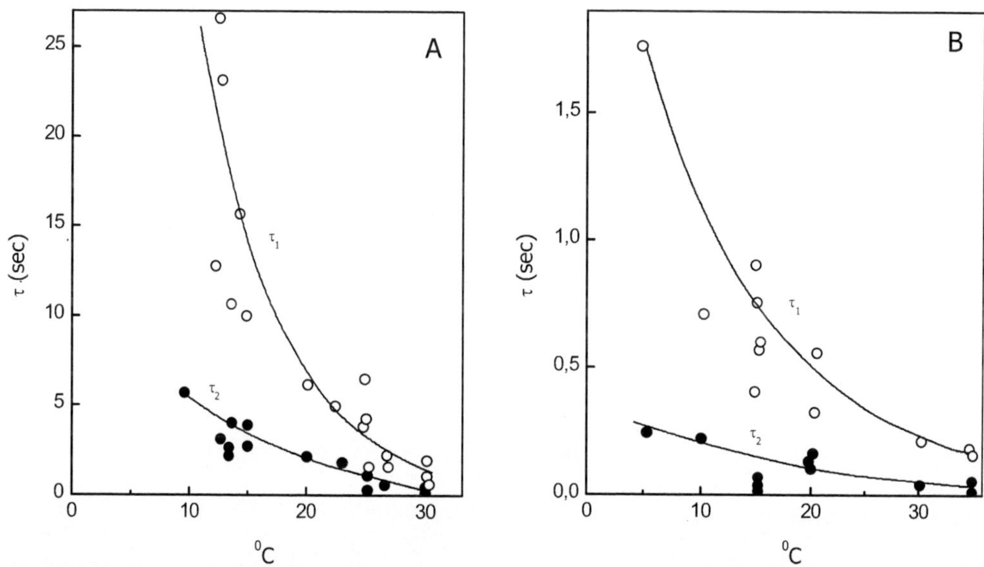

Figure 20. Temperature dependence of the time constants τ1 and τ2 determined from the stopped flow experiment on calcium (A) and magnesium (B) dissociation from cod parvalbumin.

Breen *et al.* (1985a,b) studied kinetics of dissociation of complexes of cod parvalbumin (*Gadus callarius L*) with Ca^{2+} ions at 24.8°C and pH 5.8. They also found that the fluorescence stopped-flow kinetic curves for the calcium removal by 1,2-diaminocyclohexanetetraacetic acid (DCTA) are best fitted only by a sum of two exponentials with relaxation times 0.93 and 0.17 s (0.65 and 0.095 s for Cd^{2+}; 20 and 2 s for trivalent lanthanide ions). Similar values for the relaxation times are reported in the work of White & Closset (1979) for the complex of whiting parvalbumin with Ca^{2+}. These values are somewhat lower than those obtained in the work of Permyakov *et al.*, but the difference is not very significant taking into account the differences in pH in experiments of these authors. In the work of Ogawa & Tanokura (1986a,b) kinetic curves for dissociation of the complexes of bullfrog parvalbumin with Ca^{2+} measured at 20°C and pH 6.8 were approximated by a single exponential with relaxation time 0.67 s, which is also somewhat lower than the values obtained in the work of Permyakov *et al.* At the same time, the value of relaxation time obtained in their work for Mg^{2+} (0.2 s) is in good agreement with the data of Permyakov *et al.* Potter *et al.* (1978) using a sulfhydryl-specific fluorescent label to study kinetics of Ca^{2+} dissociation from carp parvalbumin found that the relaxation time of this process is about 2 s while relaxation time for Mg^{2+} dissociation is 0.37 s, which is also in good agreement with the results of Permyakov *et al.*

White (1988) used fluorescence stopped flow method to study kinetics of the association and dissociation of calcium binding in whiting parvalbumin. In contrast to the work of Permyakov *et al.* (1987b), he did not find any intermediates in the course of spectrofluorimetric Ca^{2+}-titration and suggested that the only protein species present in significant concentrations are apo- and Ca_2-parvalbumin. It should be noted however that the isoemission point in his spectrofluorimetric EGTA-titration of parvalbumin is "smeared" and this seems to be due to not the dilution effects. The kinetics of calcium binding measured by stopped flow fluorescence were accurately single exponential from $2x10^{-7}$ to $2x10^{-4}$ M free calcium. The kinetics of calcium dissociation show a pronounced lag and are best fit by two rate constants of 1.2 and 3.0 s^{-1}, which is close to the data of Permyakov *et al.* (1987b). The lag is characteristic of a sequential two-step mechanism in which most of the spectroscopic change occurs within the second step. According to White (1988), the minimal kinetic mechanism that adequately describes the rate and equilibrium data is a branched pathway mechanism in which the rate and equilibrium constants are markedly different for each pathway:

$$\begin{array}{ccccccc} & & & \mathrm{PCa} & & & \\ 6x10^{6}\ M^{-1}s^{-1} & \nearrow\!\!\swarrow & 1.2\ s^{-1} & & 3.0\ s^{-1} & \nwarrow\!\!\searrow & 1.2x10^{8}\ M^{-1}s^{-1} \\ & \mathrm{P} & & & & \mathrm{CaPCa} & \\ \sim 10^{8}\ M^{-1}s^{-1} & \searrow\!\!\nwarrow & \sim 500\ s^{-1} & & 0.001\ s^{-1} & \nearrow\!\!\swarrow & 1.2x10^{6}\ M^{-1}s^{-1} \\ & & & \mathrm{CaP} & & & \end{array} \qquad (18)$$

At free calcium concentration lower then 2 μM the upper kinetic pathway of calcium binding predominates whereas at free calcium concentration higher then 2 μM calcium binding occurs predominantely by the lower kinetic pathway. The dissociation of calcium occurs according to the upper kinetic pathway.

Lanthanide ions dissociate from the binding sites of parvalbumin more slowly: relaxation times for the lanthanide dissociation from two sites of cod parvalbumin are 20 and 2 s (Breen *et al.*, 1985b). For carp parvalbumin pI 4.25 these values are 6.7 and 770 s (Corson et al., 1983).

The kinetic data on parvalbumin are very important from the viewpoint of elucidation of its physiological function. Gerday & Gillis (1976), Gillis & Gerday (1977), and Blum *et al.* (1977) studied the kinetics of Ca^{2+} transfer from Ca^{2+}-loaded frog parvalbumin to fragments of sarcoplasmic reticulum in a medium containing Mg-ATP and oxalate. They managed to show that the kinetics of Ca^{2+} release from parvalbumin is the same as the kinetics of Ca^{2+} removal from an aqueous solution of $CaCl_2$, i.e. the dissociation of the Ca^{2+}-parvalbumin complex is not a limiting step in this process. These authors and also Pechére *et al.* (1976, 1977) demonstrated that parvalbumin is able to take Ca^{2+} from myofibrills thereby inhibiting their ATPase activity, which can be then fully restored by addition of calcium. They concluded that there is the following Ca^{2+} gradient in muscle cells: myofibrills → parvalbumin → sarcoplasmic reticulum. They proposed that parvalbumin might serve as a shuttle for the transfer of Ca^{2+} between myofibrills and sarcoplasmic reticulum. At the same time, the presence of high concentration of parvalbumin in the sarcoplasm of muscle cell leads to a paradoxical situation: Ca^{2+} ions released in response to an electrical signal from the sarcoplasmic reticulum are forced to diffuse to the myofibrills through a powerful calcium buffer which consists of parvalbumin. This should delay the process of Ca^{2+} movement to the contractile system (Gillis & Gerday, 1977; Pechére, 1977). This contradiction can be avoided if the binding of Ca^{2+} to troponin proceeds faster than the binding to parvalbumin. Experiments of Pechére (1977) showed that an addition of Ca^{2+} to a Ca^{2+}-free system containing parvalbumin, myofibrills, and Mg-ATP causes a fast activation of the myofibrillar ATPase followed by its slow inactivation induced by the binding of Ca^{2+} to parvalbumin, i.e. the kinetics of the Ca^{2+} binding is very important in these processes.

The duration of the contraction and relaxation stages of fast skeletal muscles is 27 and 70 ms, respectively (Buller *et al.*, 1960). Since Mg^{2+} concentration in a resting muscle cell is as high as 1 to 6 mM (Brinley *et al.*, 1977) while free Ca^{2+} concentration is as low as 10^{-8} M, it was proposed that parvalbumin in such conditions must be loaded by Mg^{2+} ions (Haiech *et al.*, 1979a). When Ca^{2+} is released from sarcoplasmic reticulum, Mg^{2+} ions begin being replaced by Ca^{2+}. According to the hypothesis of Pechére *et al.* (Haiech *et al.*, 1979a), just the Mg^{2+} binding properties of parvalbumin explain why high concentrations of parvalbumin do not prevent Ca^{2+} ions from reaching the Ca^{2+}-specific sites of troponin C. A delay in the binding of Ca^{2+} caused by Mg^{2+} dissociation allows Ca^{2+} ions to reach the Ca^{2+}-specific sites of troponin C or calmodulin, which are not occupied by Mg^{2+}. Later on, the Ca^{2+},Mg^{2+}-sites of parvalbumin will successfully compete for Ca^{2+} and help to remove Ca^{2+} from the Ca^{2+}-specific sites of troponin C, which will then cause muscle relaxation. After that, Ca^{2+} ions bound to parvalbumin could be removed by the sarcoplasmic reticulum pump. So, Pechére *et al.* (Haiech *et al.*, 1979a) and other researchers (Briggs, 1975) proposed considering parvalbumin as a soluble relaxing factor in skeletal muscles.

In spite of the attractiveness of the Pechére's hypothesis that parvalbumin is a soluble relaxing factor it was disputed by many researchers. For example, according to the data of Cox *et al.* (1979) the dissociation of Mg^{2+} is too slow, so that Mg^{2+} content in parvalbumin is almost unchanged even after several successive contraction-relaxation cycles. Practically the same conclusion was obtained in the work of Potter *et al.* (Potter *et al.*, 1978, 1980, 1981; Robertson & Potter, 1980) on computer simulation of the kinetics of the Ca^{2+} and Mg^{2+} binding to various types of binding sites in muscle cell. According to their results, the mechanism of functioning of parvalbumin as a relaxing factor is not possible since Ca^{2+} and Mg^{2+} ions do not dissociate fast enough. Nevertheless they did not exclude that the Ca^{2+},Mg^{2+}- sites of parvalbumin still could exchange a large part of the bound Mg^{2+} for Ca^{2+} during repeated twitches or in tetanus.

Lehky & Stein (1979), knowing the amount of ATP hydrolyzed during the activation of myofibrills and their maximal ATPase activity in the presence of Ca^{2+}, calculated that it would take about 25 s to transfer Ca^{2+} from myofibrills to parvalbumin. This is three orders of magnitude slower than is required for the inactivation of fast skeletal muscle, which has a relaxation time of 70 ms. Lehky & Stein (1979) also believe that the exchange of Mg^{2+} by Ca^{2+} and regulation of Mg^{2+}-activated enzymes can proceed only after intensive muscle activity. The data of Moeshler *et al.* (1980) on thermodynamical and structural properties of Ca^{2+},Mg^{2+}-loaded parvalbumin are in favor of this idea.

Gillis *et al.* (1980, 1982) simulated kinetics of Ca^{2+} distribution between the Ca^{2+}-specific sites of troponin C and the Ca^{2+},Mg^{2+}-sites of parvalbumin and troponin C in response to a single Ca^{2+} impulse. They have considered the situation of the frog muscles wherein the concentration of the Ca^{2+},Mg^{2+}-sites is about 5 times higher than that of the Ca^{2+}-specific sites. Their results showed that after an impulse consisting of an increase in Ca^{2+} concentration, the Ca^{2+}-regulatory sites of troponin C are saturated very rapidly while the binding of Ca^{2+} to the Ca^{2+},Mg^{2+}-sites is rather slow. Nevertheless, Ca^{2+} is transferred gradually from the regulatory to the Ca^{2+},Mg^{2+}-sites such that the transfer is completed after 200 ms. This time is comparable with the time for a single contraction-relaxation cycle of frog *Sartorius*. However, if the concentration of the regulatory sites is comparable with the concentration of the Ca^{2+},Mg^{2+}-sites, as is the case in mammalian muscles, then the relaxing effect of the Ca^{2+},Mg^{2+}-sites will be negligibly small. The authors drew the conclusion that parvalbumins can participate in the muscle relaxation only when they are present in high concentration, i.e. when the concentration of the Ca^{2+},Mg^{2+}-sites is more than 5 times higher than the concentration of the Ca^{2+}-specific sites. It was proposed that the biological function of parvalbumin is to provide fast muscle relaxation over a wide temperature range since its relaxing effect is due to Ca^{2+} binding, a process which is less sensitive to temperature than the pumping of Ca^{2+} by sarcoplasmic reticulum which is controlled by the enzyme reaction. However, Gerday (1988) did not find any essential difference in parvalbumin content in two groups of frogs adapted to normal and low temperatures and also in tropical species in comparison with amphibians living at middle latitudes. He concluded that parvalbumins do not compensate for the loss of effectiveness in Ca^{2+} pumping by sarcoplasmic reticulum at low temperatures.

A model of calcium movement during activation of frog skeletal muscle was constructed by Cannell & Allen (1984). The model was based on the half sarcomere of a myofibril and included compartments representing the terminal cisternae, the longitudinal sarcoplasmic reticulum, the extramyofibrillar space and myofibrillar space. Troponin, parvalbumin and calsequestrin were present in appropriate locations and realistic binding kinetics. The model showed that parvalbumin gives a very important contribution to the fall of free calcium concentration in a twitch, i.e. it assists muscle relaxation.

Results of the work of Baylor *et al.* (1983) are in favor of Pechére's hypothesis as to function of parvalbumin. They measured free Ca^{2+} concentration inside the frog muscle cell during its contraction by means of the Ca^{2+}-sensitive fluorescent probe arsenazo III. The experimental data were used in calculations of concentrations of Ca^{2+} bound by Ca^{2+}-regulatory sites of troponin and Ca^{2+},Mg^{2+}-sites of parvalbumin and troponin C. The values of the association rate constants for these proteins were taken from literature. The calculations showed that two of every three Ca^{2+} ions leaving troponin are temporarily detained by parvalbumin before their return to sarcoplasmic reticulum.

Several authors proposed a multi-compartment model to estimate the time course of spread of calcium ions Ca^{2+} within a half sarcomere of a skeletal muscle fiber activated by an action potential (see for example Baylor & Hollingworth, 1998). Under the assumption that the sites of sarcoplasmic reticulum Ca^{2+} release are located radially around each myofibril at the Z line, such models can calculate the spread of released Ca^{2+} both along and into the half sarcomere. During diffusion, Ca^{2+} is assumed to react with metal-binding sites on parvalbumin (a diffusible Ca^{2+}- and Mg^{2+}-binding protein) as well as with fixed sites on troponin.

In the work of Carroll *et al.* (1997) calcium transients were calculated from fura-2 fluorescence signals (corrected for kinetic delays in the Ca^{2+}-fura-2 reaction) from single rat skeletal muscle fibres, either fully dissociated from the fast-twitch *flexor digitorum brevis* (FDB) muscle or in small bundles from the slow-twitch *soleus* muscle. Using the magnitude of the decline in the rate constant of free Ca^{2+} decay with increasing stimulation duration as an index of relative contribution of the saturable Ca^{2+} binding sites on parvalbumin, sub-populations termed 'high', 'medium' and 'low', referring to estimated parvalbumin content, were determined within each group of FDB and *soleus* fibres. In fibres assigned to the 'high' and 'medium' groups, parvalbumin was the major contributor (50-73%) to the free Ca^{2+} decay rate constant after a single action potential. In fibres in the 'low' group, parvalbumin contributed only 0-28% to the rate constant of free Ca^{2+} decay.

The fact of the good correlation of parvalbumin content in muscles with their relaxation rates (Heizmann *et al.*, 1982) is also in favor of the hypothesis of a direct participation of parvalbumin in the relaxation of fast muscles.

An analysis of the data presented shows that the contradictions therein have often arisen because of a poor knowledge of the association and dissociation rate constants for the complexes of parvalbumin with Ca^{2+} and Mg^{2+}. The data of Permyakov *et al.* (1987a,b) show that at physiological temperatures (10 to 15°C for cod-fish (Moiseev *et al.*, 1981)) Mg^{2+} dissociation times for cod parvalbumin are 900 and 100 ms. The duration of the contraction and relaxation stages for fast skeletal muscles of fishes is about several tens of milliseconds. This means that if the Mg^{2+}-Ca^{2+} exchange in parvalbumin follows the common competition scheme, the protein cannot serve as a soluble relaxing factor accelerating the removal of Ca^{2+} from troponin C during a single twitch. One can think that the Mg^{2+}-Ca^{2+} exchange in parvalbumin can be completed only after several successive twitches.

One could propose that parvalbumin still can serve as a soluble relaxation factor if the Ca^{2+}-Mg^{2+} exchange proceeds not due to simple competition, but due to the so called Ca^{2+}-activated Mg^{2+} dissociation. However, kinetic curves of the Mg^{2+}-Ca^{2+} exchange and Mg^{2+}-dissociation for polar cod parvalbumin practically coincide with each other. This suggests the absence of the Ca^{2+}-activated Mg^{2+}-dissociation in parvalbumin.

One more possibility for parvalbumin to serve as a soluble relaxing factor would be its direct interactions with troponin, but such a phenomenon has not yet been found.

Taking into consideration the high Na^+ and K^+ concentrations in muscle cell and the fact of the correspondence of equilibrium Na^+ and K^+ dissociation constants to their physiological concentrations, one could suppose that in a cell a certain part of the parvalbumin binding sites would be occupied just by these cations. If they dissociate from parvalbumin sufficiently quickly, this part of the protein would work as a soluble relaxing factor. However, this is not the case since the concentration of such protein must be negligibly small because of the very high Mg^{2+} binding constants of parvalbumin.

So, keeping in mind all the data presented we think that parvalbumin cannot serve as a soluble relaxing factor during a single contraction-relaxation cycle. The Mg^{2+}-Ca^{2+} exchange in parvalbumin can occur only after several successive twitches. At the same time, after many successive twitches parvalbumin will be saturated with Ca^{2+} and no longer contribute to relaxation which therefore will be slowed. Westerblad & Allen (1994, 1996) produced prolonged tetani in intact, single fibres of *Xenopus* frogs and single fibres from a mouse foot muscle while measuring force and the free myoplasmic calcium ($[Ca^{2+}]_i$) with indo-1. Mean rate constants of slowing of force relaxation with increasing tetanus duration ranged between 3.2 and 4.8 s^{-1}, thus, approximately similar to the Mg^{2+} off rate of parvalbumin. The authors concluded that in *Xenopus* fibres the slowing of relaxation with increasing tetanus duration can be explained by altered Ca^{2+} handling due to parvalbumin Ca^{2+} loading and impaired sarcoplasmic reticulum Ca^{2+} uptake. At the same time, this is not the case for mouse foot muscle.

EFFECTS OF METAL BINDING ON STABILITY OF PARVALBUMIN

Thermal Stability

The strong stabilizing effects of the interactions of metal cations with proteins are especially evident from the data obtained for parvalbumins. In the absence of bound cations, thermal stability of parvalbumin is not very high. According to NMR data of Williams *et al.* (1986a), apo-α-parvalbumin from rat muscle has two distinct conformations in aqueous solution. At 25°C, its highly structured form predominates. As deduced from both ^{1}H NMR and circular dichroism spectroscopy, this conformation is exceedingly similar to those of its Mg^{2+}-, Ca^{2+}-, and Lu^{3+}-bound forms. The temperature dependences of several well-resolved aromatic and upfield-shifted methyl ^{1}H NMR resonances and several circular dichroism bands indicate that the native, highly helical structure of rat apo-α-parvalbumin is unfolded according to a concerted mechanism, showing no indication of partially structured intermediates. The melting temperature of rat apo-α-parvalbumin is 35±0.5°C as calculated by both spectroscopic techniques. At 45°C, rat apo-α-parvalbumin is entirely unfolded, losing the tertiary structure that characterizes its folded form: not only are the ring-current-shifted aromatic and methyl ^{1}H NMR resonances leveled, but the 262- and 269-nm circular dichroism bands are also severely reduced. This less structured form of rat apo-α-parvalbumin shows an approximate 50% loss in apparent α-helical content compared to its folded state. The C_2H proton resonances of His26 and His48 in the folded form steadily decrease with temperature and disappear at 45°C. Two sets of ortho Phe doublet resonances are gradually reduced to zero by this increase in temperature. Thermal denaturation of the apo-form appears to eliminate all characteristic tertiary structure as well as to reduce its apparent α-helical content. The sharpness of the transition indicates that denaturation is highly cooperative.

The melting temperature of Ca^{2+}-loaded rat parvalbumin was estimated as 87°C (Williams *et al.*, 1986a). Interestingly, the transitions from structured to disordered forms not only occur at higher temperatures but are also less sharp for both Mg^{2+} and Ca^{2+} complexes. It was suggested that the denaturation process is catalyzed by charge-charge repulsions between anionic ligands in the CD and EF metal-binding loops, causing a disruption of interdomain contacts. The phenomenal stability of the tertiary structure of metal-bound parvalbumins arises due to a combination of interactions which are not directly affected by metal binding: hydrophobic contacts between the three helix-loop-helix domains; ion pair bonds such as the internal Glu81-Arg75 salt bridge; a network of paired surface charges such as Lys27-C-terminus ion pair.

Differential scanning microcalorimetry data for parvalbumin clearly show that the binding of calcium to parvalbumin shifts its thermal unfolding transition toward higher temperatures by more than 50°C. Ca^{2+}-loaded parvalbumin is very thermostable. The protein with two bound calcium ions is the most stable one. Transition enthalpy measured by means of microcalorimetry for cod parvalbumin is 432.6 kJ/mol. This is rather close to the value obtained for carp parvalbumin: 498.5 kJ/mol (Filimonov *et al.*, 1978). The heat sorption curve is asymmetrical for both parvalbumins. The calorimetric enthalpy for carp parvalbumin is 1.1 times higher than the Van't Hoff enthalpy, which does not permit one to assume that the melting process is a transition between two states. Fluorescence data for some of parvalbumins also show two stages in the thermal denaturation of calcium loaded parvalbumin (Permyakov *et al.*, 1983). The two-stepped character of the thermal transition is especially clearly seen for the protein with one bound Ca^{2+} ion. In the apo-form the two-stepped character of the transition disappears. It should be noted that the calorimetric and fluorescent curves for the protein in the presence of one equivalent of Ca^{2+} can not be represented as a sum of the corresponding curves for the apo- and $2Ca^{2+}$-protein, which is in favor of the successive scheme of the Ca^{2+} binding. The most pronounced changes of tryptophan fluorescence parameters take place in the second denaturation stage.

Thus, for parvalbumin the location and parameters of a thermal transition strongly depend on Ca^{2+} concentration. Taking into account the successive Ca^{2+}-binding scheme for parvalbumin and the two-stepped character of their thermal transition, one can suggest the following scheme of thermal denaturation of parvalbumin:

$$\begin{array}{ccccc} P & \overset{K_1}{\leftrightarrow} & PCa & \overset{K_2}{\leftrightarrow} & CaPCa \\ \alpha \downarrow\uparrow & & \downarrow\uparrow \beta & & \downarrow\uparrow \gamma \\ P^* & \overset{K_1^*}{\leftrightarrow} & P^*Ca & \overset{K_2^*}{\leftrightarrow} & CaP^*Ca \\ \alpha_1 \downarrow\uparrow & & \downarrow\uparrow \beta_1 & & \downarrow\uparrow \gamma_1 \\ P^{**} & \overset{K_1^{**}}{\leftrightarrow} & P^{**}Ca & \overset{K_2^{**}}{\leftrightarrow} & CaP^{**}Ca \end{array} \qquad (19)$$

where P is the native protein and P^* and P^{**} are the protein in two thermally changed states; K_1, K_2, K_1^*, K_2^*, K_1^{**}, and K_2^{**} are effective Ca^{2+} dissociation constants for different parvalbumin states; α, β, γ, and α_1, β_1, and γ_1 are equilibrium constants for the thermal transitions.

The two stages of the transitions $PCa \leftrightarrow P^*Ca \leftrightarrow P^{**}Ca$ and $CaPCa \leftrightarrow CaP^*Ca \leftrightarrow CaP^{**}Ca$ are clearly seen from the calorimetric and fluorimetric measurements.

The apparent constants K_1 and K_2 can be measured in the experiment on Ca^{2+} binding; the constants α, α_1 and β, β_1 could be evaluated from the corresponding experiments on the thermal denaturation, but the accuracy of the experimental methods is not sufficiently good for this. Moreover, α and α_1 cannot be obtained from the experimental data. At present it is not possible to evaluate the constants K_1^*, K_1^{**}, and K_2^*, K_2^{**}. The scheme predicts a Ca^{2+}-induced shift of the equilibrium towards more thermostable states, which actually is observed in experiments.

It should be noted that the thermal denaturation of parvalbumin does not cause formation of the random coil conformation. According to the optical rotation dispersion method, the helical content of Ca^{2+}-loaded carp parvalbumin pI 3.95 only slightly changes upon heating from 65 to 70°C. The helical content of parvalbumin measured at 25, 50 and 83°C is practically the same (Cave *et al.*, 1979b). ^{43}Ca NMR studies on carp parvalbumin pI 4.25 show that this protein binds Ca^{2+} even at temperatures above 80°C (Parello *et al.*, 1978), which also suggests the existence of rather compact protein structure at these temperatures.

Even in the absence of Ca^{2+} the helical content of parvalbumin almost does not change during the thermal denaturation process while its tertiary structure changes drastically (Cave *et al.*, 1979b): the denaturation causes the loss of residual contacts between aromatic rings of Phe residues and methyl groups of the aliphatic residues. In some cases the thermal denaturation of apo-parvalbumin is complicated by a protein aggregation (Permyakov *et al.*, 1983).

According to the data of Permyakov *et al.* (1983), Mg^{2+}-loaded parvalbumin denatures at temperatures 10 to 20°C lower than the Ca^{2+}-loaded protein. As in the case of Ca^{2+}, the transition passes through two stages, the transition temperatures differing by about 15°C.

Na^{+}- and K^{+}-saturated parvalbumins denature at about the same temperatures as the apo-protein though the monovalent cations in very large excess can increase the denaturation temperature of parvalbumin. For instance, carp parvalbumin pI 4.25 in 2 M NaCl denatures at 55°C instead of 32°C for the apo-protein (Filimonov *et al.*, 1978). Melting temperature for α isoform of rat parvalbumin in the absence of calcium and magnesium increases from 32 to 57°C when Na^{+} or K^{+} concentration increases from 0.01 to 2.0 M (Henzl *et al.*, 2000). For β isoform of rat parvalbumin such change in Na^{+} or K^{+} concentration results in a shift of the melting temperature from 40 to 66°C. Whereas the melting temperature of the α isoform is just 35 °C in 0.24 M K^{+} at pH 7.4, it shifts up to 46 °C in 0.24 M Na^{+}. The disparate effects of K^{+} and Na^{+} on α-parvalbumin stability contrast the influence of simple electrostatic screening (K^{+}) and that due to specific ion binding (Na^{+}). Moreover, the pronounced impact of Na^{+} binding in just one of the two EF-hand motifs also suggests that electrostatic destabilization is concentrated in the region of the ion-binding sites, consistent with the clustering of negatively charged side chains in the binding loops.

It is of interest that despite its higher net charge and reduced opportunities for favorable tertiary interactions, Ca^{2+}-free rat β-parvalbumin is more stable than rat α-parvalbumin (Henzl & Graham, 1999; Agah *et al.*, 2003). Using scanning calorimetry method these authors showed that under conditions wherein α-parvalbumin denatures at 45.8°C, β-parvalbumin denatures at 53.6°C in the absence of calcium ions. Extension of the F helix in the β-parvalbumin, by insertion of Ser109, has a modest stabilizing effect raising the denaturation temperature by 1.1°C. Truncation of the α-parvalbumin F helix, by removal of Glu108, has a more profound impact, lowering the denaturation temperature by 4.0°C. The homologous chicken β-parvalbumin isoform known as CPV3 also exhibits elevated stability. Individual P21A and P26A mutations lower the denaturation temperature of rat β-parvalbumin by 3.2°C, decreasing conformational stability by 0.74 kcal/mol. Simultaneous replacement of Pro21 and Pro26 essentially decreases the stability (ΔT_m = -7.6°C; $\Delta\Delta G_{conf}$ = -1.77 kcal/mol). Surprisingly, the P21A/P26A mutant displays Ca^{2+} affinity virtually indistinguishable from wild-type β-parvalbumin, implying that structural alterations in the AB domain do not necessarily influence the divalent ion affinity of the CD-EF domain.

The consequences of introducing proline at positions 21 and 26 in rat α-parvalbumin were also examined (Agah *et al.*, 2003). Whereas the H26P mutation raises the denaturation temperature by 5.6°C ($\Delta\Delta G_{conf}$ = 1.25 kcal/mol), the mutation A21P lowers the denaturation temperature by 8.5°C ($\Delta\Delta G_{conf}$ = -1.9 kcal/mol). The replacement of Ala21 by proline in an α

AB/β CD-EF chimera increases the denaturation temperature by 5.8°C ($\Delta\Delta G_{conf}$ = 0.95 kcal/mol), which seems to show that the destabilization of α-parvalbumin by Pro21 results from steric conflict with a residue in the CD-EF domain. Consistent with that hypothesis, the K80S mutation markedly stabilizes α A21P, yielding a protein with a denaturation temperature 2.0°C higher than wild-type α-parvalbumin.

Stability Against Action of Denaturants

Ca^{2+}-loaded parvalbumins are also very stable against the action of the denaturants. The helical content of Ca^{2+}-loaded parvalbumins of carp, pike and whiting does not change up to 4-5 M urea; further increase in urea concentration up to 8-9 M causes the complete denaturation of the protein with formation of a random coil structure (Closset & Gerday, 1975; Cox *et al.*, 1979; Donato & Martin, 1974). These results are in good agreement with the data of Permyakov *et al.* (1982b, 1983a) obtained by the intrinsic fluorescence method for whiting and cod parvalbumins. The increase in urea concentration causes pronounced spectral changes. The fluorescence spectrum of the protein in 7.5 M urea is shifted towards longer wavelengths by 33-34 nm in comparison with the spectrum of the native protein and its position coincides with that for free tryptophan in water. Fluorescence spectrum of the apo-parvalbumin has a maximum at 346 nm. This means that the structure of the cation-free parvalbumin is much more compact than that of the protein in 7.5 M urea.

The main part of the spectral changes upon the denaturation of Ca^{2+}-loaded whiting or cod parvalbumin occurs at urea concentrations from 4 to 6 M (Permyakov *et al.*, 1982b, 1983a). The two-stepped decrease in fluorescence quantum yield suggests the existence of an intermediate state in the course of the denaturation process at 3.2 M urea. The simplest denaturation scheme of parvalbumin taking into consideration the binding of calcium ions to different protein states looks like this:

$$\begin{array}{ccccc} P & \leftrightarrow & PCa & \leftrightarrow & CaPCa \\ \downarrow\uparrow & & \downarrow\uparrow & & \downarrow\uparrow \\ P^{*} & \leftrightarrow & P^{*}Ca & \leftrightarrow & CaP^{*}Ca \\ \downarrow\uparrow & & \downarrow\uparrow & & \downarrow\uparrow \\ P^{**} & \leftrightarrow & P^{**}Ca & \leftrightarrow & CaP^{**}Ca \end{array} \qquad (20)$$

P^{*}, $P^{*}Ca$, and $CaP^{*}Ca$ are for intermediate state of parvalbumin; P^{**}, $P^{**}Ca$, and $CaP^{**}Ca$ are for fully denatured states of parvalbumin.

Different parvalbumins have different stability against urea. For example, according to the fluorescence data, Ca^{2+}-loaded pike parvalbumin pI 4.2 starts to denature at urea concentrations above 6 M, while another pike parvalbumin pI 5.0 at the same conditions denatures at urea concentrations above 8 M (Figure 21A).

In spite of the fact that urea causes almost complete unfolding of parvalbumin, the protein continues to bind Ca^{2+} even in the presence of 7-9 M urea though with very low affinity. A gradual increase of Ca^{2+} concentration induces changes in protein structure, which are opposite to those caused by the urea denaturation. However, we can hardly reach structural parameters, which are characteristic of the Ca^{2+}-loaded protein even at very high Ca^{2+} concentrations (up to 12 mM). According to the intrinsic fluorescence data, Ca^{2+} binding constants for whiting parvalbumin in the presence of 7.5 M urea are within the range 10^2 to 10^3 M^{-1} (Permyakov *et al.*, 1982b, 1983a).

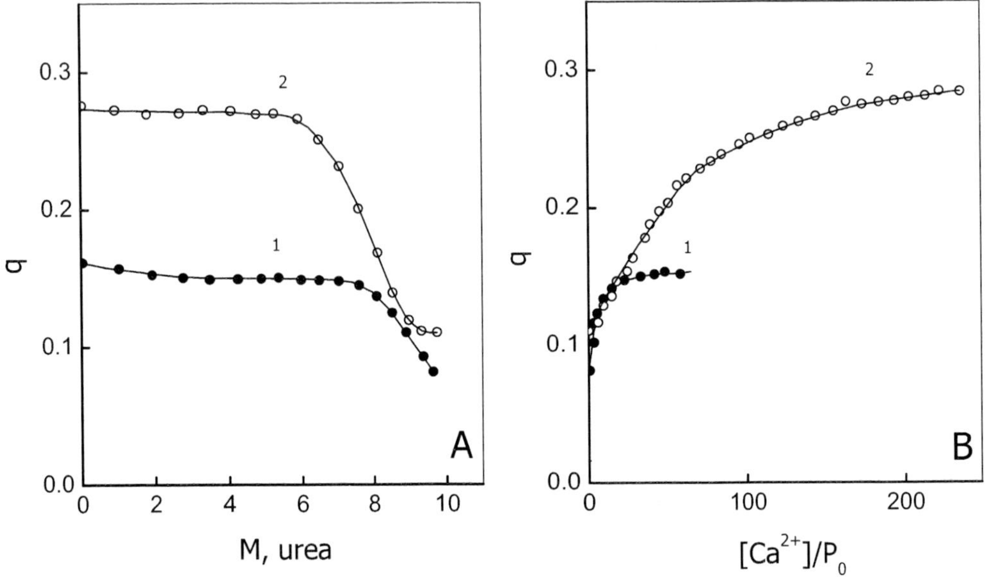

Figure 21. A - urea denaturation of pike parvalbimins pI 4.2 and 5.0 monitored by intrinsic tyrosine and phenylalanine fluorescence. B – Ca^{2+}-titration of pike parvalbimins pI 4.2 and 5.0 in the presence of 9.7 M urea.

The results of the experiment involving Ca^{2+}-titration of two pike parvalbumins in 9.7 M urea are shown in Figure 21B. The Ca^{2+}-titration of both parvalbumins results in restoration of the fluorescence yield values characteristic of the native proteins. The curve for pike parvalbumin pI 5.0 reaches a plateau faster than that for pike parvalbumin pI 4.2. The binding of Ca^{2+} seems to restore the native structure of the parvalbumins even in the presence of such high urea concentrations.

The results of the experiments described above mean that the position of the denaturation transition for parvalbumins in the scale of urea concentrations strongly depends on Ca^{2+} concentration in the solution. For example, only 20-fold molar excess of Ca^{2+} protects pike parvalbumin pI 5.0 against the denaturing action of 10 M urea. In contrast to this, Ca^{2+}-free parvalbumin denatures at rather low urea concentrations: the whole denaturation process is complete at 2 to 3 M of urea.

Ca^{2+}-loaded parvalbumins are very stable also against denaturation by guanidine hydrochloride. According to the data of fluorescence emission and triplet absorption spectroscopy, denaturation of cod parvalbumin in the presence of 5 mM Ca^{2+} occurs within the region from 3 to 5 M of guanidine hydrochloride (Sudhakar *et al.*, 1995b). The guanidine hydrochloride unfolding reaction depends upon Ca^{2+}, and ΔG values are as follows: 22.9 and 44.2 kJ/mol for no added Ca^{2+} and 5 mM Ca^{2+}, respectively. The stability toward denaturation imparted by the binding of two Ca^{2+} is about -60 kJ/mol. Circular dichroism and fluorescence data show that carp parvalbumins pI 3.95, 4.25, and 4.47 lose almost completely their ordered structure in 3.5 M guanidine hydrochloride and that the denaturation curves have midpoints at 1.8-2.2 M (Lin & Brandts, 1978). Proton NMR spectra of carp parvalbumin pI 4.25 and hake parvalbumin pI 4.36 in 6 M guanidine hydrochloride allow one to conclude that even in this case the random coil structure is not formed since the observed chemical shifts of some resonances are larger than those for proteins in the random coil state (Parello *et al.*, 1974).

The denaturation of parvalbumins by guanidine hydrochloride is completely reversible, which was checked in dilution experiments. Lin & Brandts (1978) studied kinetics of denaturation of carp parvalbumins by guanidine hydrochloride and kinetics of their renaturation. Fluorescent kinetical curves of all the parvalbumins contain a fast stage, which can be splited into two with relaxation times differing by a factor of 4 to 7. At the transition point the shortest relaxation time, τ_1, is 16 s for carp parvalbumin pI 4.47 and 25 s for carp parvalbumin pI 4.25. Kinetic curves for carp parvalbumin pI 3.95 contain an additional, the most slow stage. Its existence is connected with the presence of the proline residue in this parvalbumin, which is absent in two others. The activation enthalpy of renaturation of carp parvalbumin pI 3.95 is 67-75 kJ/mol, which corresponds to the value, expected if the denaturation and renaturation kinetics of this parvalbumin are controlled by isomerization of the proline residue. An increase in Ca^{2+} concentration accelerates the first two renaturation stages for all the parvalbumins but does not influence the slowest renaturation stage of carp parvalbumin pI 3.95.

Reversible equilibrium unfolding of carp parvalbumin III when treated with guanidine hydrochloride and the kinetics of folding and unfolding by concentration jump of the denaturant was studied by the peptide circular dichroism spectra (Kuwajima *et al.*, 1988). It turns out that the equilibrium and kinetics of the guanidine hydrochloride-induced unfolding of parvalbumin can be fully interpreted in terms of a three-state reaction scheme. In spite of the fact that the protein has no proline residues, in the kinetic refolding reaction, a transient folding intermediate was found to be rapidly accumulated within the dead time of the stopped-flow instrument (18 ms). The intermediate has 60-80% of the α-helix of the native protein, but does not have the specific structure responsible for strong Ca^{2+} binding, and it is similar to the partially unfolded state observed at acid pH. Three stages were revealed in the folding process of the parvalbumin: (i) rapid formation of the secondary structure framework; (ii) organization of a part of the specific tertiary structure including one of the two Ca^{2+}-binding domains, with this step leading to the activated state of folding; and (iii) the final stabilization associated with organization of the rest of the molecule including the other Ca^{2+}-binding domain.

Acid Transition in Parvalbumin

Potentiometric titration of carp parvalbumin pI 4.25 in the presence of 33 mM $CaCl_2$ or 8 M guanidine hydrochloride (almost completely unfolded protein) was performed by Filimonov *et al.* (1978). The amount of titratable groups in pH region from 8 and 1.5 is 21 in agreement with the amino acid content of the protein (14 Asp, 6 Glu, 1 His). The most interesting feature of the titration curve measured in the presence of $CaCl_2$ is the jump in the proton uptake beneath the isoelectric point. The titration is fully reversible in the ranges of pH from 11 to 4 and from 3 to 1.5, but at pH 3-4 slight turbidity appears.

At low pH values protons can compete with Ca^{2+} for the same carboxylate oxygens in the Ca^{2+} binding sites and displace calcium ions from the sites at very acid pH values, which results in changes of the protein conformation similar to those caused by removal of calcium at neutral pH. This is reflected, for example, in pronounced spectral effects (two-stepped red spectral shift and decrease in tryptophan fluorescence yield for tryptophan-containing parvalbumins). Increase in Ca^{2+} concentration shifts the transitions toward lower pH values. One can assume that the two-stepped character of the curves arises because of the protonation of two classes of carboxylates in the two binding sites of parvalbumin. Spectrofluorimetric pH-titration curves for whiting (Permyakov *et al.*, 1982b) and pike (Permyakov *et al.*, 1983a) parvalbumins have similar two-stepped character. The changes in ionic state of carboxyl groups in carp parvalbumin pI 4.47 upon change in pH from 4.8 to 6.5 were detected by proton NMR methods (Birdsall *et al.*, 1979).

Brandts *et al.* (1977) and Lin & Brandts (1978), using fluorescence and circular dichroism methods, studied the kinetics of conformational changes in carp parvalbumins pI 4.25 and 3.95, which were induced by pH jump in the acidic region. Kinetic curves for parvalbumin pI 4.25 display two stages with relaxation times from several milliseconds to several tens of milliseconds (depending on initial and final pH). Kinetic curves for parvalbumin pI 3.95 contain an additional slower stage with relaxation time about 5 s. The presence of this stage is connected with the presence of the proline residue in this parvalbumin. Isomerization of the proline slows down the conformational change.

Interactions with Organic Substances and Peptides

Parvalbumin is able to interact with some organic compounds and peptides. For example, Permyakov *et al.* (1982a) observed a binding of nucleotides ATP and ADP to parvalbumin. The parameters for the interactions of parvalbumin with the nucleotides were evaluated using intrinsic fluorescence method. The values of the parameters are collected in Table 8.

It may be worth taking into consideration some physiological consequences of the binding of nucleotides to parvalbumin. The affinity of parvalbumin for nucleotides is sufficiently high that a significant part of parvalbumin in a muscle cell could be in a complex with ATP, (especially in the apo- and Mg^{2+}-loaded states i.e. in a resting cell) or with ADP (especially in the Ca^{2+}-loaded state, i.e. during the relaxation stage when the ADP level is especially high). Just this change in the affinity of parvalbumin for ATP and ADP forces us to think about a physiological significance of the nucleotide binding.

Table 8. Apparent equilibrium constants for the binding of ATP and ADP to whiting parvalbumin (Permyakov *et al.*, 1982a).

State of protein	ATP		ADP	
	K_1 (M^{-1})	K_2 (M^{-1})	K_1 (M^{-1})	K_2 (M^{-1})
Metal-free	770±100	95±20 m=4-6	290±20	
Ca^{2+}-loaded	77±20		670±100	106±20 m=4-6
Mg^{2+}-loaded			190±20	

It was found that parvalbumin binds heme-CO with association constant $2x10^6$ M^{-1} in the presence of 1 mM Ca^{2+} (Leclerc *et al.*, 1993). The removal of calcium increases its affinity for heme-CO twofold and provokes further red shift of the Soret band. Parvalbumin binds four molecules of heme-CO in the absence of calcium, whereas only two molecules of heme-CO bind to parvalbumin in the presence of calcium. It is of interest that calmodulin and troponin C bind heme-CO only in their Ca^{2+}-bound state. Heme-CO is thought to be a useful probe in detecting and analyzing hydrophobic pockets in proteins.

Haiech *et al.* (1979) found that parvalbumin can bind EGTA but with low affinity (dissociation constant about 35 mM), which should be taken into account when EGTA is used as Ca^{2+}-buffer for determination of calcium binding constants.

Using tryptophan fluorescence and circular dichroism methods, Sudhakar *et al.* (1995a) found that cod parvalbumin can bind polyamine cations, spermine, spermidine, and putrescine, which occur naturally in eucaryotic cells at millimolar concentrations. The dissociation constants for the interactions at pH 6 and 7 are about 4 mM, 8 mM and >20 mM for spermine, spermidine and putrescine, respectively, indicating dependence on the charge (+4, +3, and +2, respectively) and/or length and size of the polyamine. The polyamines bind to Ca^{2+}-free or partially saturated with Ca^{2+} parvalbumin. The authors suggested that polyamines *via* parvalbumin may have a role in maintaining Ca^{2+} concentrations or that Ca^{2+} and polyamines may together play a role in keeping parvalbumin in a stable form.

While trifluoperazine induces pronounced changes in ^{113}Cd NMR spectrum of ^{113}Cd-substituted calmodulin and troponin C, it does not change the ^{113}Cd NMR spectrum of parvalbumin (Thulin *et al.*, 1980). The conclusion was that parvalbumin does not bind trifluoroperazine. Our data obtained by differential spectrophotometry method demonstrate that parvalbumin can bind some drugs such as verpamil and that this interaction is Ca^{2+}-dependent. Verpamil binding constant for Ca^{2+}-loaded pike parvalbumin pI 4.2 is about 10^5 M^{-1}.

Silver hake parvalbumin has an ability to bind the hydrophobic fluorescence probe molecule 1-anilino-naphthalene-8-sulfonate (ANS) in the absence of calcium and in the presence of magnesium ions (Zhang *et al.*, 1990). Addition of ANS to the apo-protein results in dramatic enhancement of ANS fluorescence. Dissociation constant of the 1:1 complex is 10-15 μM. Additional ANS molecules bind with weaker affinity to surface sites.

Parvalbumin can interact with some dangerous, poisoning substances. One of them is acrylonitrile, which is an important industrial chemical used in the production of plastic, synthetic fibers, and insecticides. Acrylonitrile was shown to produce a variety of toxicological effects, ranging from symptoms of cyanide poisoning acutely to tumor formation in chronic studies. A relationship between exposure to acrylonitrile and induction of tumors at several sites in the rat was demonstrated. It was shown that acrylonitrile has a long half-life in rainbow trout muscle and that [^{14}C] acrylonitrile appears to be bound to parvalbumin (Lech *et al.*, 1995). To study the reaction between parvalbumin and [^{14}C] acrylonitrile, frog parvalbumin was incubated with [^{14}C] acrylonitrile *in vitro* under various conditions (Lech *et al.*, 1996). It was found that the maximum labeling occurred at 1 nmol/nmol parvalbumin and at pH 7. The reaction occurs only in the presence of calcium ions. Amino acid analysis of the labeled protein indicated that the labeled amino acid is probably histidine, and endoproteinase Glu-C (V-8) digestion studies revealed that the ^{14}C is located in the 1-81 amino acid segment of the protein, an area that contains two histidines at positions 2 and 27.

Parvalbumin can interact with short amhiphylic peptides. Permyakov *et al.* (1989b) revealed an interaction of parvalbumin with bee venom melittin. It was found that Ca^{2+}- or Mg^{2+}-loaded pike parvalbumin pI 4.2 causes very weak changes in the tryptophan fluorescence of melittin, while in the absence of the bivalent cations parvalbumin induces 10 nm blue shift in the melittin fluorescence spectrum and an increase in fluorescence yield. Similar but less pronounced effects are induced by pike parvalbumin pI 5.0 and its fragment 38-108. The titration curves reach plateau at a protein:peptide ratio 1:1. This suggests 1:1 stoichiometry for the parvalbumin-melittin complex. The formation of the complex transfers the single tryptophan residue of melittin from aqueous to a more hydrophobic environment. The binding constant of melittin to pike parvalbumin pI 4.2 at 18°C is 10^6 M^{-1}. At 47°C, when the apo-parvalbumin is in the thermally unfolded state, the binding constant is still higher: 1.3×10^7 M^{-1}. Pike parvalbumin pI 5.0 has lower affinity to melittin: at 18°C the binding constant is 1.8×10^5 M^{-1}. The melittin binding constant of fragment 38-108 (the protein without A and B helices) is 6×10^4 M^{-1}.

INTERACTIONS WITH MEMBRANE SYSTEMS

Parvalbumins interact with bilayer membrane systems and this interaction is modulated by the binding of divalent cations. Permyakov *et al.* (1988b) revealed an interaction of parvalbumin with membranes of dipalmitoylphosphatidylcholine vesicles. According to the electron microscopy data, the diameter of the liposomes used in these works was from several hundred to several thousand Angstroms. These authors used in their work microcalorimetry, intrinsic fluorescence and gel-chromatography methods. It is well known that the narrow calorimetric peak with maximum at 41°C for pure liposomes from dipalmitoyl-phosphatidylcholine corresponds to the phase transition from the gel to the liquid crystalline state. Interactions of liposomes with various substances cause changes in this transition. In order to increase the sensitivity of the microcalorimetry method to changes in position and shape of the thermogram for liposomes in the presence of proteins, the authors measured the difference between the heat sorption curves for liposomes in the presence and in the absence of protein. The thermal denaturation of the protein was measured by microcalorimetry and intrinsic fluorescence.

The data obtained show the existence of interactions between a part of populations of parvalbumin and dipalmitoylphosphatidylcholine vesicles. The interactions change the physical properties of the protein and liposomes, increasing or decreasing their thermostability. It is of importance that these interactions are modulated by Ca^{2+} and Mg^{2+} ions.

The calorimetric and fluorimetric data are corroborated by results of gel-chromatography on Sepharose 4B. Upon chromatography of the mixture of parvalbumin with liposomes a portion of the protein is eluted with liposomes, i.e. a part of the protein is in a complex with liposomes.

The same results were obtained also in an experiment using chromatography of the mixture of the vesicles with cod parvalbumin on a Sephadex G-75 column. It was determined that about 10 % of the total protein is bound to the liposomes (about several protein molecules per liposome).

Chromatography on Sephadex G-75 in another regime, in which the column is equilibrated with protein solution of a known concentration and the liposomes are applied to it (Hummel-Dreyer method), demonstrates that the total quantity of parvalbumin bound to the liposomes is higher: roughly about 30 molecules per liposome for the Ca^{2+}-loaded protein, about 50 molecules per liposome for the Mg^{2+}-loaded protein and about 70 molecules per liposome for the apo-protein. The association constant for the complexes of liposomes with the protein evaluated from the Scatchard plots is 10^5 M^{-1} (Permyakov *et al.*, 1989a).

Different metal-bound states of parvalbumin affect the phase transition in the liposomes differently. The most pronounced effects are found in the case of the Ca^{2+}-free protein. These effects are less pronounced for the Ca^{2+}-loaded protein. Taking into consideration the opposite signs of the effects of liposomes on the thermal transition in Ca^{2+}-loaded and Mg^{2+}-loaded parvalbumin, one can assume that the interaction of these forms of the protein are of different character.

Interstingly, parvalbumin immunoreactivity was found in (or near) membraneous systems, like mitochondria and (or) microtubuli in male gonads of *Drosophila melanog*aster (Fritz-Niggli *et al.*, 1988).

Chapter 4

FUNCTIONS OF PARVALBUMIN

CALCIUM BINDING PROTEINS IN MUSCLE CONTRACTION

Calcium ions play very important role in regulation of contraction of all types of muscles. As in any other cell, the free calcium concentration in a resting muscle cell is 0.05-0.1 μM (reviewed by Konishi, 1998). However, when an external signal comes to the cell, free calcium concentration increases by one to two orders of magnitude. Depending on the type of muscle, these calcium ions are released either from intracellular stores (sarcoplasmic reticulum) or pumped up through sarcolemma from the intercellular space where the free calcium concentration is about several millimoles per liter. Free magnesium concentration in a resting muscle cell is about 1 mM (reviewed by Konishi, 1998).

In skeletal muscle, following action potential propagation along the transverse tubular membrane, calcium ions are released from sarcoplasmic reticulum to produce a transient rise of free Ca^{2+} ($[Ca]_i$). Inside sarcoplasmic reticulum calcium ions are bound by calsequestrin and calcium binding protein 55 kDa. In response to an electrical signal the membrane of sarcoplasmic reticulum becomes permeable to calcium ions and they pass to the sarcoplasm. The sarcoplasm of any cell contains calmodulin, which is a Ca^{2+}- dependent regulator of the activity of many enzyme systems in the cell. The binding of calcium to calmodulin induces changes in its conformation and in this state it is able to activate many enzyme systems. The sarcoplasm of fast muscles contains high concentrations of parvalbumin. From the knowledge of physico-chemical properties of parvalbumin, it is evident that it can modulate Ca^{2+} flows in the cell. Ca^{2+} ions reach the contractile apparatus passing through the sarcoplasm. Here we can also find several important calcium binding proteins. The main component of the thick filaments in myofibrils, myosin, is a calcium binding protein because it contains regulatory light chains characterized by high affinity to calcium. The main component of thin filaments, actin, is also high affinity calcium binding protein. Moreover, thin filaments contain troponin complex, one of the components of which, troponin C, also binds calcium. The binding of calcium to troponin C triggers acto-myosin interaction in skeletal muscles. Contractile force is a function of peak amplitude and time course of the Ca^{2+} transients. The return of calcium ions to sarcoplasmic reticulum is achieved by means of the activity of Ca^{2+},Mg^{2+}-ATPase of sarcoplasmic reticulum.

Let's consider the calcium binding proteins of these three systems (sarcoplasmic reticulum, sarcoplasm and myofibrils) in more detail.

The main protein of the sarcoplasmic reticulum membrane is Ca^{2+}-Mg^{2+}-ATPase (for review see, for example, Mooren & Kinne, 1998; Carafoli, 2002). This enzyme, tightly bound to the membrane, represents a Ca^{2+}-pump. There are at least five different isoforms with molecular masses varying from 100 to 115 kDa. The polypeptide chain of Ca^{2+}-Mg^{2+}-ATPase is supposed to span the membrane 10 times and to have 3 cytoplasmic units. The external part of the Ca^{2+}-pump protein contains one ATP binding site and two calcium binding sites with binding constant 3×10^{6} M^{-1}. The Ca^{2+} binding sites differ from the EF-hand sites. One cycle of the ATPase couples the hydrolysis of one ATP molecule to the uptake of two calcium ions. Transmembrane helices 4, 5, 6, and 8 appear to be involved in Ca^{2+} binding and form the path for Ca^{2+} across the protein. The three cytosolic domains have been termed N (nucleotide binding), P (phosphorylation), and A (actuator, or N-anchoring domain). The distance between the ATP binding site in the N domain and the catalytic aspartic acid in the P domain is about 25Å, but large conformational changes during ATP-energized calcium translocation move the N and P domains closer to each other.

The Ca^{2+} binding site I is located in the space between transmembrane domains 5 and 6 with a contribution from domain 8. The site II is formed almost entirely by carbonyl oxygens on the domain 4 and by D800 of the domain 6. One mole of the enzyme binds 2 mol Ca^{2+} cooperatively. Conformational change is needed for the cooperative binding of two Ca^{2+}, occlusion of the bound ion, and movement through a channel formed by transmembrane helices. Experimental data support the idea that a single conformational change of the ATPase from state E1 to state E2 may explain how Ca^{2+} can be pumped through the membrane: E1 and E2 conformation states correspond to high and low cation binding affinities. The equilibrium between the two states is controlled by ligand binding and ATP consumption. Active transport depends on the functional linkage between the phosphorylation and the Ca^{2+}-binding domain.

The affinity for Ca^{2+} of the pump can be greatly enhanced by phospholamban, homopentameric molecule, consisting of subunits with a molecular mass of 6 kDa. Upon phosphorylation, the inhibitory action of phospholamban on the pump is abolished and the calcium affinity of the pump is increased.

The skeletal muscle Ca^{2+} release channel, known also as the ryanodine receptor, RyR, to which the plant alkaloid ryanodine specifically binds, is composed of four polypeptides of about 5000 amino acids each. RyR channels are highly clustered square structures arranged in regular rows and that the corners of adjacent channels contact each other. Analysis of the deduced 1513 C-terminal amino acid sequence suggests the existence of a cytoplasmic Ca^{2+}-binding domain consisting of two EF-hand Ca^{2+}-binding motifs (Xiong et al., 1998). The protein binds two Ca^{2+} ions per molecule with dissociation constant 0.9 mM and Hill coefficient 1.4. It is well known that both mammalian and lobster ryanodine receptors are activated by micromolar Ca^{2+} and inhibited by millimolar Ca^{2+}. This biphasic behavior may result from at least two classes of Ca^{2+} binding sites, high-affinity activation site and a low-affinity inactivation site. So the sites found seem to be the low-affinity inactivation sites. The ryanodine receptor mediates the t-tubular depolarization-induced Ca^{2+} release from the sarcoplasmic reticulum. In skeletal muscle, activation of Ca^{2+} release from the sarcoplasmic reticulum is controlled by a voltage sensor in the transverse tubular membrane. Elementary Ca^{2+} signal events represent openings of individual RyR in the sarcoplasmic reticulum membrane and have been termed Ca^{2+} sparks and Ca^{2+} quarks. The initial Ca^{2+} release activates additional Ca^{2+} sparks by Ca^{2+}-induced Ca^{2+} release from the sarcoplasmic reticulum. It is believed today that the signal transmission from the dihydropyridine receptor to the RyR is achieved by mechanical coupling.

Two other sarcoplasmic reticulum calcium binding proteins, calsequestrin and 55 kDa calcium binding protein, are contained inside the reticulum, do not bind to the membrane, and serve for the storage of calcium ions. 60 to 120 μmol Ca per g sarcoplasmic reticulum can be stored this way (Tada *et al.*, 1978). Calsequestrin is an interesting glycoprotein with molecular mass 44,000 which possesses an enormous number of the calcium binding sites (30-50 mol/mol) with rather low affinity (binding constant about 10^3 M^{-1}) (Tada *et al.*, 1978; Heilmann & Spamer, 1996). Calsequestrin has an isoelectric point of 3.75 with more than 30% of acidic residues and binds calcium electrostatically with its C-terminus. The two forms (fast and cardiac) of calsequestrin mainly differ in their C-terminal part. The cardiac form has an extended C-terminus with 71% acidic residues. Calsequestrin contains no transmembrane segments and is therefore believed to be located within the lumen of the sarcoplasmic reticulum.

Krause *et al.* (1991) found in rabbit skeletal calsequestrin 31 Ca^{2+}-binding sites with a mean dissociation constant 0.79 mM in the absence of Mg^{2+}, and 23 sites with a dissociation constant 0.88 mM in the presence of 3 mM Mg^{2+}. These authors showed that Ca^{2+} and Mg^{2+} do not bind to the same sites on calsequestrin. The binding of calcium makes calsequestrin insoluble. In the absence of calcium calsequestrin is in almost random coil conformation with α-helical content 11 %. The binding of calcium increases helical content up to 20 % and changes protein shape from elongated (Stocks radius 45 Å) to much more compact (Stocks radius 35 Å) (Cozens & Reithmeir, 1984). Ca^{2+}, but not Mg^{2+}, blocks the binding of calsequestrin to a 26-kDa protein of the junctional sarcoplasmic reticulum. Based on this finding Mitchell *et al.* (1988) forwarded the hypothesis that calsequestrin is not only a passive Ca^{2+} buffer but is actively involved in the regulation of Ca^{2+} release from sarcoplasmic reticulum. It was concluded that the Ca^{2+}-dependent conformational change of calsequestrin causes a change in the shape of sarcoplasmic reticulum membrane proteins, including ryanodine receptor. Calsequestrin controls the ryanodine receptor channel in a phosphorylation state-dependent fashion. It is possible that calsequestrin interacts with the 95-kDa protein triadin, which contains a region of basic amino acids in a luminal domain. The luminal domain of triadin interacts with calsequestrin in a Ca^{2+}-dependent manner and therefore triadin might anchor calsequestrin to the junctional region of the sarcoplasmic reticulum (reviewed by Berchtold *et al.*, 2000).

55-kDa Ca^{2+}-binding protein, called calreticulin, is regarded as the nonmuscle calsequestrin homolog present at highest concentration in the endoplasmic reticulum (Smith & Koch, 1989). It consists of 399 amino acid residues. The protein has no EF-hand motif or a sequence found in calelectrins. The protein has two distinct types of calcium binding sites located in different structural regions of the molecule (Baksh & Michalak, 1991). The P-domain, enriched in proline residues, binds Ca^{2+} with high affinity (dissociation constant 10 μM) and low capacity (1 mol of Ca^{2+} per mol of protein). The C-domain binds Ca^{2+} with low affinity (dissociation constant 2 mM) and high capacity (18 mol of Ca^{2+} per mol of protein). Specific calreticulin binding proteins were detected in rat liver smooth and rough endoplasmic reticulum and Golgi, in canine pancreatic microsomes, and in rabbit skeletal muscle sarcoplasmic reticulum (Burns & Michalak, 1993).

Endoplasmic reticulum of some cells contains one more Ca^{2+}-binding protein, called reticulocalbin (Ozawa & Muramatsu, 1993; Ozawa, 1995). This is a luminal protein of the endoplasmic reticulum with molecular mass 44,000. Its amino acid sequence of 325 residues has six repeats of a domain containing the EF-hand motif.

One more protein of endoplasmic reticulum is calnexin (Wada *et al.*, 1991). This is a 90 kDa integral membrane protein. Calnexin binds Ca^{2+} and may function as a chaperone in the transition of proteins from the endoplasmic reticulum to the outer cellular membrane. Tjoelker *et al.* (1994) isolated a subdomain of calnexin containing four internal repeats that binds Ca^{2+} with the highest affinity. The sequence is highly conserved when compared to calreticulin and represents a conserved motif for the high-affinity binding of Ca^{2+}, which is clearly distinct from the EF hand motif. Ou *et al.* (1995) showed that the luminal domain of calnexin is responsible for binding Ca^{2+} and Mg-ATP.

One of the calcium binding proteins of the sarcoplasm of skeletal muscles is parvalbumin. Parvalbumins are mainly muscle proteins. The highest concentration of parvalbumin (up to several millimoles per liter) was found in fast muscles (mainly in skeletal, but sometimes in cardiac) (reviewed by Berchtold et al., 2000; Permyakov, 1985; 1993; Schneeberger & Heizman, 1986). Parvalbumin contents are 200-300-fold higher in fast than in slow-twitch muscles (Leberer & Pette, 1986). The concentration of parvalbumins in fast muscles is high enough to bind all the calcium in the cells. Modern literature contains many works on location and concentration of parvalbumin in various types of muscle cells.

Very high concentrations of parvalbumin up to 1.5 mM are contained in superfast muscles (reviewed by Rome, 2006). Superfast muscles of vertebrates serve to produce sounds. The fastest, the swimbladder muscle of toadfish, generates mechanical power at frequencies in excess of 200 Hz. To operate at these frequencies, the speed of relaxation has had to increase approximately 50-fold. This increase is accomplished by modifications of three kinetic characteristics: (a) a fast calcium transient due to extremely high concentration of sarcoplasmic reticulum Ca^{2+} pumps and parvalbumin, (b) fast off-rate of Ca^{2+} from troponin C due to an alteration in troponin, and (c) fast cross-bridge detachment rate constant (50 times faster than that in rabbit fast-twitch muscle) due to an alteration in myosin.

The expression of parvalbumin depends upon activity of muscles. Studies of Kobayashi *et al.* (1991) clearly showed that the content of parvalbumin in cultured rat thymic myoid cell line is significantly decreased in cells paralyzed by the addition of 1 μM tetrodotoxin M compared with controls which had continuous muscle contractions. It strongly suggests that muscle contractile activity plays an important role in the expression of parvalbumin in thymic myoid cells. Olive & Ferrer (1994) examined the modifications of parvalbumin immunocytochemistry in the *anterior tibialis* muscle of the rat at different intervals following section of the sciatic nerve. During the first 2 weeks after denervation, no changes in parvalbumin immunoreactivity were seen, although a global reduction of fibre diameter was observed. Three weeks after denervation, small angulated, strongly parvalbumin-immunoreactive fibres appeared. From the second month onwards, the pattern of parvalbumin immunohistochemistry was characterized by areas composed of small, strongly immunoreactive fibres separated by less atrophic areas displaying a normal chequerboard distribution of parvalbumin immunoreactivity. The increase of parvalbumin-immunoreactivity in denervated and reinnervated muscle indicates that important changes in parvalbumin distribution occurs in muscle fibres after denervation. These changes are probably produced in an attempt to bind the free cytosolic calcium which accumulates in denervated fibres, and further reinforces the role of parvalbumin in calcium homeostasis during denervation and reinnervation.

Huber & Pette (1996) found that chronic low-frequency stimulation suppresses parvalbumin expression in fast-twitch muscles of the rat. In *extensor digitorum longus* and *tibialis anterior* muscles, parvalbumin mRNA levels steeply declined with apparent half-lives of approximately 26 h and 45 h, respectively. Measurements of parvalbumin synthesis indicated that the reduction in mRNA was immediately transmitted to the level of translation. Relative parvalbumin synthesis rates decayed with an apparent half-life of approximately 60 h. Both the decrease in parvalbumin mRNA and synthesis considerably preceded the decay of the protein. Although parvalbumin synthesis had approached zero in 14-day-stimulated muscles, parvalbumin content started to decrease only after some delay (28-day-stimulated muscles still contained 40-50% of their normal parvalbumin content). The lag time between fully suppressed synthesis and the onset of parvalbumin decay, as well as the stability of parvalbumin against tryptic cleavage in the presence of Ca^{2+} and Mg^{2+}, indicated proteolysis as an important post-translational control of parvalbumin levels. After complete suppression, parvalbumin mRNA reached control levels 4 days after cessation of stimulation, which demonstrates the complete reversibility of the stimulation-induced parvalbumin suppression. These results show that a slow motoneuron-like impulse pattern rapidly silences the parvalbumin gene, thus overriding fast-fiber-type-specific programs of gene expression.

Wilwert *et al.* (2006) examined the effect of expression levels of parvalbumin isoforms on relaxation rate in the sheepshead, *Archosargus probatocephalus* (Pisces, *F. Sparidae*). They measured relaxation rate of each of the three fiber types, white (fast-twitch), red (slow-twitch) and pink (intermediate), from three longitudinal body positions. Sheepshead show a significant longitudinal shift in relaxation rate in red muscle, with *anterior* muscle displaying faster rates of relaxation than *posterior*, but this pattern was not significant in the pink and white muscle. The authors hypothesized that patterns of parvalbumin expression determine relaxation rate along the length of the fish. The prediction is that total parvalbumin content and the relative expression of parvalbumin isoforms will differ between the *anterior* and *posterior* red muscle, but little longitudinal variation will be observed in parvalbumin expression in white and pink muscle. Protein electrophoresis (SDS-PAGE) with Western blots was used to identify two parvalbumin isoforms in each muscle fiber type. It turned out that indeed red muscle displays a significant shift, from *anterior* to *posterior*, in the relative expression of the two isoforms, both in their relative contribution and in total parvalbumin content, but white and pink muscle did not. The red muscle of southern kingfish, *Mentictrrhus americanus* (Pisces, *F. Scianidae*) showed a pattern similar to the red muscle of sheepshead.

It was shown (Gillis *et al.*, 1979) that in frog each twitch muscle fibre contains both of the molecular species of parvalbumins found in the whole muscle. Moreover, most, if not all, the parvalbumins can diffuse out a skinned muscle. It was concluded that parvalbumins are freely soluble in the muscle sarcoplasm. Immunofluorescence data show that parvalbumins appear to be evenly distributed in carp white muscle (Benzonana et al., 1977) and move free inside muscle cells. The concentrations and diffusivity of two isoforms of frog parvalbumin, IVa and IVb, were measured using quantitative SDS polyacrylamide gel electrophoresis in single fibers from semitendinosus muscles of the frog *Rana temporaria* (Maughan & Godt, 1999, 2001). The concentrations of IVa and IVb parvalbumins were 2.9±0.3 and 4.5±0.5 g L^{-1} total fiber volume, respectively. The total concentration of parvalbumin (7.4±0.8 g L^{-1} total fiber) corresponds to a cytosolic concentration of 0.9±0.1 mmol L^{-1} myoplasmic water. Estimates for the transverse and longitudinal diffusion coefficients for parvalbumin at 4°C in this work were obtained in two ways: (1) by diffusion of parvalbumin out of skinned fibers into droplets of relaxing solution, and (2) by diffusion of parvalbumin between two juxtaposed skinned fibers under oil. The juxtaposed fiber method yielded values for transverse ($4.27 \pm 0.87 \times 10^{-7}$ cm^2s^{-1}) and longitudinal ($3.20 \pm 0.74 \times 10^{-7}$ cm^2s^{-1}) diffusion coefficients that were not significantly different, suggesting that diffusion of parvalbumin in myoplasm is essentially isotropic. The average diffusion coefficient of frog parvalbumin in myoplasm ($3.74 \pm 0.81 \times 10^{-7}$ cm^2s^{-1}; 4°C) is approximately a third of that estimated for frog parvalbumin diffusing in bulk water into and out of 3% agarose cylinders (10.6×10^{-7} cm^2s^{-1}; 4°C). The reduced translational mobility of parvalbumin in myoplasm reflects an elevated effective viscosity due to tortuosity and viscous drag imposed by the fixed proteins of the cytomatrix and the numerous diffusible particles of the cytosol. The osmotic coefficient of the parvalbumin in frog muscle fibers is approximately 3.7 mOsm mM^{-1}, i.e., roughly the same as obtained from parvalbumin-loaded agarose cylinders under comparable conditions, suggesting that the fluid interior of muscle resembles a simple salt solution as in a 4% agarose gel.

The most reasonable hypothesis for the physiological function of parvalbumin in muscle cells is that its function is connected with regulation of muscle activity. It was proposed that parvalbumin serves as a soluble relaxing factor accelerating the relaxation phase in fast muscles (see above). Parvalbumin can be detected in fast-contracting/relaxing muscle fibers of rodents starting 4 to 6 days after birth. The time period of the maximal increase in parvalbumin concentration coincides with the differentiation of the fast-twitch muscle function, indicating that parvalbumin is involved in fast-twitch muscle function.

According to the hypothesis of Pechére *et al.* (Haiech *et al.*, 1979a) the delay in the binding of Ca^{2+} caused by Mg^{2+} dissociation from the parvalbumin binding sites allows Ca^{2+} ions to reach the Ca^{2+}-specific sites of troponin C or calmodulin which are not occupied by Ca^{2+}. Later on, the Ca^{2+},Mg^{2+}-sites of parvalbumin will successfully compete for Ca^{2+} and help to remove Ca^{2+} from the Ca^{2+}-specific sites of troponin C, which will then cause muscle relaxation. However, keeping in mind all the data presented above we think that parvalbumin cannot serve as a soluble relaxing factor during a single contraction-relaxation cycle as it was proposed by Pechére *et al.* The Mg^{2+}-Ca^{2+} exchange in parvalbumin occurs only after several successive twitches. It seems to be possible that parvalbumin works as a soluble relaxation factor after several contraction-relaxation cycles (Figure 22).

There were numerous attempts to check the hyposesis of Pechére *et al.* experimentally. Some of the experiments were successful, others gave contradictory results.

Ashley & Griffiths (1983) added up to 0.5 to 1.3 mM of rabbit or perch parvalbumin into single muscle fiber of the barnacle *Balanus nubias,* which are devoid of parvalbumin. This resulted not in an increase, but in a decrease of the muscle relaxation rate, and in a decrease in the amplitude of released Ca^{2+} concentration, which is contrary to the hypothesis of participation of parvalbumin in muscle relaxation. It should be noted, however, that the addition of parvalbumin to a muscle cell devoid of its own parvalbumin is rather incorrect since calcium regulation in such muscles is different from that in the muscles, which are characterized by high parvalbumin content.

According to the data of Iaizzo (1988) obtained on isolated muscle fibers from *Rana temporaria tibialis anterior* muscles microinjected with aequorin, the disparity in the effects of temperature on the single fiber Ca^{2+} transients *vs.* the *in vitro* quenching of aequorn luminescence with parvalbumin shows that in twitch and tetanus contractions of these fibers it is unlikely that parvalbumin plays a major role in the regulation of myoplasmic calcium concentration.

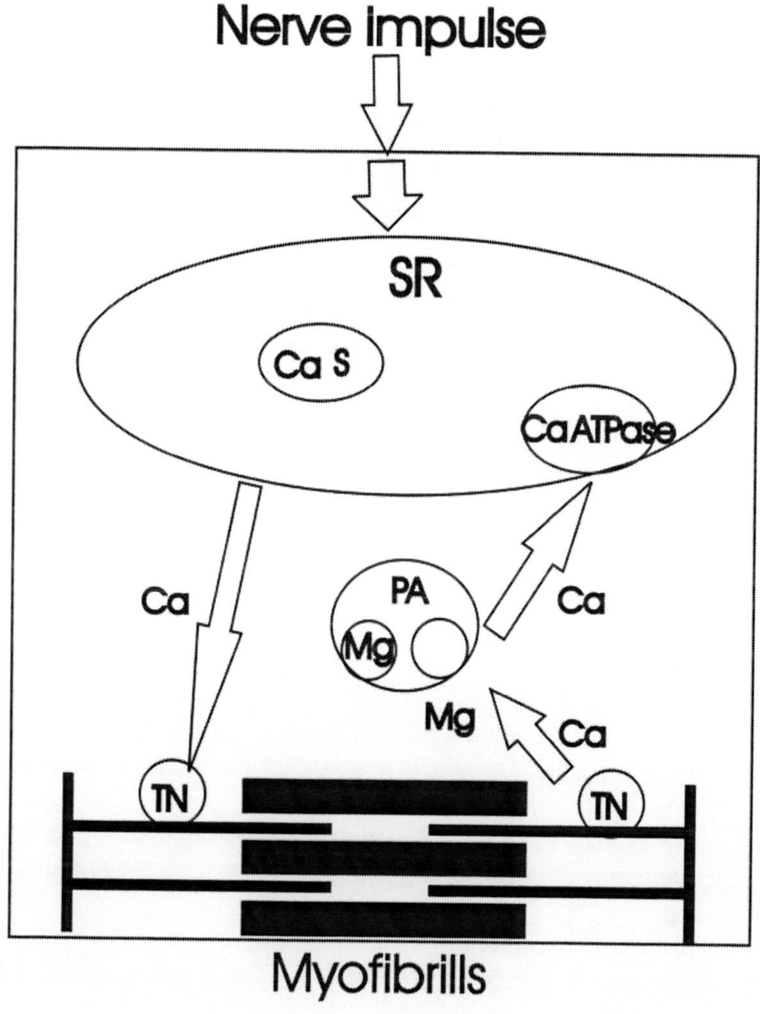

Figure 22. Participation of parvalbumin in Ca^{2+}-regulation of muscle contraction. SR – sarcoplasmic reticulum, CaS – calsequestrin, TN – troponin, PA – parvalbumin.

Hou *et al.* (1991, 1992, 1993) tested the hypothesis that parvalbumin promotes relaxation in frog skeletal muscle. Single fibers and purified parvalbumin from *R. temporaria* skeletal muscle were used to determine the relationship between Ca^{2+} and Mg^{2+} dissociation rates from parvalbumin and changes in relaxation rate as a function of isometric tetanus duration at 0°C. Relaxation rate slows as a function of tetanus duration with a rate constant of 1.18 s^{-1}. Recovery of relaxation rate after a prolonged tetanus exhibits a rate constant of 0.12 s^{-1}. Dissociation rate constants for Mg^{2+} and Ca^{2+} from purified parvalbumin, according to the data of these authors, are 0.93 s^{-1} and 0.19 s^{-1}, respectively. Thus the rates of slowing and recovery of relaxation rate may be controlled by Mg^{2+} and Ca^{2+} dissociation from parvalbumin, respectively. The influence of temperature on relaxation rate and on Ca^{2+} and Mg^{2+} dissociation rates from purified parvalbumin also was examined. The magnitude of slowing of relaxation rate with increasing tetanus duration, relative to the final, steady value of relaxation rate, is greater at 0 than at 10°C. In the 0 to 10°C range, the Q_{10} for relaxation rate increases with increasing tetanus duration. Both of these observations can be explained if the Q_{10} for Ca^{2+} uptake by the sarcoplasmic reticulum is greater than the Q_{10} for Ca^{2+} sequestration by parvalbumin during relaxation. When Ca^{2+} and Mg^{2+} dissociation rates from parvalbumin at various temperatures are compared to other proposed indicators of parvalbumin function, it is concluded that parvalbumin facilitates relaxation of frog skeletal muscle in the 0 to 20°C range.

It was shown that superfast muscles of swimbladder of toadfish (it generates high frequency sounds) contain millimoar concentrations of parvalbumin (reviewed by Rome, 2006). Toadfish call for many hours, which may suggest that parvalbumin would become quickly saturated, and hence ineffective at sequestering Ca^{2+}. However, a more detailed examination of the calling pattern suggests that parvalbumin could play an important role. Even though the toadfish can call for many hours, the duty cycle of the call [call duration/(call duration + intercall interval)] is small. A typical call is 400 ms, but the intercall interval is 5–15 s, therefore the duty cycle is approximately 2.5–7.5%. Taking into consideration this time course of calling (low duty cycle and long intercall interval) and the large parvalbumin concentration, one could suggests that parvalbumin contributes significantly to the sequestration of Ca^{2+} during calling. During the actual call, much of the released Ca^{2+} bind to parvalbumin after Mg^{2+} dissociation. This Ca^{2+} is subsequently pumped back into the sarcoplasmic reticulum during the relatively long intercall intervals. Furthermore, the high concentration of Ca^{2+} binding sites associated with parvalbumin (3 mM) seems to be sufficient to sequester much of the Ca^{2+} released during a contraction. Hence, toadfish seem to have both the need and the opportunity to utilize parvalbumin. Thus the swimbladder muscle represents a compelling model with which to examine the role of parvalbumin during *in vivo* function.

According to the model calculations of Konishi (1998), about 70% of magnesium ions released from parvalbumin and troponin in exchange with Ca^{2+} is buffered by creatine phosphate and ATP.

Jiang *et al.* (1996) found that inhibition of sarcoplasmic reticulum Ca^{2+}-ATPase with 2,5-di-(tert-butyl)-1,4-benzohydroquinone (TBQ) in frog skeletal muscle fibers at 10°C prolonged the half time of the fall of the Ca^{2+} transient by 62% and twitch force by 100% and increased peak force by 120% without increasing the amplitude of the Ca^{2+} signal. In the presence of TBQ the rate of relaxation and the rate of fall of Ca^{2+} became progressively slower in a series of twitches until relaxation failed. It is of importance that relaxation rate decreased with a time course (approximately 2 s^{-1}) similar to the Mg^{2+} off rate from purified parvalbumin. TBQ slowed the rate of fall of Ca^{2+} (5-fold) and force (8-fold) in a 0.3-s tetanus so that the rate of fall of Ca^{2+} (approximately 2.5 s^{-1}) was similar to the Mg^{2+} off rate from parvalbumin. TBQ caused a near total failure of both Ca^{2+} sequestration and relaxation in a 1.1-s tetanus, during which parvalbumin would be saturated with Ca^{2+} and could not contribute to relaxation. Thus, when the Ca^{2+}-ATPase of sarcoplasmic reticulum is inhibited, Mg^{2+}-parvalbumin can sequester Ca^{2+} and produce relaxation at a rate that is defined by the Mg^{2+} off rate from parvalbumin.

In the work of Jacquemond & Schneider (1992) the effects of low intracellular free Mg^{2+} concentration on the myoplasmic calcium removal properties of skeletal muscle were studied in voltage-clamped frog skeletal muscle fibers by analyzing the changes in intracellular free calcium and magnesium after membrane depolarization under various conditions of internal free Mg^{2+} concentration. In these experiments batches of muscle fibers were internally equilibrated with cut end solutions containing two calcium indicators, antipyrylazo III and fura-2, and different concentrations of free Mg^{2+} (25 μM - 1 mM) obtained by adding appropriate amounts of ATP and magnesium to the solutions. Changes in antipyrylazo III absorbance were used to monitor [Ca^{2+}] and [Mg^{2+}] transients, whereas fura-2 fluorescence was mostly used to monitor resting [Ca^{2+}] concentration. Shortly after applying an internal solution containing less than 60 μM free Mg^{2+} to the cut ends of depolarized fibers most of the fibers exhibited spontaneous repetitive movements, oscillations, suggesting that free internal Mg^{2+} might affect the activity of the sarcoplasmic reticulum calcium channels at rest. The spontaneous contractions generally subsided. In polarized fibers the maximal amplitude of the calcium transient induced by a depolarizing pulse was about the same whatever the internal Mg^{2+} concentration, but its decay after the end of the pulse slower at low Mg^{2+} concentration. A model characterizing the main calcium removal properties of a frog skeletal muscle fiber, including the sarcoplasmic reticulum pump and the Ca-Mg sites on parvalbumin, was fitted to the decay of the calcium transients. Results of the fits show that in low internal Mg^{2+} concentration the slowing of the decay of the calcium transient can be well predicted by both a decreased rate of sarcoplasmic reticulum calcium uptake and an expected decreased resting magnesium occupancy of parvalbumin leading to a reduced contribution of parvalbumin to the overall rate of calcium removal. These results are thus consistent with the known properties of parvalbumin as a Ca-Mg buffer.

In the work of Liu et al. (1997) skeletal muscle fibers enzymatically dissociated from adult mouse *flexor digitorum brevis* muscles were maintained in culture for up to 8 days. After various times in culture, Ca^{2+} transients for trains of 1, 5, and 10 action potentials (100 Hz) triggered by external electrical stimulation were calculated from fluorescence ratio records corrected for noninstantaneous reaction of fura 2 with Ca^{2+}. The decay rate constants of Ca^{2+} transients decreased with increasing stimulation duration, indicating a slowing of the Ca^{2+}-removal properties with increased stimulation duration. After 6 days in culture, Ca^{2+} decay rate constants decreased dramatically for all stimulation durations and the differences in decay rate constants among 1, 5, and 10 pulses became smaller. Intracellular parvalbumin content measured by single-fiber immunofluorescence decreased with time in culture in parallel with the decrease in the decay rate constant of Ca^{2+} transients. These results were interpreted as manifestation of a good correlation between parvalbumin content and the decay rate constant of the Ca^{2+} transient.

In the work of Müntener *et al.* (1987) the fast *extensor digitorium longus* and slow *soleus* muscles were cross-reinervated in both directions in the rat. The combined biochemical, histochemical and physiological study showed that the amount of parvalbumin decreased in the fast to slow and increased in the slow to fast transformation. It was found that after the altered nervous input, a slow contracting/slow relaxing muscle may even contain more parvalbumin than fast contracting/fast relaxing one.

Modern gene transfer technologies were used to reveal physiological function of parvalbumin in muscles. To test whether parvalbumin could act as a relaxing factor, direct gene transfer method was applied in normal and regenerating rat *soleus* muscles that do not synthesize detectable amounts of parvalbumin (Müntener *et al.*, 1995). Two weeks after *in vivo* transfection with parvalbumin cDNA considerable levels of parvalbumin mRNA and protein were found in uninjured and regenerating muscles. Half-relaxation time was significantly shorter in transfected than in non-transfected muscles. The relaxing role of parvalbumin in fast-contracting/relaxing muscles was directly demonstrated in mice lacking parvalbumin due to gene knock out (Schwaller *et al.*, 1999). The decrease of Ca^{2+} concentration seen after a 20-ms stimulation of the isolated *exterior digitorum longus* was slower (33% lower rate constant of Ca^{2+} decay) in knockout mice compared with wild-type animals. The knockout experiment shows that the presence of parvalbumin may also shorten the time needed to obtain peak twitch tension.

It is of interest that Chen *et al.* (2001) found that parvalbumin-deficient fast-twitch muscles of mice are significantly more resistant to fatigue than are the wild type. No upregulation of another cytosolic Ca^{2+}-binding protein was found. Mitochondria are thought to play a physiological role during muscle relaxation and were thus analyzed. The fractional volume of mitochondria in the fast-twitch muscle *extensor digitorum longus* is almost doubled in pavalbumin knockout mice, and this was reflected in an increase of cytochrome *c* oxidase. A faster decrease in intracellular Ca^{2+} concentration ($[Ca^{2+}]_i$) 200–700 ms after fast-twitch muscle stimulation observed in pavalbumin knockout muscles supports the role for mitochondria in late $[Ca^{2+}]_i$ removal. Thus alterations in the dynamics of Ca^{2+} transients detected in fast-twitch muscles of pavalbumin knockout mice might be linked to the increase in mitochondria volume and capillary density, which contribute to the greater fatigue resistance of these muscles. Quantitative analysis (Racay *et al.*, 2006) of selected mitochondrial proteins, mitochondrial DNA-encoded cytochrome oxidase c subunit I and nuclear DNA-encoded cytochrome oxidase c subunit Vb and F1-ATPase subunit β revealed the parvalbumin-/- *tibialis anterior* mitochondria composition to be almost identical to that in wild-type *soleus*, but not in wild-type fast-twitch muscles. Besides the function in energy metabolism, mitochondria in both fast- and slow-twitch muscles act as temporary Ca^{2+} stores and are thus involved in the shaping of Ca^{2+} transients in these cells. The observed altered spatio-temporal aspects of Ca^{2+} transients in parvalbumin-/- muscles are sufficient to up-regulate mitochondria biogenesis through the probable involvement of both calcineurin- and Ca^{2+}/calmodulin-dependent kinase II-dependent pathways. Racay *et al.* (2006) proposed that 'slow-twitch type' mitochondria in parvalbumin-/- fast muscles are aimed to functionally replace the slow-onset buffer parvalbumin based on similar kinetic properties of Ca^{2+} removal.

The purpose of the work of Chin et al. (2003) was to determine whether induced expression of parvalbumin in slow-twitch fibres would lead to alterations in physiological, biochemical and molecular properties reflective of a fast fibre phenotype. Transgenic mice were generated that overexpressed parvalbumin in slow (type I) muscle fibres. In *soleus* muscle (58 % type I fibres) total parvalbumin expression was 2- to 6-fold higher in transgenic mice compared to wild-type mice. Maximum twitch and tetanic tensions were similar in the wild type and transgenic mice but force at sub-tetanic frequencies (30 and 50 Hz) was reduced in trangenic soleus muscle. Twitch time-to-peak tension and half-relaxation time were significantly decreased in trangenic soleus muscle (time-to-peak tension: 39.3±2.6 vs. 55.1±4.7 ms; half-relaxation time: 42.1±3.5 vs. 68.1±9.6 ms). Since parvalbumin is a slow Ca^{2+} buffer, the rise time of free Ca^{2+} would be reduced and the amount of Ca^{2+} bound to troponin C reduced at any given time during a semi-fused tetanus. According to these authors, this would explain why force output at sub-tetanic frequencies was lower in the transgenic mice. The alterations in relaxation time and sub-tetanic force were probably a direct consequence of the increased buffering of cytosolic Ca^{2+} by parvalbumin. These data show that overexpression of parvalbumin, resulting also in decreased calcineurin activity, can alter the functional and metabolic profile of muscle and influence the expression of key marker genes in a predominantly slow-twitch muscle with minimal effects on the expression of muscle contractile proteins.

Experiments with parvalbumin-transduced myocytes at different time points after gene transfer and mathematical modeling showed that there is an optimal parvalbumin range for which relaxation rate is increased with little effect on contractile amplitude and that parvalbumin effectiveness decreases as the stimulus frequency increases (Coutu & Metzger, 2002). At low parvalbumin concentrations (approximately 0.01 mM), contractile parameters were unchanged; at intermediate parvalbumin concentrations, relaxation rate of the mechanical contraction and the decay rate of the calcium transient increased with little effects on amplitude; and at high parvalbumin concentrations (approximately 0.1 mM), relaxation rate was further increased, but the amplitudes of the mechanical contraction and the calcium transient were diminished when compared with control myocytes. The sarcomere length shortening *versus* frequency relationship exhibited a biphasic response to increasing stimulus frequency in controls (decrease in amplitude and re-lengthening time from 0.2 to 1.0 Hz followed by an increase in these parameters from 2.0 to 4.0 Hz). The effect of parvalbumin was to flatten this frequency response. This flattening effect was partly explained by a reduction in the variation in fractional binding of parvalbumin to calcium during beats at high frequency. These results clearly demonstrate that parvalbumin plays an important role in fast and phasic muscle contraction.

Coutu & Metzger (2005a) applied two genetic experimental approaches, *de novo* expression of parvalbumin and overexpression of sarco(endo)plasmic reticulum Ca^{2+}-ATPase (SERCA2a), to increase relaxation rates in myocardial tissue. They used gene transfer in isolated rat adult cardiac myocytes. When parvalbumin is expressed in elevated concentration (>0.1 mM), the transduced myocytes showed a reduction in sarcomere-shortening amplitude: 129±17, 81±8, and 149±14 nm for control, parvalbumin, and SERCA2a, respectively. In SERCA2a myocytes, the increase in shortening was slightly less than in parvalbumin or control myocytes: 108±14, 169±39, and 34±12% for control, parvalbumin, and SERCA2a, respectively. In another test set, parvalbumin myocytes had the strongest early postrest potentiation among all groups studied (rest time = 2-10 s), and SERCA2a myocytes were the least sensitive to variations in stimulation rhythm. To replicate the deficient Ca^{2+} removal observed in heart failure, 150 nM thapsigargin was used. Under these conditions, control myocytes exhibited slowed relaxation, whereas parvalbumin myocytes retained their rapid kinetics, showing that parvalbumin is still able to control relaxation, even when SERCA2a function is impaired. A mathematical model (Coutu & Metzger, 2005a) was utilized to predict whether parvalbumin metal-binding characteristics might be modified to improve diastolic and systolic functions and whether parvalbumin or SERCA2a might affect diastolic Ca^{2+} levels and myocyte energetics. One outcome of the model was to demonstrate a higher peak and total ATP consumption in SERCA2a myocytes and more even distribution of ATP throughout the cardiac cycle in parvalbumin myocytes. This finding may have implications for failing hearts that are energetically compromised.

Raymackers *et al.* (2000) studied the effects of tetanus duration on the relaxation rate of *extensor digitorum longus* and *flexor digitorum brevis* muscles in normal (wild-type) and parvalbumin-deficient mice, at 20°C. In *extensor digitorum longus* of parvalbumin-deficient mice, the relaxation rate was low and unaffected by tetanus duration (< 3.2 s). In contrast, the relaxation rate of wild type muscles decreased when tetanus duration increased from 0.2 to 3.2 s. Moreover, in wild type muscles, fast relaxation recovered as the rest interval increased. Specific effect of parvalbumin was asserted by calculating the difference in relaxation rate between wild type and parvalbumin-deficient mice muscles. For *extensor digitorum longus*, the rate constant of relaxation slowing was 1.10 s^{-1} of tetanization; the rate constant of relaxation recovery was 0.05 s^{-1} of rest. In *flexor digitorum brevis*, the effects of tetanus duration on wild type and parvalbumin-deficient mice muscles were qualitatively similar to those observed in *extensor digitorum longus*. Relaxation slowing as tetanus duration increases, was interpreted as the progressive saturation of parvalbumin by Ca^{2+}, while recovery as rest interval increases as the return to Ca^{2+}-free parvalbumin. At all tetanus durations, relaxation rate still remained slightly faster in wild type muscles. This shows that parvalbumin facilitates calcium traffic from myofibrils to the sarcoplasmic reticulum. The results of this work clearly demonstate that *the physiological function of parvalbumin in skeletal muscles is connected with acceleration of calcium transfer from myofibrils to sarcoplasmic reticulum not after single contraction-relaxation cycle but mostly after tetanus.* This effect may be further enhanced if parvalbumin interacts with the sarcoplasmic reticulum and stimulates calcium uptake as proposed by Ushio & Watabe (1994). Most probably, such calcium traffic, in a parvalbumin bound state, would largely escape detection by the usual calcium indicators and it would be rate limited by the value of the diffusion coefficient of parvalbumin (Raymackers *et al.*, 2000). Indeed, direct measurements of the diffusion coefficient of $^{45}Ca^{2+}$ injected into frog fibres (thus containing parvalbumin) (Kushmerick & Podolsky, 1969) and of parvalbumin are identical (Maughan & Godt, 1999).

It is of interest that cooling increases the twitch force of frog skeletal muscle (*Rana temporaria; Rana pipiens*), but decreases the twitch force of tropical toad muscle (*Leptodactylus insularis*) (Caputo *et al.*, 1998). Action potentials and intramembranous charge movement in frog and toad fibers were slowed identically by cooling. Cooling increased the integral of twitch Ca^{2+} detected by aequorin in frog fibers (1.4-fold), while also decreasing the peak and slowing the rate of decay. Conversely, cooling decreased the integral (0.6-fold) and the peak of twitch Ca^{2+} in toad fibers, without affecting the rate of decay. The difference in entire Ca^{2+} transients may account for cold-induced twitch potentiation in frogs and twitch paralysis in toads. Parvalbumins are thought to promote relaxation of frog muscle in the cold. The unique parvalbumin isoforms in toad muscle apparently lack this property.

Johnson *et al.* (1999) used EDTA as an "artificial" parvalbumin in that it exhibits similar rate constants for Mg^{2+} (3 s^{-1}) and Ca^{2+} (0.7 s^{-1}) dissociation at 10°C. When introduced into frog skeletal muscle, EDTA increases the relaxation rate by approximately 2.7-fold, and with increasing tetanus duration, EDTA looses its ability to contribute to relaxation (and Ca^{2+} sequestration) at its Mg^{2+} off-rate. Intracellular EDTA recovers its ability to contribute to muscle relaxation and Ca^{2+} sequestration at its Ca^{2+} off-rate. Like parvalbumin, EDTA's contribution to muscle relaxation and Ca^{2+} sequestration is more clearly observed when the sarcoplasmic reticulum Ca-ATPase is inhibited. Introduction of EDTA into rat *soleus* muscle, which has low parvalbumin concentration, increases the relaxation rate in a manner that is analogous to the way in which parvalbumin facilitates relaxation of frog skeletal muscle. The authors believe that intracellular EDTA serves as an effective mimic of parvalbumin, and its use should add in our understanding of PA's function in muscle and nerve.

It is of interest that parvalbumin isotypes isolated even from the same cell can differ in their metal binding properties. For example, pike parvalbumin pI 5.0 possesses two equivalent Ca^{2+},Mg^{2+}-binding sites, while pike parvalbumin pI 4.2 has two Ca^{2+},Mg^{2+}-sites with different affinities to Ca^{2+} (Mg^{2+}). The polymorphic pattern of parvalbumins in white muscles of pike is mainly constant but in old pikes the content of parvalbumin pI 4.2 is often lowered (Gosselin-Rey, 1974).

The parvalbumin isotype pattern was found species specific in fish. It constitutes thus a convenient and effective tool for studying within family and especially between-family relationships. Eleven parvalbumin isotypes expressed during the development of clariids *Heterobranchus longifilis* and *Clarias gariepinus* and claroteid *Chrysichthys auratus* were purified and electrophoresed on sodium-dodecyl-sulfate polyacrylamide gels in the work of Huriaux et al. (2002). Antibodies raised against *H. longifilis* parvalbumin I (larval-juvenile isotype) and against *C. gariepinus* parvalbumin IIIa (juvenile-adult isotype) cross-reacted to a rather similar extent despite a weaker cross-reaction of anti-parvalbumin IIIa with larval-juvenile isotypes. On the other hand, antibodies raised against *H. longifilis* parvalbumin IV (an exclusively adult isotype) recognized markedly only *H. longifilis* parvalbumin IV and *C. gariepinus* parvalbumin IIIb. These two adult isotypes most likely belong to the α lineage, and all the others to the β lineage. These results show that parvalbumin isotypes synthesized at different stages of fish growth differ structurally, and that the most marked difference is between larval-juvenile and adult clariid isotypes.

In the work of Focant *et al.* (2003) developmental changes in parvalbumin isoform composition were investigated in the myotomal muscle of the flatfish *Solea solea*, characterized by a very brief metamorphic stage. Results were compared with the data on another pleuronectiform teleost, the turbot (*Scophthalmus maximus*), displaying prolonged metamorphosis. Non-denaturing polyacrylamide gel electrophoresis revealed different parvalbumin isoforms in the sole and turbot: in the former, predominance of a single parvalbumin isoform (parvalbumin III), and in the latter, a major larval parvalbumin II and a minor adult parvalbumin V. Analysis of parvalbumin synthesis during sole growth has yielded some unusual findings. One peculiarity is the early appearance of adult parvalbumin III (on day 50), which quickly becomes the predominant component. Larval parvalbumins (parvalbumin I and parvalbumin II) were detected only in trace amounts, appearing on day 35 and decreasing thereafter. The early predominance of adult parvalbumin III in the sole and the fast disappearance of larval parvalbumin I and parvalbumin II may be related to the brevity of metamorphosis. During the longer turbot metamorphosis, the early appearing larval parvalbumin IIb isoform and the late appearing larval parvalbumin IIa isoform were successively synthesized, and parvalbumin IIa remained a major isoform until the young adult stage. The authors suggested that the turbot's late isoform transition is probably involved in metamorphosis-linked physiological modifications of the muscle machinery. Noteworthy is the fact that sole parvalbumin isoforms were detected on gels later than the myofibrillar proteins.

Until now, the distinct physiological role of each parvalbumin isotype is unknown. But the fact that these isotypes appear at different developmental stages must probably implicate specific physiological adaptations of the propulsive musculature related to fish size and performance (locomotion, feeding) (Huriaux *et al.*, 2002).

Parvalbumins are a powerful Ca^{2+}- and Mg^{2+}-buffer in muscle cells and the presence of several different parvalbumins inside the same cell should be considered as a widening of the range of its Ca^{2+}- and Mg^{2+}-capacity. Perhaps, such a widening of the Ca^{2+} and Mg^{2+}

concentration regions buffered by parvalbumins has some physiological significance especially for animals living in water with high concentrations of divalent metal cations.

It is possible that studies of interactions of parvalbumin with membrane systems or nucleotides and peptides will shed some light on the problem of parvalbumin function. Ushio & Watabe (1994) found a direct interaction between carp parvalbumin and sarcoplasmic reticulum. Immunoblotting using an anti-parvalbumin antibody revealed that parvalbumin binds to the light sarcoplasmic reticulum isolated from carp skeletal muscle in the presence of Ca^{2+}. In their experiments parvalbumin enhanced Ca^{2+} uptake activity of the light sarcoplasmic reticulum. Using a photoreactive crosslinking, they detected a protein 130 kDa in the light sarcoplasmic reticulum, which binds to parvalbumin in a Ca^{2+}-dependent manner. The authors suggested that parvalbumin may directly interact with the light sarcoplasmic reticulum in contraction-relaxation cycle of fast skeletal muscle.

Annoh et al. (1995) tried to clarify the parvalbumin localization within the cytoplasm of single muscle fibers by immunohistochemical methods using the confocal laser scanning microscope and transmission electron microscope. Fluorescent immunohistochemical study has clearly shown that both parvalbumin and troponin are located intimately in the I-band of the skeletal muscle fibers. The finding by the immunofluorescent study correlated well with those which have been seen at the ultrastructural level.

The sarcoplasm of any muscle cell contains calmodulin. Ca^{2+}-loaded calmodulin is able to interact with many enzymes activating them (reviewed, for example, by Allan & Hepler, 1989). Among these enzymes are such as cyclic nucleotide phosphodiesterase, adenylatcyclase, and kinases of various specificity including myosin light chain kinase. The binding of Ca^{2+} to calmodulin seems to follow the successive scheme and each state of calmodulin interacts with target proteins. Activation of some enzymes requires the binding of three or even four Ca^{2+} ions per molecule. At present the scheme has been most studied for myosin light chain kinase.

The complex of myosin light chain kinase with Ca^{2+}-calmodulin can phosphorylate myosin light chains. The functional significance of this process in skeletal and cardiac muscles is not clear now. In smooth muscles the phosphorylation of myosin light chains is one of the regulatory mechanisms of muscle contraction (Bremel, 1974; Sommerville & Hartshorne, 1986). Another mechanism of the regulation of smooth muscle contraction involving the participation of calmodulin is the interaction of Ca^{2+}-calmodulin with caldesmon, a protein, which is located on the thin filaments and interacts with actin and tropomyosin (Sommerville & Hartshorne, 1986; Marston *et al.*, 1988). The interaction results in elimination of the inhibition of the actomyosin interaction. So, in smooth muscles may exist at least two mechanisms of Ca^{2+}-regulation of contraction, one *via* thick and one *via* thin filaments, and calmodulin is a key element in both of them.

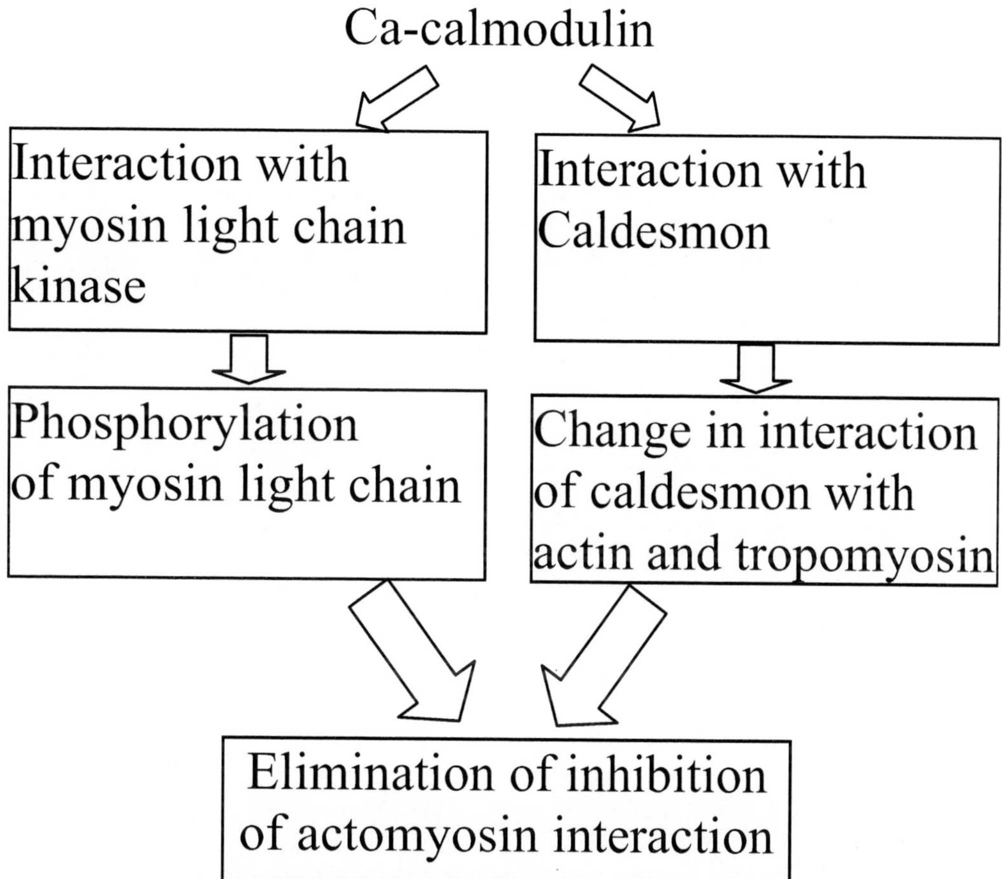

Caldesmon is a basic protein of molecular mass of around 120,000 found in association with actin filaments in all smooth muscles and also in many non-muscle motile cells (for review see for example Marston *et al.*, 1988 and Sobue & Sellers, 1991). It is a highly extended flexible protein having a contour length of about 146 nm and a secondary structure composed primarily of random coil (Lynch *et al.*, 1987; Marston & Redwood, 1991). It can readily undergo intra and intermolecular disulfide bond formation under oxidizing conditions. Pure caldesmon is an F-actin-binding protein, which typically has an affinity of greater than 10^6 M^{-1}. The binding affinity is enhanced about five-fold when tropomyosin is also bound to the F-actin. Caldesmon binds to actin at a stoichiometry of 1 molecule per 28 actin monomers. In the presence of tropomyosin the bound caldesmon inhibits actin activation of myosin ATPase by up to 95%. Neither the binding nor the inhibition is Ca^{2+} sensitive. In the presence of Ca^{2+}, calmodulin binds to caldesmon and weakens its interaction with actin-tropomyosin as a result of which there may be a partial dissociation of Ca^{2+}-calmodulin-caldesmon from the actin-tropomyosin. This Ca^{2+}-calmodulin binding releases caldesmon's inhibition of actomyosin MgATPase activity. Ca^{2+}-calmodulin binds to caldesmon with association constant about 10^6-10^7 M^{-1} (Smith *et al.*, 1987; Shirinsky *et al.*, 1988; Czurylo *et al.*, 1991). Chymotriptic cleavage experiments have localized the actin- and calmodulin-binding sites to a C-terminal third of the caldesmon molecule. It is of interest that the natively unfolded C-terminal domain of caldesmon remains substantially unstructered after the effective binding to calmodulin (Permyakov *et al.*, 2003). The 34,000-dalton fragment containing these sites was shown to inhibit actomyosin ATPase to roughly the same extent, as did intact caldesmon. The binding of calmodulin to the 34,000 fragment is of two orders of magnitude weaker than that to intact caldesmon. It was found that this fragment contains also a tropomyosin-binding site. Binding constants of tropomyosin to caldesmon and its 34,000 fragment are similar (3×10^{5} M^{-1} according to Czurylo *et al.*, 1991).

It is worth noting that Ca^{2+}-S100 protein, like Ca^{2+}-calmodulin, inhibits the binding of caldesmon to F-actin due to a direct interaction between S100 and caldesmon (Fujii *et al.* 1990). Ca^{2+}-sensitivity curve of the inhibition has a midpoint at 3×10^{-5} M for S100 and at 8×10^{-6} M for calmodulin. Mani *et al.* (1992) showed that one more Ca^{2+}-binding protein, 11-kDa smooth muscle calcium-binding protein named caltropin, could release the inhibitory effect of caldesmon in the presence of calcium. Complete reversal is obtained when 1 mol of caltropin is added per mol of caldesmon. In reversing the inhibitory effect of caldesmon caltropin is as potent as calmodulin, if not better. The authors suspect that caltropin is in fact a calcyclin. The affinity of Ca^{2+}-caltropin to caldesmon is very high (association constant 5.0×10^{6} - 1.25×10^{7} M^{-1}).

It turned out that caldesmon binds stoichiometrically to smooth muscle myosin in an ATP-dependent manner and Ca^{2+}-calmodulin reduces the binding of caldesmon to myosin with the same effectiveness as it does the binding of caldesmon to actin (Hemric *et al.*, 1993). It was shown that Ca-calmodulin binding to the C-terminal region of caldesmon is responsible for the reversal of binding to myosin. Phosphorylation of the N-terminal region of caldesmon by the co-purifying kinase weakens but does not eliminate the binding of caldesmon to myosin.

It is usually assumed in most models for calmodulin regulation of smooth and non-muscle contractivity that calmodulin is freely diffusible at resting intracellular concentrations of free Ca^{2+}. However, data of Luby-Phelps *et al.* (1995) obtained from fluorescence recovery after photobleaching measurements of three different fluorescent analogs of calmodulin in cultured bovine tracheal smooth muscle cells suggest that concentration of free calmodulin may be very limited in unstimulated cells. According to their results, at most 5% of total endogenous calmodulin molecules in resting smooth muscle cells are unbound even at intracellular Ca^{2+} concentrations as low as 17 nM. When Ca^{2+} is elevated to between 450 nM and 3 μM, some of the bound calmodulin is released. At higher Ca^{2+} concentrations, calmodulin becomes increasingly immobilized. In about 50% of the cell population, clamping Ca^{2+} at micromolar levels results in translocation of cytoplasmic calmodulin to the nucleus.

Apart from parvalbumin and calmodulin, the sarcoplasm of any muscle cell contains many other calcium binding proteins. These are, for example, calpains, non-lysosomal proteases, which act only in the presence of calcium. Calpains are a family of cytosolic cysteine proteinases whose enzymatic activities depend on Ca^{2+} (papain-like cysteine protease) (reviewed by Sorimachi & Suzuki, 2001). Members of the calpain family are believed to participate in various biological processes, including integrin-mediated cell migration, cytoskeletal remodeling, cell differentiation and apoptosis. μ-Calpain requires micromolar levels of Ca^{2+}, while m-calpain requires millimolar concentrations of Ca^{2+} (Nishimura & Goll, 1991; Pontremoli & Melloni, 1986). Currently, at least 12 different calpains have been identified in mammals. The m- and μ-calpains are the best characterized members of this family. Both proteins are heterodimers composed of a large 78-80 kDa catalytic subunit and a common small 29 kDa regulatory subunit. The large subunit consists of four domains (dI-dIV), while the small subunit has two domains (dV and dVI) (Imajoh *et al.*, 1988). Crystal structure of the full-length human m-calpain in the Ca^{2+}-free form was determined (Strobl *et al.*, 2000). Domain dI comprises a single α-helix placed in a cavity of dVI domain. Domain dII contains the catalytic site and can be divided into two sub-domains, IIa and IIb containing the active site Cys105 and His262/Asn286, respectively. Domain dIII consists of eight β-strands with topology similar to the C_2 domain found in some proteins including protein kinase C and phospholipase C, which is known to interact with Ca^{2+} and phospholipids (Rizo & Sudhof, 1998). It was suggested that this domain is responsible for the Ca^{2+}-dependent translocation of calpain to membranes. Domains dIV and dVI are Ca^{2+}-binding domains, each containing five EF-hand motifs. These domains provide heterodimerization of the large and small subunits through a unique interaction between their fifth EF-hands.

Activated calpains are very unstable. Suzuki *et al.* (1995) proposed that the first step in activation of calpains is a Ca^{2+}-dependent translocation to the cell membrane in the heterodimeric form. Calpain can then be activated by phospholipids, which leads to autolysis at the cell membrane with 76 kDa and 18 kDa subunits as active forms. The dissociated large subunit is enzymatically fully active and exhibits Ca^{2+} sensitivity identical to the sensitivity of the activated form of calpain that is twofold higher than the one of nondissociated calpain. The small subunit has a stabilizing rather than an activating effect on the large subunit, which contains the catalytic activity. The activity of calpains is regulated by the specific inhibitor calpstatin, a protein with molecular mass 110 kDa. The enzyme and the inhibitor were found in sarcolemma. However, the exact physiological role of calpain inhibition by calpstatin is not understood so far.

Calpains initiate turnover of myofibrillar proteins by specifically cleaving them and releasing large peptide fragments from the myofiblils (Goll *et al.*, 1992), which seems to be required for muscle growth. Moreover, calpain-mediated cleavage of cytoskeletal proteins such as filamin might be important for myogenic differentiation.

Sorcin, a calpain-like protein of 22 kDa, has been reported to be abundant in skeletal, cardiac, and smooth muscles (reviewed by Meyers, 1996). The protein consists of eight α-helices (A-H) organized in five calcium binding motifs (EF1-EF5). Sorcin has been found to be associated with ryanodine receptor, therefore it was suggested that the function of sorcin is a Ca^{2+} release from intracellular stores.

Another calcium binding protein, which is contained in muscle cell sarcoplasm, is calcycline, a protein with molecular mass 10,500 (Kuznicki *et al.*, 1989). This protein is thought to take part in regulation of cell proliferation (Calabretta *et al.*, 1986).

Cardiac and smooth muscles contain several proteins, which interact with either cellular components or hydrophobic matrices in a calcium-dependent manner. These proteins with molecular masses of 67,000; 35,000; 33,000 and 30,000 were called calcimedins (reviewed by Dedman, 1986). Calcimedins do not substitute for calmodulin in several assays including cAMP phosphodiesterase, adenylate cyclase, and myosin light chain kinase. Most of these proteins are now refer to the family of annexins. These proteins interact with membranes in a calcium-dependent manner. Annexins have been frequently suggested to function as connecting elements between plasma membrane and intracellular components, such as cytoskeleton (F-actin, spectrin), endosomes, lisosomes, and other vesicular structures (reviewed by Gerke & Moss, 2002).

Low molecular mass sarcoplasmic Ca^{2+}-binding proteins have been isolated from muscle extracts of various invertebrate species. They have been found in considerable amounts (≈2.5 g/kg) in fast striated muscle of crustacea and protochordates. They form a group less homogeneous than parvalbumins. Crayfish sarcoplasmic Ca^{2+}-binding protein is a dimer of identical 22,000-dalton subunits and contains six high affinity Ca^{2+}-binding sites (Cox *et al.*, 1976; Wnuk *et al.*, 1979). The binding of Ca^{2+} displays positive cooperativity, which is enhanced in the presence of Mg^{2+}. A 20,000-dalton sarcoplasmic Ca^{2+}-binding protein was isolated from annelid *Nereis diversicolor* (sandworm) (Cox & Stein, 1981; Collins et al., 1988) and *Nereis virens* (Gerday et al., 1981). Similar 22,000-dalton sarcoplasmic Ca^{2+}-binding protein was also isolated from scallop striated muscles (Collins *et al.*, 1983). An EF-hand Ca^{2+}-binding protein with similarity to calcyphosine was isolated from abdominal muscle of crustaceans (Sauter *et al.*, 1995). Since all these proteins interact neither with target proteins nor with hydrophobic matrices, their physiological function is still unknown.

Very little is known about the Ca^{2+}-binding proteins of nuclei, although there are indications that Ca^{2+} may play a role in the fusion and differentiation of muscle cells. Schibeci & Martonosi (1980) found Ca^{2+}-binding proteins in nonhistone chromosomal protein fractions and in the insoluble residue of nuclei isolated from skeletal muscle of 15-day-old chick embryos, adult chickens and rabbits. They suggested that the interaction of Ca^{2+}-binding proteins with chromatin might be of importance in the regulation of the gene expression in response to changes in cytoplasmic and nucleoplasmic free Ca^{2+} concentration.

Engelkamp et al. (1992) found three members of the S100 protein family (S100 , CACY, and CAPL) in human heart. All three proteins are expressed at high levels. Their function in muscle cells is unknown. S100A1 is especially interesting with respect to the muscle physiology, since it has been shown to be present at high concentration in slow-twitch muscle fibers; it co-locolizes with the sarcoplasmic reticulum in skeletal muscle, whereas in the heart it was found in sarcolemma (Zimmer & Landar, 1995). It was shown that S100A1 activates ryanodine binding activity of the ryanodine receptor at nanomolar calcium concentrations (Treves *et al.*, 1997). S100A1 was found also to bind and specifically activate the protein kinase twitchin in a Ca^{2+}-dependent manner (Heierhorst *et al.*, 1996).

It has been known for a long time that mitochondria possessed the necessary machinery to transport Ca^{2+} into and out of their matrix (reviewed by Gailly, 2002): Ca^{2+} entry is promoted by a very large electrochemical gradient: Ca^{2+} efflux occurs *via* an electroneutral exchange with 2 Na^{+} or 2 H^{+} ions. Initially, these processes were proposed to contribute to free Ca^{2+} homeostasis. Later on, it was also suggested that Ca^{2+} ions might play a role in the regulation of mitochondrial metabolism. It was suggested that endoplasmic reticulum and mitochondria form structural and functional units that allow mitochondria to sense free Ca^{2+} oscillations more than slow and global increases of internal free Ca^{2+}, the frequency of these oscillations controlling the activation of calcium-sensitive dehydrogenases. But it is now also clear that mitochondria play a role in $[Ca^{2+}]_i$ homeostasis in skeletal muscles.

Let us consider calcium binding proteins of the contractile system. The main protein of thick filaments, myosin, comprises lights chains, the amino acid sequence of which is homologous to the sequences of troponin C, calmodulin and parvalbumin (Kendrick-Jones & Jakes, 1977; Weeds & McLachlan, 1974). In spite of the homology, the so called essential (enzymatic) light chains do not bind Ca^{2+}, while the regulatory chains bind one Ca^{2+} per molecule and this does not depend upon their association with the heavy chains (Werber, 1978). The Ca^{2+} binding constant of regulatory light chains is 10^5 to 10^6 M^{-1} (Werber, 1978). The binding of Ca^{2+} to myosin light chains seems to trigger the contraction of scallop muscles (Kendrick-Jones & Jakes, 1977; Scholey *et al.*, 1981). The role of the myosin light chain in these muscles is simply to switch on and off the binding of the cross-bridge to actin in response to the changing calcium levels within the muscle. Of the four EF hand-like domains in molluscan essential light chains, only domain III has amino acid sequence predicted to be capable of binding Ca^{2+}, nevertheless the data of Fromherz & Szent-Gyorgyi (1995) indicate that the unique and required contribution of molluscan essential light chains to Ca^{2+} regulation of molluscan myosin resides exclusively in domain I. In striated muscles of vertebrata myosin light chains also bind Ca^{2+}, but at present we have no direct evidence of their participation in the regulation of contraction of these muscles.

In both smooth and skeletal muscles, there exist myosin light chain kinases that catalyze the rapid incorporation of phosphate into a specific serine residue of some light chains (reviewed by Stull *et al.*, 1980, for example). Dephosphorylation of these light chains is catalyzed by another enzyme, myosin light chain phosphatase. This class of light chains is referred to as the phosphorylatable or P-light chains. A number of *in vitro* studies showed that the phosphorylation of the myosin P-light chains is an effective mechanism for control of contraction in smooth muscles. The Ca^{2+} sensitivity of the myosin light chain kinase is mediated by calmodulin. Ca^{2+} first binds to calmodulin to form a complex, which, in turn, binds to the inactive myosin light chain kinase to form a catalytically active phosphotransferase. Experimental data on smooth (intestinal or arterial) muscles demonstrate a close correlation between the degree of phosphorylation of the myosin light chains and tension development. The inhibition of the light chain kinase results in dephosphorylation of the myosin light chains and inactivation of tension in the presence of Ca^{2+}. No such effects were revealed for striated muscle fibers. (Kerrick *et al.*, 1980).

Another Ca^{2+}-binding protein of myofibrills is the main protein of thin filaments, actin. It has one strong Ca^{2+}-binding site with a very high Ca^{2+} affinity (binding constant 10^8 to 10^9 M^{-1}) (Gershman *et al.*, 1986, 1991; Estes *et al.*, 1987; Novak *et al.*, 1988; Carlier, 1991). The analysis of the actin primary structure aimed at identification of sequences resembling the known Ca^{2+}-binding patterns has revealed the absence of an EF -hand Ca^{2+}-binding site. The best match was obtained between the sequence of the 292-301 segment and that of the Ca^{2+}-site in lectins. The high affinity Ca^{2+} binding site in actin involves sequentially distant residues from the N- and C-terminal portions of the polypeptide chain (Strzelecka-Golaszewska *et al.*, 1989). The association rate constants for Ca^{2+} and Mg^{2+} for the high affinity site are 1.9×10^7 M^{-1} s^{-1} and 2.3×10^5 $M^{-1}s^{-1}$, respectively (Gershman *et al.*, 1991). Apart from the strong site, actin possesses weaker binding sites for Ca^{2+}, Mg^{2+}, and K^+. The filling of these sites promotes actin polymerization by reducing the net negative charge of the monomers and thus weakening their repulsion. Myosin subfragment 1 inhibits dissociation of calcium from G-actin (Kasprzak, 1993). The physiological role of the Ca^{2+} binding to actin is still unclear.

The family of structurally related proteins that sever actin filaments includes gelsolin, villin, severin, fragmin, adseverin, scinderin, and fragmin 60 (for review see Janmey, 1994). All known actin-severing proteins require Ca^{2+} under physiological conditions. There exist different classes of end-blocking proteins called F-actin capping proteins and some of them are regulated by Ca^{2+} (gCap39/MCP, for example).

Gelsolin (90 kDa) takes part in reorganization of the actin cytoskeleton. It consists of six structurally similar domains, G1-G6 (Burtnick *et al.*, 1997). Elevation of Ca^{2+} concentration releases latches within the constrained structure and leads to large shifts in the relative location of the domains, permitting gelsolin to bind to and sever actin filaments. Crystallographic data of Choe *et al.* (2002) revealed two classes of Ca^{2+}-binding sites in gelsolin: type 1 sites share coordination of calcium with actin, while type 2 sites are wholly contained within gelsolin. In total, gelsolin has the potential to bind 8 calcium ions, two type 1 and six type 2. The function of the type 2 sites is to facilitate structural rearrangements within gelsolin as a part of the activation and actin-binding and severing processes. The type 2 site in G6 seems to be the critical site that starts overall activation of gelsolin by releasing the tail latch that locks calcium-free gelsolin in conformation unable to bind actin.

α-Actinin is an EF-hand protein associated with the actin filaments in muscles. It is present in the Z-disks and thought to connect the ends of the parallel and antiparallel arrays of actin filaments (reviewed by Noegel, 1996). The binding of Ca^{2+} regulates activity of non-muscle α-actinin, whereas α-actinin cross-linking in muscles is Ca^{2+}-independent.

The next Ca^{2+}-binding protein of myofibrills is troponin C, the closest relative of calmodulin. Troponin C is a component of the troponin complex consisting of troponins C, I, and T. Within the complex troponin C interacts with both troponin I and troponin T. Inclusion of troponin C in the complex increases its affinity to Ca^{2+} by almost an order of magnitude (Potter & Gergely, 1975). Some authors believe that only Ca^{2+}-specific sites take part in the regulation of muscle contraction (Potter & Gergely, 1975; Holroyde *et al.*, 1980) other authors do not rule out the possibility of participation of the Ca^{2+},Mg^{2+}-sites in it (Iio & Kondo, 1980a,b). Literature data on effects of Mg^{2+} on the ATPase activity of actomyosin are contradictory.

The binding of Ca^{2+} to troponin C causes changes in its structure, and these changes in their turn alter the structure of troponins I and T, which induces changes in the location of tropomyosin on the actin filament and opens the sites of actomyosin interaction. The data of Potter *et al.* (1995) suggest that the binding of Ca^{2+} to troponin C would have a dual role: release of the ATPase inhibition by troponin I and activation of the ATPase through interaction with troponin T. The results of Tripet *et al.* (1997) demonstrate that Ca^{2+} binding to the regulatory sites of troponin C located in the N-domain alters the binding affinity between the N-terminus and C-terminus of troponin I for troponin C, i.e. a Ca^{2+}-dependent switch between these two sites of troponin I for the C domain of troponin C.

Li & Fajer (1994) used EPR to study orientational changes of troponin C accompanying muscle activation by Ca^{2+}. Rabbit skeletal muscle troponin C was labeled with maleimide spin label at Cys98 and reconstituted into an oriented skinned muscle fiber. The orientational analysis revealed a bimodal orientational distribution of troponin C in the absence of Ca^{2+} and attached myosin heads. One of the components is well-ordered with its probe axis inclined at 22° to the fiber axis, while the other one is more disordered and inclined at 58°. Ca^{2+} and/or cross-bridge binding significantly disorder the labeled domain and increase the average probe axis angle by 20-30° away from the fiber axis. Thus, troponin C exists in many different orientational conformations depending on which ligand is bound.

Results of Morimoto & Ohtsuki (1994), obtained on isolated porcine cardiac myofibrils, provide direct evidence that some feedback mechanism exists between myosin crossbridge attachment and the Ca^{2+}-binding to the regulatory site II of cardiac troponin C, which thus may confer positive cooperativity on the Ca^{2+} activation of myofibrillar ATPase activity.

They showed that the activation of ATPase is in direct proportion to Ca^{2+} binding to site II of troponin C and that the myosin crossbridge interaction with actin in the presence of ATP confers cooperativity on the Ca^{2+}-binding to site II of troponin C and, therefore, causes the cooperative activation of myofibrillar ATPase.

Site-directed mutagenesis was used to convert Asp to Ala at the first coordinating position in each of the four Ca^{2+}-binding sites of chicken skeletal muscle troponin C (Sorenson *et al.*, 1995). The functional effects of each mutation in the reconstitution assays (their ability to associate with other components of the troponin-tropomyosin regulatory complex and to regulate thin filamens) were largely confined to the domain in which it occurs, where the unmutated site was unable to compensate for the defect. The mutants of sites I and II bind to the regulatory complex but are impaired in ability to regulate tension and actomyosin ATPase activity, whereas the mutants of sites III and IV regulate activity but are unable to remain bound to thin filaments unless Ca^{2+} is present.

An *in vitro* motility assay was used to study troponin regulation of individual actin-tropomyosin filaments moving over immobilized skeletal muscle heavy meromyosin (Fraser & Marston, 1995). The most striking observation is that the actin-tropomyosin filament appears to be regulated as a single unit. At pCa 9.0 addition of up to 4 nM troponin causes the proportion of filaments motile to decrease from >85% to 20% with no dissociation of the filaments from the heavy meromyosin surface or change in velocity. Increasing Ca^{2+} concentration causes the filaments to be switched back. This is an "all or none" process in which an entire filament, up to 15 μm long, switches rapidly as a single cooperative unit. The work of Zhao *et al.* (1995) is in line with this conclusion. Their results indicate that the troponin complex is relatively rigid in relaxed muscle, but becomes more flexible when Ca^{2+} binds to regulatory sites in troponin C. The increased flexibility may be propagated to the whole thin filament, releasing the inhibition of actomyosin ATPase activity and allowing the muscle to contract.

According to the data of Ishikawa & Wakabayashi (1999), without Ca^{2+}, the three-dimensional map from electron cryomicrographs of (troponin C - troponin I) on actin-tropomyosin reveals the extra-density region due to troponin(C+I), which extends perpendicularly to the helix axis and covers the N-terminal and C-terminal regions of actin. In the presence of Ca^{2+}, the C-terminal region of actin becomes more exposed, and troponin(C+I) becomes V-shaped with one arm extending towards the pointed end of the actin filament. At the same time, fluorescence resonance energy transfer results on reconstituted skeletal muscle thin fibers suggest that the C-terminal domain of troponin I moves to the outer domain of actin during inhibition, while the C-terminal domain of troponin does not move much (Miki *et al.*, 1998). Using EPR method, Li & Fajer (1998) found that troponin C in thin filaments feels structural changes in myosin during force generation, implying that there is a structural coupling between actomyosin and troponin C.

As was mentioned above, most researchers suppose that only two Ca^{2+}-specific sites of troponin C take part in the regulation of muscle contraction. At the same time, the dependence of muscle fiber tension on free Ca^{2+} concentration is very steep (Hill's coefficient 5 to 6) (Gillis, 1985). This means that the two sites are not enough to regulate the actomyosin interaction. Several models were proposed to solve this contradiction. Some of them are based on the assumption that the Ca^{2+} affinity of the binding sites in troponin C increases with the rise in myofibrillar tension (see for example Brandt *et al.*, 1980). Butters *et al.* (1997) found that the thin filament-myosin S1 MgATPase cycle requires calcium binding to adjacent troponin molecules and that this binding is cooperatively promoted by a single cycling cross bridge. They think that this mechanism is a potential explanation for Ca^{2+}-mediated regulation of cross-bridge kinetics in muscle fibers.

It is interesting that even some troponins T are metal-binding proteins. For example, a repeating metal-binding ($Cu^{2+}>Ni^{2+}>Zn^{2+}\approx Co^{2+}$) sequence $(HE/AEAH)_4$ has been found in troponin T isoforms expressed in the breast but not leg muscles of all *Galliformes* and *Graciformes* (Jin & Smillie, 1994). Concentration of the metal-binding sites is adequate to affect free metal levels in the muscle cell.

From the data presented one can see that almost all the main muscle proteins are Ca^{2+}-binding ones. Some of them are in tight complexes with each other and with other proteins, and some are in a free state and form complexes with other proteins only temporarily. Some of them are incorporated to membranes. All these proteins can bind Ca^{2+}, Mg^{2+}, Na^{+}, and K^{+}, and some of them interact with ATP and ADP. Many details of their functioning are known, but in order to reach a quantitative understanding of the principles of regulation of muscle contraction further studies in this field are needed.

Parvalbumin and Muscle Diseases

Some muscle deseases are reflected in changes of parvalbumin level inside muscle cells and even in serum. Employing Sandwich ELISA method, Jockusch *et al.* (1990) showed that parvalbumin is present in the serum of normal mice and that its level is indicative of the disease status of muscle. Elevated parvalbumin levels were found in mice with X-linked dystrophy (*mdx*) and reduced levels in myotonic (ADR) mice. Because myotonic mouse muscle is characterized by strongly reduced parvalbumin content, the reduced parvalbumin serum level in ADR mice indicates that serum parvalbumin is derived from skeletal muscle. It was suggested that serum parvalbumin in *mdx* mice, in which muscle parvalbumin content is close to normal, is a measure of the necrosis of fast muscle fibers.

Muscles from dystrophic (*mdx*) mice and from Duchenne dystrophy patients are deficient in dystrophin, a cytoskeletal protein associated with the plasma membrane. Recent experimental results led to the hypothesis that dystrophin-lacking fibres suffer from a chronic Ca^{2+} overload, resulting in the activation of Ca^{2+}-dependent proteases. Indeed, the total Ca^{2+} content of *mdx* muscle is about doubled, but as far as the cytosolic Ca^{2+} concentration is concerned, reports are contradictory. Gailly et al. (1993) hypothesized that in muscle from *mdx* mice, parvalbumin prevents a significant rise in cytosolic Ca^{2+} concentration and the subsequent proteases activation so that dystrophy remains benign. Parvalbumin mRNA was assayed by Northern blot analysis in muscles from normal and dystrophic (*mdx*) mice and its content was found to be specifically higher in *mdx* fast muscles than in control preparations. This suggests an increased expression of the protein in dystrophin-lacking fast fibres.

Raymackers *et al.* (2003) tested the hypothesis whether the mild dystrophy in *mdx* mice could result from the contribution of parvalbumin in maintaining a normal cytosolic calcium concentration $[Ca^{2+}]_i$, in spite of an increased passive Ca^{2+} influx. By crossing *mdx* mice with parvalbumin-deficient mice, double mutant mice, lacking both dystrophin and parvalbumin, were obtained. Though resting cytosolic $[Ca^{2+}]_i$ and total calcium content were similar to that of *mdx* muscles, this new animal model presented a slightly more severe phenotype than the *mdx* mouse. Such mice suffer from a more pronounced muscle degeneration associated with a worsening of some of its mechanical defects and of interstitial fibrosis. However, regeneration is not impaired and eventually overcomes these defects. Notwithstanding, it is clear that, even in the absence of parvalbumin, the *mdx* myopathy remained mild compared with the severity of the affection in *xmd* dogs and Duchenne dystrophy patients. It was concluded that the lack of muscle parvalbumin in dogs and humans does not seem an essential factor in determining the severity of their dystrophinopathy.

It is of interest that parvalbumin now is tested as a pharmacological remedy to treat heart disfunctions. Diastolic dysfunction is a characteristic feature of the aged mammalian heart. The reason of the diastolic dysfunction is an impaired ventricular relaxation and it is an important component of human heart failure. In an estimated 40% of patients who suffer from heart failure, disease progression results specifically from a slowing of myocardial relaxation (diastolic dysfunction). In these patients, the heart relaxes too slowly after each contraction, thus compromising the refilling of the cardiac chambers with blood for the next beat. At present, abnormal relaxation of the heart, termed diastolic dysfunction, is a significant and growing problem that is a major cause of heart failure in the aged population. No specific treatments for diastolic dysfunction in human patients currently exist. Genetic modification of intracellular calcium-handling proteins may help to correct the diastolic dysfunction. A viral gene transfer is usually used in such studies. Since parvalbumin acts as a Ca^{2+} sink and enhances relaxation in skeletal muscle, overexpression of parvalbumin in myocardium should increase cardiac relaxation *in vitro* as well as *in vivo*.

Indeed, the expression of parvalbumin dramatically increases the rate of Ca^{2+} sequestration and the relaxation rate in normal cardiac myocytes. Importantly, parvalbumin fully restores the relaxation rate in diseased cardiac myocytes isolated from an animal model of human diastolic dysfunction (Wahr *et al.*, 1999). These authors concluded that gene therapy of parvalbumin may address the impaired Ca^{2+} homeostasis and diastolic dysfunction without an increase in energy expenditure.

Szatkowski *et al.* (2001) showed that parvalbumin gene transfer to the heart *in vivo* produces levels of parvalbumin characteristic of fast skeletal muscles, causes a physiologically relevant acceleration of heart relaxation performance in normal hearts, and enhances relaxation performance in an animal model of slowed cardiac muscle relaxation. They concluded that parvalbumin may offer the unique potential to correct defective relaxation in energetically compromised failing hearts because the relaxation-enhancement effect of parvalbumin arises from an ATP-independent mechanism. They also concluded that parvalbumin gene transfer may provide a new therapeutic approach to correct cellular disturbances in calcium signaling pathways that cause abnormal growth or damage in the heart or other organs.

Similar study was carried out by Michele et al. (2004): the potential of gene transfer of parvalbumin to improve diastolic function in the aged myocardium *in vivo* was evaluated. It turned out that parvalbumin gene transfer and expression *in vivo* were sufficient to improve time constant for relaxation, a load-independent indicator of diastolic function, assessed using catheter-based micromanometry in the aged myocardium. These results suggest that expression of parvalbumin may represent an effective approach to functional correction of the failing heart in the aging. Parvalbumin significantly improved relaxation parameters in senescent myocytes: both the rate of calcium transient decay and the rate of myocyte relengthening were dramatically increased in senescent cardiac myocytes transduced with parvalbumin compared with nontransduced and GFP-expressing controls, with no effect on myocyte shortening (Huq *et al.*, 2004). Since abnormalities of myocyte relaxation underlie diastolic dysfunction in a large proportion of elderly patients with heart failure, gene transfer of parvalbumin may thus be a novel approach to target diastolic dysfunction in senescent myocardium.

Schmidt *et al.* (2005) used adenovirus to transfer parvalbumin into two different rat models of aging: the Fischer 344 (F344) and the Fischer 344xBrown Norway F1 hybrid (F344xBN). It turned out that *in vivo* overexpression of parvalbumin in both rat aging models had no effect on systolic parameters but reduced left ventricular diastolic pressure and the time course of pressure decline. Overexpression of parvalbumin also improved the force-frequency relationship in senescent rats.

Gene transfer of the sarco(endo)plasmic reticulum calcium-ATPase (SERCA2a) and parvalbumin restored cardiac myocyte relaxation in a dose-dependent manner under baseline conditions (Hirsch *et al.*, 2004). In the presence of high parvalbumin concentrations, sarcomere shortening was depressed. In contrast, during β-adrenergic stimulation, the expected enhancement of myocyte contraction (inotropy) was abrogated by SERCA2a but not by parvalbumin. The mechanism of this effect is unknown, but it could relate to the uncoupling of SERCA2a/phospholamban in SERCA2a myocytes.

Hypertrophic cardiomyopathy is an inherited disease characterized by an increase in ventricular wall thickness, impairment in relaxation properties (diastolic dysfunction), and an increased risk of sudden death. At least eight hypertrophic cardiomyopathy causing mutations in α-tropomyosin have been reported. Hypertrophic cardiomyopathy mutations A63V and E180G in α-tropomyosin were shown to cause slow cardiac muscle relaxation. Coutu *et al.* (2004) used two complementary genetic strategies, gene transfer in isolated rat myocytes and transgenesis in mice, to correct the diastolic dysfunction caused by these mutations. Dual gene transfer of α-tropomyosin A63V and parvalbumin causes a marked decrease in time from peak sarcomere shortening to 50% relengthening, T50R: 29.8±1.0 ms vs. 36.8±1.0 ms in controls. However, this increase in relaxation rate is accompanied with a decrease in shortening amplitude (114.6±4.4 nm in A63V+parvalbumin, 137.8±5.3 nm in controls). Using an asynchronous gene transfer strategy, parvalbumin expression was reduced (from approximately 0.12 to approximately 0.016 mmol/L), slow relaxation redressed, and shortening amplitude maintained (T50R=33.9±1.6 ms, sarcomere shortening amplitude=132.2±7.0 nm in A63V+parvalbumin delayed; n=56).

The issues of application of parvalbumin for correction of diastolic disfunction are reviewed in the work of Coutu *et al.* (2003). According to these authors, additional issues to consider in the future include the effects of β-adrenergic stimulation on parvalbumin expressing myocytes. What effect, if any, would parvalbumin have on ischemic myocardium? Are there any long-term adaptations of the heart to the presence of parvalbumin? How does parvalbumin compare with other potential strategies, for example modification of SERCA2a/phospholamban? Is it possible to engineer parvalbumin metal binding sites with Ca^{2+}/Mg^{2+} properties that optimally correct diastolic dysfunction? These questions and others will need to be answered to fully understand and appreciate the potential beneficial effects and limitations of parvalbumin in the context of the failing heart.

CALCIUM BINDING PROTEINS IN NERVOUS SYSTEM

Changes in intracellular Ca^{2+} concentration play a central role in functioning of neurons by acting as a trigger for neurotransmitter release. Moreover, many aspects of neuronal activity, from rapid modulations of channel function to long term switches in gene expression, are controlled by changes in cytosolic concentration of Ca^{2+}.

The central nervous system contains many various calcium binding proteins. The most well described calcium binding proteins include parvalbumin, calbindin-D_{28k}, calretinin, calmodulin, calcineurin, synaptotagmins and the protein S100 family. Most these proteins are contained in neurons. Calbindin and at least one of the S100 proteins, S100h, have also been shown to be present in glia. It is assumed that in the central nervous system, parvalbumin, calbindin and calretinin play a vital role in calcium homeostasis and are generally thought of as calcium buffering rather than calcium regulatory proteins.

Most neuronal Ca^{2+}-binding proteins can be classified into two groups based on the type of Ca^{2+}-binding domains: EF-hand domain proteins such as parvalbumin and calmodulin and C_2 domain proteins such as synaptotagmins which are generally membrane bound and have overall lower calcium affinities. In contrast to EF-hands, C_2-domains are characterized by more complex Ca^{2+}-binding domains in which multiple Ca^{2+}-binding sites are formed by discontinuous sequences in a β-sandwich structure (reviewed by Rizo & Südhof, 1998). The best characterized C_2-domain protein is synaptotagmin I, a synaptic vesicle protein that binds multiple Ca^{2+} ions by means of two C_2-domains and is required for Ca^{2+}-triggering of exocytosis (reviewed by Bai & Chapman, 2004). Synaptotagmin I might operate as a Ca^{2+}-sensor that triggers rapid exocytosis. It is anchored to the membrane of secretory vesicles *via* a single transmembrane domain.

Sometimes neuronal calcium sensor proteins, such as calmodulin, are classified into a separate group. Calmodulin is expressed ubiquitously in all the cells and tissues of eukaryotic organisms. Calmodulin has the unique structure with two globular domains, each containing a pair of EF-hand motifs connected by a central helix (see above). Upon Ca^{2+} binding, opening of the interfaces between α-helices within these domains allows calmodulin to associate with a wide range of target proteins (reviewed by Hoeflich & Ikura, 2002; Kortvely & Gulya, 2004). In the absence of Ca^{2+} calmodulin can interact with another set of target proteins. These less geometrically restricted hydrophobic interactions and conformational plasticity allow calmodulin to interact with proteins that share very little homology.

Each brain cell contains calmodulin, the well known neuronal calcium sensor protein, with its numerous functions. Although calmodulin is ubiquitous, it is nevertheless highly enriched in brain in neurons. The characterized neuronal calmodulin binding proteins include adenyl cyclase VII, calcineurin A, Ca^{2+}/calmodulin kinase II, Ca^{2+}/calmodulin kinase kinase, neuronal nitric oxide synthase and various calcium ion channels. Calmodulin is involved in controlling synaptic vesicle recruitment *via* activation of Ca^{2+}-calmodilin-dependent protein kinases and the phosphorylation of the synapsin proteins on the synaptic vesicle (reviewed by Burgoyne & Weiss, 2001). The phosphorylation leads to dissociation of synapsins from the vesicle, release of synaptic vesicles from a cytoskeletal attachment and vesicle movement to the presynaptic plasma membrane. Calmodulin-dependent kinases are abundant in the brain and thymus: there exists a cascade that consists of three Ca^{2+}/calmodulin-dependent kinases (Soderling, 1996): calmodulin kinase I (CaMKI), calmodulin kinase IV (CaMKIV), and calmodulin kinase kinase (CaMKK). CaMKK acts as an upstream activator of CaMKI and CaMKIV by phosphorilating Thr177 in CaMKI and Thr196 in CaMKIV. It results in an increase in efficiency of CaMKI and CaMKIV in phosphorylation of various protein substrates including MAP kinases. CaMKI and CaMKIV are predominantly localized in the cytoplasm, CaMKIV and CaMKK can also be localized in the nucleus. It is of importance that one of the well-characterized functions of nuclear CaMKIV involves the regulation of transcription through phosphorylation of CREB, the camp-regulated transcriptional activator that activates transcription of target genes in part through direct interactions with the coactivator calcium-binding protein p300.

Calmodulin takes part in many events of Ca^{2+}-dependent modulation of neuronal function including post-synaptic changes during synaptic plasticity, alternation of gene expression, Ca^{2+}-dependent inactivation of voltage-gated Ca^{2+} channels, activation of Ca^{2+}-dependent K^{+} channels, modulation of glutamate receptor and metabolic function (reviewed by Burgoyne & Weiss, 2001).

Other neuronal calcium sensor proteins represent at least five sub-families (reviewed by Burgoyne & Weiss, 2001; Haeseleer *et al.*, 2002): frequenin (NCS-1), visinin-like protein (VILIP), recoverin, guanilate cyclase-activating protein (GCAP), K_v-channel-interacting protein (KChIP). Recoverin and guanilate cyclase-activating protein are expressed mostly in retinal photoreceptors and take part in phototransduction. Frequenin, visin-like protein and K_v-channel-interacting protein are widely expressed in the nervous system and seem to regulate neurotransmitter release, the A-type K^{+} channels and the binding of mRNAs to cytoskeleton. The gene encoding K_v-channel-interacting protein also produces calsenilin, a protein which interacts with a membrane protein presenilin that is genetically linked to Alzheimer disease.

All neuronal calcium sensor proteins have four EF-hand motifs, of which some might not be functional and most are N-terminally myristoylated. Various members of the neuronal calcium sensor proteins family show 25-35% identity with calmodulin. Their calcium dissociation constants are within the 0.1 – 2 μM region. Myristoylation may be responsible for membrane targeting of the neuronal calcium sensor proteins, which, in the case of recoverin, has been shown to be due to exposure of the myristoyl group following Ca^{2+} binding (Ames *et al.*, 1997). Other neuronal calcium sensor proteins can be membrane-associated, however, even at low Ca^{2+} concentration. The most studied neuronal calcium sensor protein is the bovine photoreceptor protein recoverin. This protein has only two functional EF-hands (EF2 and EF3). The Ca^{2+}-free form of recoverin has a compact structure with the myristoyl group buried within a hydrophobic pocket formed from residues within EF1 and hydrophobic residues contributed by other helices. Comparison of the Ca^{2+}-free form with the Ca^{2+}-bound form of recoverin (Ames *et al.*, 1997) demonstrated extensive conformational changes due to Ca^{2+} binding. In the Ca^{2+}-bound form the myristoyl group has 'flipped out' into the aqueous solution, leaving an exposed hydrophobic surface potentially able to interact with target proteins. In addition, the N-terminal domain is rotated by 45° relative to the C-terminal domain in the Ca^{2+}-bound form. The C-terminal domain is relatively unaffected, apart from small changes to accommodate the bound Ca^{2+} in EF3. It was suggested that similar conformational changes should occur in other members of the family following Ca^{2+} binding (Ames *et al.*, 1997).

The structural studies on recoverin led to the concept of the so-called "Ca^{2+}-myristoyl switch". The idea is that Ca^{2+} binding would allow extrusion of the myristoyl group, which would, in turn, allow membrane association of recoverin as the myristoyl group inserted into the lipid bilayer. The idea is corroborated by the fact that recoverin only binds to membranes in a Ca^{2+}-dependent manner. Other neuronal calcium sensor proteins show similar properties.

Sugita *et al.* (2002) identified a new family of proteins called NECABs (neuronal Ca^{2+}-binding proteins). NECABs are characterized by a single N-terminal EF-hand domain that binds Ca^{2+}, but different from many other neuronal EF-hand Ca^{2+}-binding proteins. At the C-terminus, NECABs include a DUF176 motif, a bacterial domain of unknown function that was previously not observed in eukaryotes. In rat at least three closely related NECAB genes are expressed either primarily in brain (NECABs 1 and 2) or in brain and muscle (NECAB 3). Immunocytochemical methods revealed that NECAB 1 is restricted to subsets of neurons. In cerebral cortex, NECAB 1 is highly and uniformly expressed only in layer 4 pyramidal neurons, whereas in hippocampus only inhibitory interneurons and CA2 pyramidal cells contain NECAB 1. In these neurons, NECAB 1 fills the entire cytoplasm similar to other EF-hand Ca^{2+}-binding proteins, and is not concentrated in any particular sub-cellular compartment.

Some of the calcium binding proteins in brain, like parvalbumin, calbindin, and calretinin, proved to be useful neuronal markers for a variety of functional brain systems and their circuitries. The exact functions of these proteins are still unknown, nevertheless their major role is assumed to be buffering, transport of Ca^{2+}, and regulation of various enzyme systems (reviewed by Heizmann, 1993). One more function ascribed to these proteins is a protection of neurons against calcium overloard, which often accompanies cellular degeneration. It is assumed that neurons containing certain intracellular Ca^{2+}-binding proteins may have a greater capacity to buffer Ca^{2+} and therefore would be more resistant to degeneration.

About 25% of cortical neurons utilize the inhibitory neurotransmitter, γ-aminobutyric acid (GABA). Sub-classes of GABA neurons differ in their morphological, biochemical, and functional characteristics, which indicates that they likely play distinct roles in regulating cortical circuitry. Some GABA neurons provide inhibitory synapses to the cell bodies of nearby excitatory pyramidal neurons and to their initial axon segments. These neurons appear to be positioned to regulate the firing of pyramidal neurons. Other GABA neurons provide inhibitory synapses to the distal portions of pyramidal neuron dendrites, where they modulate neighboring excitatory inputs to pyramidal neurons. The synapses of a third group of GABA neurons provide inhibitory inputs to other GABA neurons, resulting in a downstream "disinhibition" of pyramidal neurons. It is of interest that these three sub-classes of GABA neurons express different calcium-binding proteins (Heizmann & Celio, 1987; Heizmann, 1988; Hof *et al.*, 1999). The first type of GABA neurons expresses parvalbumin, whereas the second and the third types of GABA neurons express calretinin. The cell type-specific expression of parvalbumin and calretinin make it possible to use the expression of the messenger RNA (mRNA) encoding each protein to study the involvement of different sub-classes of GABA neurons in brain diseases.

Parvalbumin is thought to regulate calcium-dependent metabolic and electric processes within the population of the GABA-ergic neurons (Celio, 1986). Its intracellular concentration in interneurones was reported to be of the order of 50 μM (6 μM in the caudatoputamen and 45 μM in the cerebellum) (Plogmann & Celio, 1993). The data of electron microscopy show that in the brain regions of zebra finch responsible for songs (Zuschratter *et al.*, 1985) and in neurons of the optic cortex of cat (Stichel *et al.*, 1987) parvalbumin is concentrated in amorphous material, dendrites and axons, in most nuclei and in association with microtubules, postsynaptic densities and intracellular membranes. Parvalbumin-containing GABAergic interneurons are a key neuronal cell population that can significantly regulate input-output functions in some brain regions.

In 1977 Potter *et al.* found that carp parvalbumin can activate rat brain phosphodiesterase in a Ca^{2+}-dependent manner. The concentration of Ca^{2+} required for half-maximal stimulation by parvalbumin is $1.4x10^{-7}$ M, whereas calmodulin requires $1.2x10^{-6}$ M. Later on, LeDonne & Coffee (1979) showed that Ca^{2+}-sensitive phosphodiesterase-stimulating activity associated with parvalbumin preparations is in fact due to contaminating (<0.1%) amounts of carp muscle calmodulin-like protein. This protein can be separated from parvalbumin by Sephadex G-75 chromatography and has many characteristics of calmodulin. They found as well that parvalbumin itself causes a non-specific stimulation of phosphodiesterase at all calcium concentrations. In the presence of calmodulin, it can result in an apparent shift to lower concentrations of the calcium level required for half-maximal stimulation.

The rat brain parvalbumin is indistinguishable from its muscle counterpart by its molecular mass, isoelectric point, chromatographic behavior, calcium content, amino acid composition, tryptic peptide maps, and immunological properties (Berchtold *et al.*, 1982b).

Schmidt *et al.* (2003b) quantified the diffusion of the endogenous parvalbumin in spiny dendrites of cerebellar Purkinje neurons with two-photon fluorescence recovery after photobleaching. Fluorescently labeled by Alexa Fluor 488 rat parvalbumin diffused readily between spines and dendrites with a median time constant of 49 ms. Based on published data on spine geometry, these authors concluded that this value corresponds to an apparent diffusion coefficient of 43 $\mu m^2\ s^{-1}$, which is similar to the 43 and 32 $\mu m^2\ s^{-1}$ values reported by Maughan & Godt (1999) for transverse and longitudinal parvalbumin diffusion in frog myoplasm. The absence of large or immobile binding partners for parvalbumin was confirmed in parvalbumin null-mutant mice. These data show that parvalbumin is a highly mobile endogenous Ca^{2+} buffer that diffuses readily between spines and dendrites on the timescale of synaptic Ca^{2+} transients. Thus, in cells that express large concentrations of paralbumin, such as Purkinje neurons, diffusion of parvalbumin-buffered Ca^{2+} will seriously affect the spatio-temporal extent of dendritic Ca^{2+} signals.

As a consequence of its high mobility, parvalbumin will be readily washed out during whole-cell patch-clamp recordings. Assuming an access resistance of 10 MΩ, the concentration of endogenous parvalbumin in a round cell with a diameter of 15 µm, for example, will drop with a time constant of about 170 s (Pusch & Neher, 1988). In neuronal dendrites washout will occur over tens of minutes. The parvalbumin concentration in terminal dendrites (150–200 µm distant from the soma) of Purkinje neurons will drop with a time constant of 15–25 min (Schmidt *et al.*, 2003a).

The function of parvalbumin, calbindin and calretinin in brain is not fully understood, although strong evidence supports their prominent role in physiological calcium regulation. These proteins regulate and are regulated by intracellular calcium level. For example, they may directly or indirectly enable sensitization or desensitization of calcium channels, and may further block calcium entry into the cells, like the calcium-sensor proteins. The absence of calcium buffer proteins results in marked abnormalities in cell firing with alterations in simple and complex spikes or transformation of depressing synapses into facilitating synapses.

Calcium-binding protein implication in resistance to degeneration is a controversial issue. Neurons rich in calcium-binding proteins, especially calbindin-D_{28k} and parvalbumin, seem to be relatively resistant to degeneration in a variety of acute and chronic disorders. However other data support that an absence of calcium-binding proteins may also have a neuroprotective effect. It is not unlikely that neurons may have a dual action mechanism where a decrease in calcium-binding proteins has a first short-term beneficial effect while it becomes detrimental for the cell over the long term.

Chard *et al.* (1993) examined the ability of calbindin D_{28k} and paravalbumin to modulate increases in the intracellular free Ca^{2+} concentration ($[Ca^{2+}]_i$), produced by brief depolarizations, in rat dorsal root ganglion neurones. Calbindin D_{28k} causes an 8-fold decrease in the rate of rise in $[Ca^{2+}]_i$ and alters the kinetics of decay of $[Ca^{2+}]_i$ to a single slow component. Parvalbumin also slows the rate of rise in $[Ca^{2+}]_i$. Parvalbumin selectively increases a fast component in the decay of the Ca^{2+} signal. These data demonstrate that both calbindin D_{28k} and paravalbumin effectively buffer Ca^{2+} in a cellular environment and may therefore regulate Ca^{2+}-dependent aspects of neuronal function.

It was proposed that efficient Ca^{2+} buffering by parvalbumin and its high concentration in parvalbumin-expressing nonpyramidal cells serve for the proficient inhibition of cortical networks (DeFelipe, 1997). To check this hypothesis, Schwaller *et al.* (1999) used mice lacking parvalbumin (parvalbumin-/-), which show no obvious abnormalities when maintained under standard housing conditions. In parvalbumin-/- mice, changes in the contraction/relaxation cycle of fast-twitch muscles, which in wild type animals contain significant amounts of parvalbumin, can be directly correlated with the absence of parvalbumin. A similar effect of parvalbumin on the kinetics of Ca^{2+} transients is detected in parvalbumin-injected chromaffin cells (Lee *et al.*, 2000b) and more important in a sub-population of inhibitory hippocampal neurons (Lee *et al.*, 2000a).

Ca^{2+} dynamics in parvalbumin (parvalbumin-/-) and parvalbumin/calbindin D_{28k} null-mutant (parvalbumin/calbindin-/-) mice were compared with responses in wild-type animals (Schmidt *et al.*, 2003a). In the wild type mice, Ca^{2+} transients in dendritic shafts were characterised by double exponential decay kinetics while Ca^{2+} transients in parvalbumin-/- Purkinje cells reached the same peak amplitude as in the wild type but the biphasic nature of the decay was less pronounced, an effect that could be attributed to parvalbumin's slow binding kinetics. In contrast, peak amplitudes in parvalbumin/calbindin-/- Purkinje cells were about two times higher than in the wild type animals and the decay became nearly monophasic. It was shown in this work that in spiny dendrites of Purkinje cells, parvalbumin acts as a slow buffer that does not affect the peak amplitude of Ca^{2+} transients but accelerates the decay during the first 300 ms. This effect was not attributed to saturation of parvalbumin but to its peculiar Ca^{2+} and Mg^{2+} sensitivity. The buffering of Ca^{2+} transients by parvalbumin has to be preceded by Mg^{2+} dissociation, which is an inherently slow process (see above). This delayed Ca^{2+} buffering explains why, despite the high parvalbumin concentration in Purkinje cells (50–100 μM), parvalbumin only slightly affects the rapid kinetics of synaptically evoked Ca^{2+} transients. Neuronal integration in Purkinje neurons involves many forms of Ca^{2+} signaling (reviewed by Hartmann & Konnerth, 2005). The resulting intracellular Ca^{2+} signals seem to be shaped by the Ca^{2+} buffers calbindin and parvalbumin.

Collin *et al.* (2005) showed that expression of parvalbumin in cerebellar interneurons is cell specific and developmentally regulated, leading to characteristic changes in presynaptic Ca^{2+} dynamics ($[Ca^{2+}]_i$). Biochemical and immunocytochemical analysis showed parallel changes in the expression levels and cellular distribution of parvalbumin. By comparing wild-type and parvalbumin(-/-) mice, parvalbumin was shown to accelerate the initial decay of action potential-evoked $[Ca^{2+}]_i$ signals in single varicosities and to introduce an additional slow phase that summates during bursts of action potentials. The slow decay component is responsible for a pronounced, parvalbumin-dependent, delayed transmitter release that occurs at interneuron-interneuron synapses after presynaptic bursts of action potentials. Overall, parvalbumin looks as a major contributor to presynaptic $[Ca^{2+}]_i$ signals and synaptic integration in the cerebellar cortex.

The immediate consequences of past neuronal activity on synaptic strength were examined by measuring the ratio (called paired-pulse ratio) between the mean synaptic current in response to a test stimulation over that obtained with a conditioning stimulus. If the paired-pulse ratio is larger than 1, the synapse is considered as facilitating, whereas values smaller than 1 are characteristic of depressing synapses. Current hypotheses link facilitation to the fact that some of the Ca^{2+} ions entering the presynaptic terminal during the first stimulus are still present when the second stimulus is delivered. Caillard *et al.* (2000) tested the possibility that parvalbumin may affect the amplitude and time course of intracellular Ca^{2+} transients in terminals after an action potential, and hence may regulate short-term synaptic plasticity. They applied paired-pulse stimulations (with 30- to 300-ms intervals) at GABAergic synapses between interneurons and Purkinje cells, both in wild-type (parvalbumin+/+) mice and in parvalbumin knockout (parvalbumin-/-) mice. They found paired-pulse depression in parvalbumin+/+ mice, but paired-pulse facilitation in parvalbumin-/- mice. In paired recordings of connected interneuron-Purkinje cells, dialysis of the presynaptic interneuron with Ca^{2+} buffer EGTA (1 mM) rescues paired-pulse depression in parvalbumin-/- mice. The authors concluded that parvalbumin potently modulates short-term synaptic plasticity.

In modern literature there exist a lot of publications about distribution of parvalbumin, calbindin, and calretinin in various parts of brain.

Although parvalbumin, calbindin, and calretinin share in common their ability to bind calcium, they belong to different sub-families. They present, in general, specific developmental and distribution patterns (reviewed by Hof *et al.*, 1999; Schwaller *et al.*, 2002; Bastianelli, 2003). Most Purkinje cells express calbindin-D_{28k} and parvalbumin, whereas basket, stellate and Golgi cells parvalbumin only. They are, almost all, calretinin negative. Calbindin-D_{28k} and parvalbumin are present throughout the axon, soma, dendrites and spines of Purkinje cells. Granule, lugaro and unipolar brush cells present an opposite immunoreactivity profile, most of them being calretinin positive while lacking calbindin-D_{28k} and parvalbumin.

Parvalbumin, calbindin, and calretinin are found in morphologically distinct classes of inhibitory interneurons as well as in some pyramidal neurons in the mammalian neocortex (see Hof *et al*, 1999 for a review). Although there is a wide variability in the qualitative and quantitative characteristics of the neocortical sub-populations of calcium binding protein-immunoreactive neurons in mammals, most of the available data show that there is a fundamental similarity among the mammalian species investigated so far, in terms of the distribution of parvalbumin, calbindin, and calretinin across the depth of the neocortex. Thus, calbindin- and calretinin-immunoreactive neurons are predominant in layers II and III, but are present across all cortical layers, whereas parvalbumin-immunoreactive neurons are more prevalent in the middle and lower cortical layers. These different neuronal populations have well defined regional and laminar distribution, neurochemical characteristics and synaptic connections, and each of these cell types displays a particular developmental sequence.

The developmental pattern of appearance of these proteins seems to follow the maturation of neurons. Calbindin-D_{28k} appears early, shortly after cessation of mitosis when neurons become ready to start migration and differentiation while parvalbumin is expressed later in parallel with an increase in neuronal activity. The other proteins are generally detected later. During development, some of these proteins, like calretinin, are transiently expressed in specific cellular sub-populations. Analysis of different brain regions suggests that these proteins are involved in regulating calcium pools critical for synaptic plasticity. Surprisingly, a major role of any of these three calcium-binding proteins as an endogenous neuroprotectant is not generally supported.

The distribution of calbindin, calretinin and parvalbumin during the development of the mouse main olfactory bulb was studied using immunohistochemistry techniques (Qin *et al.*, 2005). Parvalbumin-immunoreactive profiles are mainly located in the external plexiform layer (except for P10 mice); weakly stained parvalbumin-immunoreactive profiles are present in the glomerular layer at all stages; and no parvalbumin was detected in the mitral cell layer at any stage.

The presence and distribution of three cytoplasmic calcium binding proteins, calbindin, calretinin, and parvalbumin, were investigated in the projection neurons of the cochlear nucleus complex in adult rats by using immunohistochemistry in free-floating slices (Por *et al.*, 2005). The data obtained in these experiments demonstrated the presence of parvalbumin in pyramidal neurons and globular and spherical bushy cells of rat cochlear nucleus, whereas octopus and giant cells did not contain parvalbumin. According to the double immunolabeling co-localization experiments, the pyramidal neurons, Purkinje-like cells, globular bushy cells, and octopus cells express two different calcium binding proteins in their cytoplasm (although in different combinations) whereas giant cells and spherical bushy cells contain solely calbindin and parvalbumin, respectively. The immunolabeling of the fibers and axonal endings of the acoustic nerve in the ventral part of the cochlear nucleus indicated that these structures are also parvalbumin positive.

Gonzalez-Soriano *et al.* (2000) studied the expression pattern of calbindin D_{28k} and parvalbumin in the superior colliculus of the adult rabbit, as well as the morphology of the immunoreactive cells. It was found that calbindin neurons and neuropil form three main tiers: the first located within the *stratum zonale* and the upper part of the *stratum griseum superficiale*, the second located within the *stratum griseum intermedium*, and the third, located within the medial and central areas of the *stratum griseum profundum*. In contrast to this layer labeling, almost no calbindin-positivity is found within the other collicular layers. On the other hand, the densest concentration of parvalbumin labeled cells and terminals is found within a single dense tier that spanned the ventral part of the *startum griseum superficiale* and the dorsal part of the *stratum opticum*. Anti-parvalbumin neurons are also scattered through the deeper layers below the dense tier. In contrast, almost no anti-parvalbumin labeled neurons or neuropil are found within the *stratum zonale* and upper *stratum griseum superficiale*. The results of this study indicate that both calbindin and parvalbumin are present in a variety of neurons, which present a number of homologies between mammals, but have a different location and/or distribution, according to the different species.

Todtenkopf *et al.* (2004) quantified parvalbumin-immunoreactive cells in sub-territories of the striatum and *nucleus accumbens* in animals behaviorally sensitized to cocaine. Rats received a sensitization-inducing regimen of cocaine (twice-daily injections of 15 mg/kg i.p. for 5 consecutive days). Repeated cocaine administration resulted in robust sensitization that correlated with transient increases in the number of parvalbumin immunoreactive neurons in the ventrolateral, dorsolateral and dorsomedial striatum.

Mize (1999) used electron microscope immunocytochemistry to study directly the synaptic inputs to neurons containing calbindin and parvalbumin. Calbindin immunoreactive dendrites were usually of small to medium size and were found to receive synaptic input from retinal terminals. These retinal terminals were all small profiles forming a single synaptic contact with asymmetric densifications. Calbindin immunoreactive profiles also received other synaptic input, including from terminals with dark mitochondria that contained flattened synaptic vesicles (F profiles). No calbindin immunoreactive dendrites were found to receive cortical terminals input even though degenerating cortical terminals were found in the vicinity of calbindin immunoreactive profiles. By contrast, both retinal terminals and cortical terminals were found to contact parvalbumin immunoreactive dendrites. Retinal terminals contacting parvalbumin immunoreactive dendrites were both small and larger profiles with round synaptic vesicles and asymmetric synaptic densifications.

The connections between the cortex and the striatum are critically involved in control and execution of voluntary movements. Jinno & Kosaka (2004) studied the expression of parvalbumin in the corticostriatal pathway. The parvalbumin-positive corticostriatal projection neurons were mainly found in layer V, but occasionally seen in layers II, III, and VI. Some of the parvalbumin-positive cells showed distinct apical dendrites and were considered pyramidal cells. The main target of parvalbumin-positive cortical afferents was the caudal striatum. Unexpectedly, it was found that the majority of parvalbumin-positive corticostriatal projection neurons were glutamic acid decarboxylase (GAD)-negative, while some of them were GAD-positive. These results provide anatomical evidence for expression of parvalbumin in glutamatergic and GABAergic corticostriatal pathway in mice and suggest that a sub-set of cortical afferents may exert some inhibitory influence on striatal activity.

Parvalbumin and calbindin immunoreactivities were studied in nucleus *robustus archistriatalis* (RA) of male and female zebra finches, together with retrograde labelling of RA neurons (Wild *et al.*, 2001). The results of double and triple labelling experiments suggested that, in males, moderately and faintly parvalbumin-positive neurons were projection neurons, but that all intensely parvalbumin-positive cells were not projection neurons. The latter, which are presumably interneurons, were also intensely calbindin-positive, and may correspond to the GABAergic inhibitory interneurons identified by others. In addition, the complete RA pathway and its terminal fields in the respiratory-vocal nuclei of the brainstem were strongly parvalbumin-positive. It is of interest that in female zebra finches, which do not sing, no evidence was found that parvalbumin-positive RA cells were projection neurons, yet the pattern of projections of RA neurons was very similar to that of RA in males. Parvalbumin immunoreactivity was also present in RA and its projections in males of several other songbird species (northern cardinal, brown headed cowbird, canary) and in the female cardinal, which sings to some extent, but the labelling was not as intense as that in male zebra finches.

The study of Wu *et al.* (2000) was aimed at evaluating the relative number and comparing the pattern of distribution of interneurons containing calretinin, parvalbumin or

NADPH-d in the striatum of rats, squirrel monkeys and humans. In primates, the most abundant interneurons were those expressing calretinin. The ratio of calretinin-positive/parvalbumin-positive neurons was approximately 2-3:1. In contrast, the most frequently encountered interneurons in the rat striatum were those expressing parvalbumin. In rodents, all three interneurons were more abundant rostrally than caudally, but calretinin-positive neurons displayed a particularly striking rostrocaudal decreasing gradient. In monkeys and humans, the three striatal interneurons were distributed rather uniformly rostrocaudally, but calretinin-positive and parvalbumin-positive interneurons were significantly more numerous in the caudate nucleus than in the putamen in humans. In monkeys, only parvalbumin-positive neurons were more abundant in the caudate nucleus than in putamen. Overall, the density of the three striatal interneurons was much higher in monkeys than in rats and humans.

An immunohistochemical study in the rat was carried out to evaluate the distributions of parvalbumin in the striatum and extended amygdale (Zahm *et al.*, 2003). Parvalbumins occupy all structures currently regarded as having a striatal composition, including the *caudate-putamen*, *nucleus accumbens*, and *olfactory tubercle*, as well as structures that receive outputs from these, including the *globus pallidus*, *ventral pallidum*, *entopeduncular nucleus* and *substantia nigra reticulata*. These authors found that the density of larger neostriatal parvalbumins with extensive and densely immunoreactive dendritic and local axonal arbors is greatest laterally, particularly in striatal districts with slight calbindin-immunoreactivity, including the striatal patch compartment. In contrast to the situation in striatum, few parvalbumins were observed in the central and medial divisions of the extended amygdala, including the bed nucleus of stria terminalis, interstitial nucleus of the posterior limb of the anterior commissure and central and medial nuclei of the amygdala, or in mesopontine, peribrachial and medullary structures that receive extended amygdala output. The paucity of parvalbumins may be a characteristic feature distinguishing extended amygdala and its projection areas from striatopallidum, as well as the general character of neural processing that occurs in each.

Real *et al.* (2003) studied the distribution of calbindin D_{28k}, calretinin, and parvalbumin in the mouse dorsal *claustrum* and *endopiriform nucleus*. The three calcium-binding proteins were distinctly expressed in structures of both the *claustrum* and the *endopiriform nucleus*. Calbindin was the calcium-binding protein showing the highest expression in the claustrum and the endopiriform nucleus. In contrast, calretinin-immunoreactive structures, particularly cell bodies, were very scarce in these regions. Both calbindin-immunoreactive and parvalbumin-immunoreactive neurons were more abundant in the *claustrum* than in the *endopiriform nucleus*, and more in rostral than in caudal levels. Nevertheless, calcium-binding protein immunoreactive neurons constitute a minority population of claustral neurons. The co-localization study of calbindin and parvalbumin immunoreactivities has demonstrated that both calcium-binding proteins are mostly expressed by separate claustral neurons in the mouse.

The central nucleus of the inferior colliculus but not the surrounding regions contains parvalbumin-positive neuronal somata and fibres (Tardif *et al.*, 2003). Calbindin-positive neurons and fibres are concentrated in the dorsal aspect of the central nucleus and in structures surrounding it: the dorsal cortex, the lateral lemniscus, the ventrolateral nucleus, and the intercollicular region. In the dorsal cortex, labelling of calbindin and calretinin revealed four distinct layers. Thus, calcium-binding protein reactivity reveals in the human inferior colliculus distinct neuronal populations that are anatomically segregated. The different calcium-binding protein-defined sub-divisions may belong to parallel auditory pathways that were previously demonstrated in non-human primates, and they may constitute a first indication of parallel processing in human sub-cortical auditory structures.

Using monoclonal antibodies against parvalbumin and calbindin, and a polyclonal antiserum against calretinin, the expression patterns of these proteins in the retina of the tench and rainbow trout were studied at light microscopic level in in toto preparations and radial sections (Weruaga *et al.*, 2000). Parvalbumin was found in sub-populations of small amacrine cells in both species, but these cells were more abundant and had a clear centre-periphery gradient distribution in the tench. Calbindin was observed in glial cells such as Muller cells, astrocytes in the nerve fibre layer, and sparse large cells close to the entrance of the optic nerve in both species. Moreover, it was revealed in H1 horizontal cells and their thick axon terminals in the tench retina. Calretinin was expressed in different types of ganglion cells and numerous neurones located in the inner plexiform layer in both species, but was more abundant and more strongly stained in the trout retina, where some bipolar cells were easily distinguishable.

The distribution of parvalbumin immunoreactivity in the developing cerebellum of the rainbow trout was studied by using a specific monoclonal antibody and the avidin-biotin peroxidase method (Porteros *et al.*, 1998). Parvalbumin immunoreactivity was absent during the embryonic development of the cerebellum. The first immunoreactive elements, identified by their localization and posterior morphological evolution as immature Purkinje cells, appeared at 6 days posthatching in the presumptive *corpus cerebelli* and *lobus vestibulolateralis*. The labeling extended throughout the cerebellum following a caudorostral gradient, and in 21 days alevins, parvalbumin immunoreactive Purkinje cells were also observed in the *valvula cerebelli*. The appearance of parvalbumin-immunostaining in the Purkinje cells was not simultaneous; the labeling was observed initially in the cell body, extending gradually to the dendritic branches and finally to the axon. From 1 year onwards, parvalbumin immunoreactive terminal puncta from the Purkinje cell axons were observed surrounding the cell bodies of eurydendroid cells, that were parvalbumin immunonegative in all developmental stages studied. The spatio-temporal pattern of parvalbumin immunoreactivity in the rainbow trout cerebellum is different to previous observations in the cerebellum of amniotes.

Due to its strategic position, the thalamic reticular nucleus (TRN) plays an important role within the thalamo-cortical circuits. The perireticular thalamic nucleus (PRN) is a smaller group of cells, which is associated with the TRN and lies among the fibres of the internal capsule. In the study of Contreras-Rodriguez *et al.* (2002), sections from rabbits at different ages (prenatal, postnatal and adult) were examined to determine the parvalbumin expression in the developing TRN and PRN. In the TRN, there is one wave of parvalbumin expression during development, from caudal parts of the nucleus towards the rostral pole. At E22 stage there is already an incipient parvalbumin expression. In the adult stage, the TRN is completely positive to parvalbumin. This study clearly indicates the presence of the PRN in the developing rabbit. The first parvalbumin positive cells were visible at E24 stage, meanwhile the immunoreactivity was at its maximum at early postnatal stages (P0-P8).

Bu *et al.* (2003) investigated age-related changes in calcium binding proteins in the human brain by studying the status of calbindin-, calretinin-, and parvalbumin-positive neurons in 17 cortical areas. They found a trend toward a decrease in the number of calbindin-immunoreactive neurons in all areas studied. However, this trend reached significance in only 4 areas in which the loss of calbindin-positive neurons ranged between 20 and 46%. Immunoreactivity for calretinin was also decreased in many areas and this difference reached significance in three regions (26-37%). It is of interest that cortical neurons displaying parvalbumin immunoreactivity did not show an age-related change. These observations indicate that loss of calcium binding proteins-positive neurons occurs in restricted cortical regions and is not a specific change as other neurochemically specific neurons also display restricted age-related changes.

The entorhinal cortex is an essential component in the organization of the human hippocampal formation related to cortical activity. It transfers neocortical information (ultimately distributed to the dentate gyrus and hippocampus) and receives most of the hippocampal output directed to neocortex. At birth, the human entorhinal cortex presents similar layer organization as in adults, although layer II (cell islands) and upper layer III have a protracted maturation. Grateron *et al.* (2003) reviewed works on postnatal development of calcium-binding proteins immunoreactivity (parvalbumin, calbindin, calretinin) in the human entorhinal cortex. The presence of interneurons expressing parvalbumin, calbindin-D_{28k} and calretinin is well documented in the adult human entorhinal cortex. In many of them the calcium binding is co-localized with GABA. Parvalbumin-immunoreactive cells and fibers are virtually absent at birth, their presence increasing gradually in deep layer III, mostly in the lateral and caudal portions of the entorhinal cortex from the 5th month onwards. Calbindin immunoreactive cells and fibers are present at birth, mainly in layers II and upper III; mostly at rostral and lateral portions of the entorhinal cortex, increasing in number and extending to deep layers from the 5th month onwards. Calretinin immunoreactivity is present at birth, homogeneously distributed over layers I, II and upper V, throughout the entorhinal cortex. A substantial increase in the number of calretinin neurons in layer V is observed at the 5th month. The postnatal development of parvalbumin, calbindin and calretinin may have an important role in the functional maturation of the entorhinal cortex through the control of hippocampal, cortical and sub-cortical information.

Parvalbumin is contained even in retinal cells. The study of Chiquet *et al.* (2005) was aimed to characterize the distribution of calbindin, calretinin, parvalbumin and recoverin in relation to retinal cell types in a strepsirhine primate (mouse lemur, *Microcebus*) in comparison with primate species of the three main haplorhine lineages (marmoset, macaque and human), as well as a rodent (gerbil, *Taterillus*). It was found that whereas the recoverin antibody labels both rod and cone photoreceptors in all species, calbindin consistently labels cones, but not rods, in the haplorhine primates marmoset, macaque and human, but none of the photoreceptors in the mouse lemur. Marmoset and macaque also show a distinct label of cone outer segments with calretinin. Depending on the species, bipolar cells express calbindin and/or recoverin, while amacrine, horizontal and ganglion cells are labeled to varying degrees with calbindin, calretinin and parvalbumin. Haplorhine and strepsirhine primates clearly differ in the expression of calcium-binding protein expression in horizontal cells. In all haplorhine species, horizontal cells are densely labeled with parvalbumin whereas in mouse lemur horizontal cells express calbindin but not parvalbumin. Several characteristics of the calcium-binding immunostaining in the retina of the mouse lemur are similar to those observed in the rodent, and distinguish this species from the diurnal haphorhine primates. These differences may be related to adaptations of retinal structure and function to the nocturnal niche, since nocturnal strepsirhine and haphorhine (*Tarsius* and *Aotus*) primates share some features of calcium-binding expression.

Ross & Porter (2002) and Porter *et al.* (1999) studied the effects of dopamine and 17 β-estradiol upon parvalbumin expression in rodent frontal cortex during development. Under control conditions, parvalbumin immunoreactive somata and fibers were primarily found in the deep laminae. In comparison, slices in all treatment groups exhibited a pattern of parvalbumin expression that was significantly different than controls. Specifically, dopamine treatment increased the percentage of parvalbumin immunoreactive somata, dendritic length and density in the deep cortical layers, but not in the superficial cortical layers. Both 17 β-estradiol and dopamine+17 β-estradiol treatments induced similar changes in both the deep and the superficial cortical layers.

Some experimental works were devoted to the studies of effects of parvalbumin on cytosolic free calcium concentration. For example, the effect of parvalbumin on $[Ca^{2+}]_i$ transients was investigated by perfusing adrenal chromaffin cells with fura-2 and fluorescein isothiocyanate (FITC) labeled parvalbumin (Lee *et al.*, 2000). As parvalbumin diffused into cells, the decay of $[Ca^{2+}]_i$ transients was transformed from monophasic into biphasic. The proportion of the initial fast decay phase increased in parallel with the fluorescence intensity of FITC, indicating that parvalbumin is responsible for the initial fast decay phase. From these results the authors concluded that Ca^{2+} buffers with slow kinetics, such as parvalbumin, may cause biexponential decays in $[Ca^{2+}]_i$ transients, thereby complicating the analysis of endogenous Ca^{2+} binding ratios (k_S) based on time constants. Nevertheless, estimates of k_S based on Ca^{2+} increments provide reasonable estimates for Ca^{2+} binding ratios before equilibration with parvalbumin.

Dreessen *et al.* (1996) investigated whether the expression of human α-parvalbumin affects depolarization-induced elevations of the cytosolic free calcium concentration ($[Ca^{2+}]_i$) in human neuroblastoma SKNBE2 cells. A full length human parvalbumin cDNA was transiently transfected into SKNBE2 cells. In parvalbumin-expressing SKNBE2 cells, parvalbumin concentration amounted to 0.42 mM. In a fraction of cells, depolarization by 50 mM K^+ induced a transient elevation in $[Ca^{2+}]_i$. Cells with a significant parvalbumin immunofluorescence responded to depolarization with smaller elevations in $[Ca^{2+}]_i$ than non-parvalbumin-expressing cells. Resting $[Ca^{2+}]_i$ did not differ between parvalbumin-expressing and control cells. These observations indicate that large depolarization-induced transient elevations of $[Ca^{2+}]_i$ in neuroblastoma cells can be suppressed by parvalbumin.

Neurons that express calcium binding proteins are characterized by some specific features. Apoptosis is attenuated in cells that contain calbindin (Dowd *et al.*, 1992; Dowd, 1995). Hippocampal neurons in culture containing calbindin were shown to be more effective in reducing intracellular Ca^{2+} concentrations compared with non-calbindin-containing neurons (Mattson *et al.*, 1991) and this was proposed as a mechanism for increased seizure resistance in these cells (Scharfman & Schwartzkroin, 1989). Cortical neurons in culture containing calbindin are selectively resistant to excitotoxins (Lukas & Jones, 1994), and motor neurons and hippocampal neurons transfected with cDNA for calbindin were shown to have increased ability to buffer calcium and increased survival after sclerotic, hypoglycemic or excitotoxic induced injury (Lukas & Jones, 1994; Ho *et al.*, 1996). Calbindin was also shown to protect neurons from oxidative stress (Dowd, 1995). Hippocampal progenitor cells transfected with calbindin cDNA are much less susceptible to nicotine-induced apoptosis (Berger *et al.*, 1998). There is also evidence that calcium binding proteins are neuroprotective *in vivo*. Hippocampal or motor neurons containing calretinin are resistant to Ca^{2+}-induced excitotoxicity while those without this protein were vulnerable (Mockel & Fischer, 1994; Terro et al., 1998). In tissue from patients with either motor neuron disease or temporal lobe epilepsy, neurons containing calbindin or parvalbumin survived while those without these proteins degenerated (Ince *et al*., 1993; Sloviter, 1989). The association between calcium binding proteins and ability to survive neuronal stress is not consistent in all regions and is a subject of debate (Andressen *et al*., 1993; Freund *et al*., 1990; Heizmann & Braun, 1992; Airaksinen *et al*., 1997).

Andressen *et al*. (1995) transfected parvalbumin cDNA into a human ovarian adenocarcinoma cell line, which normally does not express this protein. The induced expression of parvalbumin under the control of three different promoters causes: (1) changes in the morphology of the cells from epitheloid to fusiform, (2) an increase in motility of whole cell clusters, and (3) a decrease in the mitotic rate. Transfection with a mutated cDNA of rat parvalbumin incapable of binding Ca^{2+} had no effect on these three parameters. These results indicate that ectopic expression of parvalbumin influences cell shape and motility by modulating intracellular Ca^{2+} handling. The authors suggested that this may be a basic function of parvalbumin when it is intrinsically expressed in differentiated non-muscle cells.

Indirect evidences support a protective role of parvalbumin against calcium-induced neurotoxicity. For example, in the work of Muller *et al*. (1996) after cloning the parvalbumin cDNA into an expression vector, teratocarcinoma cells (PCC7) were transfected. The results show that parvalbumin-transfected PCC7 cells had much better calcium buffering capacity than control cells.

Excitotoxic effects leading to neuronal cell degeneration are often accompanied by a prolonged increase in the intracellular level of Ca^{2+} ions and L-glutamate-induced toxicity is assumed to be mediated *via* a Ca^{2+}-dependent mechanism. Due to their buffering properties, EF-hand Ca^{2+}-binding proteins (calretinin, calbindin D_{28k} and parvalbumin) can affect intracellular Ca^{2+} homeostasis. D'Orlando *et al*. (2002) transfected N18-RE 105 neuroblastoma-retina hybrid cells with the cDNAs for the three calcium binding proteins and investigated the effect of these proteins on the L-glutamate-induced, Ca^{2+}-dependent cytotoxicity. It was found that in untransfected and parvalbumin-transfected cells, lactate dehydrogenase released into the medium progressively increased (starting from the 20th hour) reaching maximum levels after 28-30 h of glutamate application. In contrast, lactate dehydrogenase release in both, calretinin and calbindin D-28k-transfected clones, was not significantly different from unstimulated transfected or untransfected cells over the same period of time. The results indicate that the 'fast' Ca^{2+}-buffers calretinin and calbindin D-28k, but not the 'slow' buffer parvalbumin can protect N18-RE 105 cells from this type of Ca^{2+}-dependent L-glutamate-induced delayed cytotoxicity.

Similar conclusion was obtained in the work of Maetzler *et al*. (2004). They injected the glutamate agonist ibotenic acid into the striatum of adult mice ectopically expressing parvalbumin in neurons. Striatal ibotenic acid injection results in local nerve cell loss and reactive astrogliosis. Light microscopic evaluation, carried out after a delay of 2 and 4 weeks, reveals an enlarged and accelerated neurodegenerative process in mice ectopically expressing neuronal parvalbumin. These authors concluded that parvalbumin is not neuroprotective, it rather enhances nerve cell death.

Van Den Bosch *et al*. (2002) investigated whether increasing the motor neuron's cytosolic Ca^{2+}-buffering capacity protects them from excitotoxic death. To obtain motor neurons with increased Ca^{2+} buffering capacity, they generated transgenic mice overexpressing parvalbumin. These mice have no apparent phenotype. Parvalbumin overexpression was present in the central nervous system, kidney, thymus, and spleen. Motor neurons from these transgenic mice expressed parvalbumin in culture and were partially protected from kainic acid-induced death as compared to those isolated from non-transgenic littermates. Parvalbumin overexpression also attenuated kainic acid-induced Ca^{2+} transients, but not those induced by depolarization. The authors concluded that overexpression of a high-affinity Ca^{2+} buffer such as parvalbumin protects the motor neuron from excitotoxicity and this protective effect depends upon the mode of Ca^{2+} entry into the cell.

The effects of kainic acid, which induces *status epilepticus*, on the expressions of calbindin D_{28k}, parvalbumin and calretinin was examined in the rat striatum by immunohistochemistry and microdensitometry by Lee *et al.* (2002). At 1, 3 and 6 days after kainic acid-induced seizure, the number of calretinin-positive neurons in the striatum was significantly lower than in control rats. However, no significant difference was observed in the number of calbindin D_{28k}- and parvalbumin-positive neurons in control and seizure rats. At 1, 3 and 6 days after seizure the optical densities of calretinin- and parvalbumin-positive neurons in the striatum were significantly lower than in control rats. These data suggests that calcium binding proteins in the striatum have differential vulnerabilities to kainic acid-induced seizure.

Recently it was suggested that parvalbumin plays a role in the generation of synchronous spikes in a network of GABAergic neurons in the cerebral cortex. Galarreta & Hestrin (2002) used transgenic mice expressing the enhanced green fluorescent protein in GABAergic interneurons containing parvalbumin to study the synaptic connectivity among fast-spiking cells in slices from adult animals (2-7 months old). They found that the majority of such cells are electrically coupled (61%, 14 of 23 pairs). In addition, 78% of the pairs are connected *via* GABAergic chemical synapses, often reciprocally. Their results indicate that such cells are highly interconnected in the adult cerebral cortex by both electrical and chemical synapses, establishing networks that can have important implications for coordinating activity in cortical circuits. Caillard *et al.* (2000) demonstrated that the lack of parvalbumin presynaptically in the axo-axonic (chandelier) and basket cells of the cerebellum affects the paired-pulse modulation at this synapse to the Purkinje cell. In the hippocampus, parvalbumin deficiency facilitated repetitive inhibitory postsynaptic currents at frequencies >20 Hz and the power of related inhibition-based gamma oscillations was increased (Vreugdenhil *et al.*, 2003). Thus, parvalbumin is able to affect short-term modulation that likely affects temporal aspects of the entire network containing parvalbumin-expressing neurons. It was suggested that parvalbumin deficiency, due to an increased short-term facilitation of GABA release, enhances inhibition by high-frequency burst-firing parvalbumin-expressing interneurons and may affect the higher cognitive functions associated with gamma oscillations.

In spite to enormous number of publications on parvalbumin in brain, the physiological role of parvalbumin in this organ is still unclear. The results obtained in these studies are very interesting but often contradictory. This seems to be explained by complexity of this system. More direct and more clear experiments are needed to elucidate the functions of parvalbumin and other calcium binding proteins in brain.

Parvalbumin and Brain Diseases

Brain deseases can influence expression of calcium binding proteins in neurons. Moreover, there exists the calcium hypothesis of some diseases, such as Alzheimer's desease, which invokes the disruption of calcium signaling as the underlying cause of neuronal dysfunction and ultimately apoptosis (reviewed by O'Day & Myre, 2004).

The distribution of parvalbumin, calretinin, and calbindin immunoreactive neurons was studied with the help of an image analysis system (Vidas/Zeiss) in the primary visual area 17 and associative area 18 (Brodmann) of Alzheimer and control brains (Leuba *et al.*, 1998). In neither of these areas was there a significant difference between Alzheimer and control groups in the mean number of parvalbumin, calretinin, and calbindin immunoreactive neuronal profiles, counted in a cortical column going from pia to white matter. Significant differences in the mean densities (numbers per square millimeter of cortex) of parvalbumin, calretinin, and calbindin immunoreactive neuronal profiles were not observed either between groups or areas, but only between superficial, middle, and deep layers within areas 17 and 18. The frequency distribution of neuronal areas indicated significant differences between parvalbumin, calretinin, and calbindin immunoreactive neuronal profiles in both areas 17 and 18, with more large parvalbumin than calretinin, and calbindin positive profiles. There were also significantly more small and less large parvalbumin and calretinin immunoreactive neuronal profiles in Alzheimer than in controls. These data show that, although the brain pathology is moderate to severe, there is no prominent decrease of parvalbumin, calretinin, and calbindin positive neurons in the visual cortex of Alzheimer brains, but only selective changes in neuronal perikarya.

The entorhinal cortex, which is involved in neural systems related to memory, is selectively degenerated in early Alzheimer's disease. Mikkonen *et al.* (1999) examined neuropathological changes in the eight entorhinal sub-fields in post mortem Alzheimer's disease subjects using Thionin and Bielschowsky stains and parvalbumin, calretinin and calbindin-D_{28k} immunohistochemistry. Immunohistochemical staining showed that neurons and fibres that contain calcium-binding proteins were more frequently altered in these sub-fields than in the rostromedial sub-fields. Detailed analysis further revealed that non-principal cells containing parvalbumin or calbindin-D_{28k} showed morphological alterations early in the entorhinal pathology of Alzheimer's disease, whereas non-principal neurons containing calretinin were better preserved even in Alzheimer's disease patients with severe entorhinal pathology. The degeneration of parvalbumin-immunoreactive neurons and basket-like networks and calbindin-positive non-principal neurons was observed mainly in layer II, where the calretinin-positive non-principal neurons formed aggregates especially at late stages of the disease. These findings indicate that specific sub-fields of the entorhinal cortex involving neurons that contain distinct calcium-binding proteins are differentially vulnerable in Alzheimer's disease.

At the same time, the data of Capper-Loup *et al.* (2005) demonstrate that parvalbumin expression is specifically modified in the primary motor cortex of parkinsonian rats. Since GABAergic cell function is related to parvalbumin expression, these changes might be linked with a change in cortical inhibition. These results suggest that dopamine deficiency induces reversible changes in GABAergic cortical cells, which might be linked with parkinsonian symptoms.

Carretta *et al.* (2004) quantified and compared the distribution of parvalbumin- and calbindin-positive neurons in the motor, somatosensory, visual, and anterior cingulate cortices of wild-type and *mdx* mice. They were able to reveal, in *mdx* mice, changes involving both the above populations of interneurons. These changes were evident in the motor and anterior cingulate cortices but not in the somatosensory and visual cortices. In addition, the changes of coefficients of variation were area-specific in the cortex of *mdx* mice. The values increased in the motor cortex and decreased in the anterior cingulate cortex with respect to the corresponding values of wild-type animals.

Neuropathological studies demonstrated deficits of GABAergic interneurons in the hippocampus in schizophrenia and selective deficits in some GABAergic sub-populations defined by calcium-binding proteins were reported in the cortex in schizophrenia. Eyles *et al.* (2002) reviewed the role of calcium binding proteins in schizophrenia. At present there exists a suggestion that malfunction within the prefrontal cortex is a feature of schizophrenia. The crucial role played by GABAergic inhibitory interneurons in regulating prefrontal function is well accepted. In the cerebral cortex, calcium binding proteins are discrete markers of some sub-classes of these neurons (e.g. parvalbumin is found in chandelier and basket cells, calbindin in double-bouquet neurons, calretinin in bipolar and bitufted neurons (Conde *et al.*, 1994; Lund & Lewis, 1993; Lewis, 1998). Collectively, calcium binding proteins-containing interneurons make up 90% of all GABAergic neurons in this region (Lund & Lewis, 1993). In addition to the considerable heterogeneity in both the morphology and afferent control of pyramidal output from the frontal cortex, the receptor profiles for each sub-class appear highly specific.

There are many other examples of differential distribution of receptors between the calcium binding proteins-expressing interneurons of the cortex. Both D1 and D2 receptors on cortical interneurons appear to be selectively found on parvalbumin-containing cells while 5HT2A and 5HT3 receptors are differentially distributed between parvalbumin- and calbindin- and calretinin-immunoreactive neurons (Le Moine & Gaspar, 1998; Jakab & Goldman-Rakic, 2000). The strategic role these interneurons play in regulating prefrontal cortex output and the complexity of their relationship to pyramidal output neurons have made them a focus for neuropathological investigation of calcium binding proteins abnormalities in schizophrenia.

The study of prefrontal areas 9 and 46 in prefrontal cortex showed a 50–70% increased density of calbindin-positive neurons in layers III and V/VI but no group difference between schizophrenics and controls for calretinin (Daviss & Lewis, 1995). In another study, parvalbumin was found to be reduced in laminae III and IV of prefrontal area 10 in schizophrenics (Beasley & Reynolds, 1997). No change in calretinin could be found in this tissue in a later study; however, a reanalysis of the previously reported parvalbumin findings showed deficits were more dramatic in patients without the commonly observed ventricular enlargement associated with the disease (Reynolds & Beasley, 2001).

An important finding is that the distinctive axonal fibre network from prefrontal chandelier interneurons was shown to be reduced in schizophrenics (Woo *et al.*, 1998). These predominantly parvalbumin-containing interneurons provide direct inhibitory control over pyramidal cells within the prefrontal cortex. In a qualitative investigation, calbindin was shown to be unaltered in GABAergic interneurons in prefrontal area 9 but decreased in pyramidal cells from the CA2 region of the hippocampus (Iritani *et al.*, 1999). The second finding from this study was an apparent random orientation of calbindin-containing neuronal fibres in schizophrenics compared with a more ordered array in non-psychiatric controls.

There is substantial evidence for the importance of hippocampus in the pathophysiology of schizophrenia. It was shown that in hippocampal sub-fields cell number is unchanged but parvalbumin content is increased in anterior cingulate GABAergic interneurons from schizophrenics (Kalus *et al.*, 1997). At the same time, in the work of Zhang & Reynolds (2002) it was shown that, relative to normal controls, schizophrenic patients showed a significant and profound deficit in the relative density of parvalbumin-immunoreactive neurons in all hippocampal sub-fields. These reductions were more apparent in male than female schizophrenic patients, and were unrelated to antipsychotic drug treatment, age or duration of illness.

The main purpose of the review of Eyles *et al.* (2002) on calcium binding proteins and schizophrenia was to alert the reader that the so-called non-regulatory ''marker or buffer'' calbindin, calretinin and parvalbumin may themselves be functionally involved in the pathogenesis of schizophrenia. The fact that only certain calcium binding proteins-containing cells appear to be affected, i.e. parvalbumin and to a lesser extent calbindin; the implication that in some cases, neuronal calcium binding proteins content appears to be affected rather than the number of cells containing calcium binding proteins; the extensive evidence both *in vitro* and *in vivo* that neuronal function and even viability may be compromised when intracellular calcium binding proteins content is diminished; and the potential impact on prefrontal and hippocampal output when GABAergic inhibitory function is impaired, suggest that calcium binding proteins warrant closer examination.

It is well known that intracellular calcium is increased in vulnerable spinal motoneurons in immune-mediated as well as transgenic models of amyotrophic lateral sclerosis. Beers *et al.* (2001) developed transgenic mice overexpressing parvalbumin in spinal motoneurons. It is of importance that amyotrophic lateral sclerosis immunoglobulins increased intracellular calcium and spontaneous transmitter release at motoneuron terminals in control animals, but not in parvalbumin overexpressing transgenic mice. Parvalbumin transgenic mice interbred with mutant SOD1 (mSOD1) transgenic mice, an animal model of familial ALS, had significantly reduced motoneuron loss, and had delayed disease onset (17%) and prolonged survival (11%) when compared with mice with only the mSOD1 transgene. These results affirm the importance of parvalbumin in altering calcium homeostasis in motoneurons.

It is well known also that following nerve injury in neonatal rats, a large proportion of motoneurons die, possibly as a consequence of an increase in vulnerability to the excitotoxic effects of glutamate. Calcium-dependent glutamate excitotoxicity is thought to play a significant role not only in injury-induced motoneuron death, but also in motoneuron degeneration in diseases such as amyotrophic lateral sclerosis. Motoneurons are particularly vulnerable to calcium influx following glutamate receptor activation, as they lack a number of calcium binding proteins, such as calbindin-D_{28k} and parvalbumin. Therefore, it is possible that increasing the ability of motoneurons to buffer intracellular calcium may protect them from injury-induced cell death. Dekkers *et al.* (2004) tested this possibility by examining the effect of neonatal axotomy on motoneuron survival and muscle force production in normal and transgenic mice that over-express parvalbumin in their motoneurons. Following nerve injury in wild type mice, only 20.2% of injured motoneurons survive long term compared with 47.2% in parvalbumin over-expressing mice. Surprisingly, this dramatic increase in motoneuron survival was not reflected in a significant improvement in muscle function. Thus, inducing spinal motoneurons to express parvalbumin protects a large proportion of motoneurons from injury-induced cell death, but this is not sufficient to restore muscle function.

A finely tuned balance between excitatory and inhibitory activity in a neuronal circuit is required for appropriate brain function. Epilepsy, one of the most common neurological disorders, is characterized by massive hypersynchronous discharges from large assemblies of neurons. Networks of GABAergic interneurons are of utmost importance in generating and promoting synchronous activity and are involved in producing coherent oscillations. Alteration of their inhibitory activity has been proposed as a major mechanism leading to epileptic seizures. In the work of Schwaller *et al.* (2004) the role of parvalbumin in maintaining the stability of neuronal networks was assessed in knockout (parvalbumin-/-) mice. Pentylenetetrazole induced generalized tonic-clonic seizures in all genotypes, but the severity of seizures was significantly greater in parvalbumin-/- than in parvalbumin+/+ animals. Extracellular single-unit activity recorded from over 1000 neurons *in vivo* in the temporal cortex revealed an increase of units firing regularly and a decrease of cells firing in bursts. In the hippocampus, parvalbumin deficiency facilitated the GABA(A)ergic current reversal induced by high-frequency stimulation, a mechanism implied in the generation of epileptic activity. The authors postulated that parvalbumin plays a key role in the regulation of local inhibitory effects exerted by GABAergic interneurons on pyramidal neurons. It was suggested that, through an increase in inhibition, the absence of parvalbumin facilitates synchronous activity in the cortex and facilitates hypersynchrony through the depolarizing action of GABA in the hippocampus.

PARVALBUMIN AS ALLERGEN

Almost 4% of the population on Eath suffers from food allergy, which is an adverse immunological reaction to food. Allergic reactions to fish are frequent, especially in countries with a developed fish industry. After eating of fish, some of the following clinical symptoms are rapidly induced in sensitized patients: urticaria, angioedema, respiratory symptoms (asthma and rhinoconjunctivitis), gastrointestinal symptoms (diarrhea and vomiting) and, in severe cases, fatal anaphylaxis. Extensive studies of cod *Gadus callarias* first identified parvalbumin as a fish allergen (Elsayed & Aas, 1971; Elsayed & Bennich, 1974, 1975). It was shown that patients who produce IgE Abs against one parvalbumin will cross-react with the homologous proteins from other fish species, which demonstrates the importance of parvalbumins as cross-reactive fish allergens and explains why allergic individuals exhibit clinical symptoms upon contact with various fish species. Parvalbumins from different fish species, for instance cod, tuna, salmon were shown to contain cross-reactive as well as species specific epitopes when tested with the sera of fish allergic patients (Bugajska-Schretter *et al.*, 1998). After identification of parvalbumin as allergen, calcium-binding allergens were discovered in pollens of trees, grasses and weeds and, recently, as autoallergens in man (reviewed by Valenta *et al.*, 1998).

Parvalbumin from fish is a very stable allergen: extremes in pH, temperature, or high concentrations of denaturing agents do not significantly alter its allergenicity (Elsayed & Aas,

1971, 1975). Due to the very high stability of calcium loaded parvalbumin, it can sensitise patients despite cooking and exposure to the gastrointestinal tract.

In the work of Bugajska-Schretter *et al.* (2000) allergenic activity of carp parvalbumin pI 4.7 was analysed by immunoglobulin E binding and basophil histamine release tests. Purified carp parvalbumin reacted with immunoglobulin E of more than 95% of individuals allergic to fish, induced dose-dependent basophil histamine release and contained, on average, 83% of the immunoglobulin E epitopes present in other fish species. It is of interest that calcium depletion reduces the immunoglobulin E binding capacity of parvalbumin, which may be due to conformation-dependent immunoglobulin E recognition (Swoboda *et al.*, 2002).

Three species of mackerels (*Scomber japonicus, S. australasicus* and *S. scombrus*) are widely consumed and considered to be most frequently involved in incidents of immunoglobulin E-mediated fish allergy in Japan. For this reason processed food products containing them as raw materials are recommended to have labels indicating their use on packages or bottles. Purified preparations of possible candidate for the major allergen from three species gave a single band of about 11 kDa and were clearly identified as parvalbumins by analyses of their partial amino acid sequences. In ELISA experiments, four of five sera from fish-allergic patients reacted to all the purified parvalbumins, demonstrating that parvalbumin is the major allergen in common with the mackerels (Hamada *et al.*, 2003). Antigenic cross-reactivity among the mackerel parvalbumins was also established by ELISA inhibition experiments.

Van Do *et al.* (2005) determined the allergenic cross-reactivity between 9 commonly edible fish: cod, salmon, pollack, mackerel, tuna, herring, wolffish, halibut, and flounder. It was found that cod (Gad c 1), salmon (Sal s 1), pollack (The c 1), herring, and wolffish share antigenic and allergenic determinants as shown by immunoblots and IgE ELISA, whereas halibut, flounder, tuna, and mackerel displayed lowest cross-reactivities. The highest mean IgE ELISA inhibition percent of 10 sera was obtained in the case of Gad c 1, followed by The c 1, herring, Sal s 1, wolffish, halibut, flounder, tuna, and mackerel with the least inhibition. Nine of the 10 patients showed positive skin prick test to cod, salmon, and pollack; 8 patients reacted to recombinant (r) Sal s 1. Positive skin prick tests to rGad c 1 and rThe c 1 were demonstrated in one patient. It was concluded that Gad c 1, Sal s 1, The c 1, herring, and wolffish contained the most potent cross-reacting allergens, whereas halibut, flounder, tuna, and mackerel were the least allergenic in the study. Fish-hypersensitive patients can probably tolerate some fish species while being allergic to others.

Immunoglobulin E-mediated allergic reactions to frogs or frog material were the topic of a number of case reports. The symptoms and situations in which these allergic reactions occur fall into two categories: airborne and contact-mediated reactions provoked by catching or handling frogs, and food-induced anaphylactic reactions following consumption of frog meat. The frog material and the patient's serum from a case of severe food-induced anaphylaxis were used in the work of Hilger *et al.* (2002) to define the implicated allergen at the protein and DNA level. These authors determined the nucleotide sequence of the allergen from the frog of Indonesian origin that was consumed by the patient, and the homologous cDNA from *Rana esculenta*. Protein microsequencing revealed that the implicated frog allergen belonges to the parvalbumin family. cDNAs coding for α- and β-parvalbumin of *R. esculenta* and *Rana species* were cloned. The patient's serum immunoglobulin E antibodies recognized parvalbumin prepared from frog muscle and recombinant α-parvalbumin from *R. species* but not from *R. esculenta*. Recombinant β-parvalbumin was not recognized by the immunoglobulin E antibodies. It is of interest that cod parvalbumin, a major cross-reactive allergen among different fish species, shares immunoglobulin E binding epitopes with frog parvalbumin (Hilger *et al.*, 2004). This *in vitro* cross-reactivity seems to be also clinically relevant.

CONCLUDING REMARKS

Metalloproteins are one of the most diverse classes of proteins. Examination of the Protein Data Bank shows that almost one fourth of the entries contain a metal atom bound to a protein, with Zn^{2+}, Fe^{2+}, Mg^{2+} and Ca^{2+} most abundant. Any cell contains a set of metal binding proteins. The activity of some of these proteins is regulated by the binding of metal cations, other metal binding proteins serve as modulators of metal ions flows. Some metal cations compete for the same protein binding sites, others have their own distinct protein binding sites. Modern metalloproteomics studies all these complicated interactions between physiologically significant metal cations and various proteins inside various cell compartments at different stages of cell life. Such studies need good knowledge of physico-chemical properties of the metal binding proteins.

The detailed studies of calcium binding proteins, including parvalbumin, are very important for both fundamental biology and practical medicine. The structural and physico-chemical properties of parvalbumin are well known now and this knowledge is an essential contribution to our understanding of protein physics. The EF-hand domain, first found in parvalbumin, was revealed in many other calcium binding proteins. At the same time, physiological functions of parvalbumin are far from being absolutely clear. While its role as a soluble relaxing factor in fast muscles is widely accepted, its functional role in neurons and other cells and tissues is still unclear. Nevertheless, even at this stage there appeared attempts of practical use of parvalbumin as a new therapeutic tool to correct cellular disturbances in calcium signaling pathways.

This book mostly contains general information about structure, properties of parvalbumin but I hope that it will be helpful for future studies of structural, physico-chemical and functional properties of not only parvalbumin but also another metal binding proteins.

REFERENCES

Aas, K., and Elsayed, S. Physico-chemical properties and specific activity of a purified allergen (codfish). *Dev. Biol. Scand.* (1975) 29, 90–98.

Agah, S., Larson, J.D., and Henzl, M.T. Impact of proline residues on parvalbumin stability. *Biochemistry* (2003) 42, 37, 10886-10895.

Ahlström, P., Teleman, O., Jonsson, B., and Forsen, S. Molecular dynamics simulation of parvalbumin in aqueous solution. *J. Amer. Chem. Soc.* (1987) 109, 5, 1541-1551.

Ahmed, F.R., Przybylska, M., Rose, D.R., Birnbaum, G.I., Pippy, M.E., and MacManus, J.P., Structure of refined at 1.85 resolution. An example of extensive molecular aggregation via Ca^{2+}. *J. Mol. Biol.* (1990) 216, 1, 127-140.

Ahmed, F.R., Rose, D.R,, Evans, S.V., Pippy, M.E., and To, R. Refinement of recombinant oncomodulin at 1.30 Å resolution. *J. Mol. Biol.* (1993) 230, 4, 1216-1224.

Airaksinen, M.S., Thoenen, H., and Meyer, M. Vulnerability of midbrain dopaminergic neurons in calbindin-D28k-deficient mice: lack of evidence for a neuroprotective role of endogenenous calbindin in MPTP-treated and weaver mice. *Eur. J. Neurosci.* (1997) 9, 120-127.

Allan, E., and Hepler, P.K. Calmodulin and calcium binding proteins. In *Biochem. Plants: Comprehensive Treatise*, San Diego etc. (1989) 15, 455-484.

Allbritton, N.L., Meyer, T., and Stryer, L. Range of messenger action of calcium ion and inositol 1,4,5-triphosphate. *Science* (1992) 258, 1812-1815.

Allouche, D., Parello, J., and Sanejouand, Y.H. Ca^{2+}/Mg^{2+} exchange in parvalbumin and other EF-hand proteins. A theoretical study. *J. Mol. Biol.* (1999) 285, 857-873.

Ames, J. B., Ishima, R., Tanaka, T., Gordon, J. I., Stryer, L. and Ikura, M. Molecular mechanics of calcium-myristoyl switches. *Nature (London)* (1997) 389, 198-202.

Ames, J.B., Tanaka, T., Ikura, M., and Stryer, L. Nuclear magnetic resonance for Ca^{2+}-induced extrusion of the myristoyl group of recoverin. *J. Biol. Chem.* (1995) 270, 52, 30909-30913.

Ames, J.B., Tanaka, T., Stryer, L., and Ikura, M. Secondary structure of myristoylated recoverin determined by three-dimensional heteronuclear NMR: implications for the calcium-myristoyl switch. *Biochemistry* (1994) 33, 35, 10743-10753.

Ames, J.B., Tanaka, T., Ikura, M., and Stryer, L. Nuclear magnetic resonance for Ca^{2+}-induced extrusion of the myristoyl group of recoverin. *J. Biol. Chem.* (1995) 270, 52, 30909-30913.

Andressen, C., Blumcke, I., and Celio, M.R. Calcium-binding proteins: selective markers of nerve cells. *Cell Tissue Res.* (1993) 271, 181-208.

Andressen, C., Gotzos, V., Berchtold, M.W., Pauls, T.L., Schwaller, B., Fellay, B., and Celio, M.R. Changes in shape and motility of cells transfected with parvalbumin cDNA. *Exp. Cell Res.* (1995) 219, 2, 420-426.

Angiolillo, P.J., and Vanderkooi, J.M. Hydrogen atoms are produced when tryptophan within a protein is irradiated with ultraviolet light. *Photochem. Photobiol.* (1996) 64, 3, 492-495.

Annoh, H., Inokuchi, T., Ohta, K., Wakimoto, M., and Ueda, T. Immunohistochemical investigations of parvalbumin localization in the skeletal muscle fibers of rats. *Okajimas Folia Anat. Jpn.* (1995) 72, 4, 221-226.

Ashley, C.C., and Griffiths, P.J. The effect of injection of parvalbumin into single muscle fiber from the barnacle *Balanus nubilus*. *J. Physiol.* (1983) 345, 105P.

Atoji, Y., Yamamoto, Y., and Suzuki, Y. Parvalbumin in cortical epithelial cells of the pigeon thymus. *J. Anat.* (2000) 196, Pt 3, 305-311.

Babini, E., Bertini, I., Capozzi, F., Del Bianco, C., Hollender, D., Kiss, T., Luchinat, C., and Quattrone, A. Solution structure of human β-parvalbumin and structural comparison with its paralog α-parvalbumin and with their rat orthologs. *Biochemistry* (2004) 43, 51, 16076-16085.

Babor, M., Greenblatt, H.M., Edelman, M., and Sobolev, V. Flexibility of metal binding sites in proteins on a database scale. *Proteins: Structure, Function, and Bioinformatics.* (2005) 59, 221-230.

Babu, Y.S., Sack, J.S., Greenhough, T.J., Bugg, C.E., Means, A.R., and Cook, W.J. Three-dimensional structure of calmodulin. *Nature* (1985) 315, 6014, 37-40.

Bai, J., and Chapman, E.R. The C2 domains of synaptotagmin – partners in exocytosis. *TRENDS Biochem. Sci.* (2004) 29, 3, 143-151.

Baig, I., Bertini, I., Del Bianco, C., Gupta, Y.K., Lee, Y.M., Luchinat, C., and Quattrone, A. Paramagnetism-based refinement strategy for the solution structure of human α-parvalbumin. *Biochemistry* (2004) 43, 18, 5562-5573.

Baksh, S., and Michalak, M. Expression of calreticulin in *Escherichia coli* and identification of its Ca^{2+} binding domains. *J. Biol. Chem.* (1991) 266, 32, 21458-21465.

Baldellon, C., Alattia, J.R., Strub, M.P., Berchtold, M.W., Cave, A., and Padilla, A. ^{13}N NMR relaxation studies of calcium-loaded parvalbumin show tight dynamics compared to those of other EF-hand proteins. *Biochemistry* (1998) 37, 28, 9964-9975.

Baldellon, C., Padilla, A., and Cave, A. Kinetics of amide proton exchange in parvalbumin studied by ^{1}H 2-D NMR. A comparison of the calcium and magnesium loaded forms. *Biochimie* (1992) 74, 9-10, 837-844.

Banville, D., Rotaru, M., and Boie, Y. The intracisternal A particle derived solo LTR promoter of the rat oncomodulin gene is not present in the mouse gene. *Genetica* (1992) 86, 85-97.

Baron, G., Demaille, J., and Dutruge, E. The distribution of parvalbumin in muscle and in other tissues. *FEBS Letters* (1975) 56, 156-160.

Bastianelli, E. Distribution of calcium-binding proteins in the cerebellum. *Cerebellum* (2003) 2, 4, 242-262.

Baylor, S.M., Chandler, W.K., and Marshall M.W. Sarcoplasmic reticulum calcium release in frog skeletal muscle fibers estimated from arsenaso III calcium transients. *J. Physiol.* (1983) 344, 625-666.

Baylor, S.M., and Hollingworth, S. Model of sarcomeric Ca^{2+} movements, including ATP Ca^{2+} binding and diffusion, during activation of frog skeletal muscle. *J. Gen. Physiol.* (1998) 112, 3, 297-316.

Beasley, C.L., and Reynolds, G.P. Parvalbumin-immunoreactive neurons are reduced in the prefrontal cortex of schizophrenics. *Schizophr. Res.* (1997) 24, 349-355.

Beers, D.R., Ho, B.K., Siklos, L., Alexianu, M.E., Mosier, D.R., Mohamed, A.H., Otsuka, Y., Kozovska, M.E., McAlhany, R.E., Smith, R.G., and Appel, S.H. Parvalbumin overexpression alters immune-mediated increases in intracellular calcium, and delays disease onset in a transgenic model of familial amyotrophic lateral sclerosis. *J. Neurochem.* (2001) 79, 3, 499-509.

Benzonana, G., Capony, J.P., and Pechére, J.F. The binding of calcium to muscular parvalbumins. *Biochim. Biophys. Acta* (1972) 278, 1, 110-116.

Benzonana, G., Wnuk, W., Cox, J.A., and Gabbiani, G. Cellular distribution of sarcoplasmic calcium-binding proteins by immunofluorescence. *Histochem.* (1977) 51, 335-341.

Berchtold, M.W. Parvalbumin genes from human and rat are identical in intron/exon organization and contain highly homologous regulatory elements and coding sequences. *J. Mol. Biol.* (1989) 210, 417-427.

Berchtold, M.W., Brinkmeier, H., and Muntener, M. Calcium ion in skeletal muscle: its structural role for muscle function, plasticity, and disease. *Physiol. Reviws* (2000) 80, 3, 1215-1265.

Berchtold, M.W., Heizmann, C.W., and Wilson, K.J. Ca^{2+}-binding proteins: a comparative study of their behavior during high performance liquid chromatography using gradient elution on reverse-phase supports. *Analyt. Biochem.* (1983) 129, 120-131.

Berchtold, M.W., Heizmann, C.W., and Wilson, K.J. Primary structure of parvalbumin from rat skeletal muscle. *Eur. J. Biochem.* (1982a) 127, 381-389.

Berchtold, M.W., and Means, A.R. The Ca^{2+}-binding protein parvalbumin: molecular cloning and developmental regulation of mRNA abundance. *Proc. Natl. Acad. Sci. USA* (1985) 82, 1414-1418.

Berchtold, M.W., Wilson, K.J., and Heizmann, C.W. Isolation of neuronal parvalbumin by high-performance liquid chromatography. Characterization and comparison with muscle parvalbumin. *Biochemistry* (1982b) 21, 6552-6557.

Berger, F., Gage, F.H., and Vijayaraghaven, S. Nicotinic receptorinduced apoptotic cell death of hippocampal progenitor cells. *J. Neurosci.* (1998) 18, 6871–6881.

Bhushana Rao, K.S.P., Focant, B., Gerday, C., and Hamoir, G. Low molecular weight proteins of pike (Esox lucius) white muscles. II. Chemical and physical properties. *Comp. Biochem. Physiol.* (1969) 44B, 1113-1125.

Birdsall, W.J., Levine, B.A., Williams, R.J.P., Demaille, J.G., Haiech, J., and Pechére, J.F. Calcium and magnesium binding by parvalbumin: a proton magnetic resonance spectral study. *Biochimie* (1979) 67, 1, 741-750.

Blancuzzi, Y., Padilla, A., Parello, J., and Cave, A. Symmetrical rearrangement of the cation-binding sites of parvalbumin upon Ca^{2+}/Mg^{2+} exchange. A study by ^{1}H ^{2}D NMR. *Biochemistry* (1993) 32, 1302-1309.

Blum, H.E., Lehky, P., Kohler, L., Stein, E.A., and Fisher, E.H. Comparative properties of vertebrate parvalbumins. *J. Biol. Chem.* (1977) 252, 9, 2834-2838.

Blum, J.K., Berchtold, M.W. Calmodulin-like effect of oncomodulin on cell proliferation. *J. Cell Physiol.* (1994) 160, 3, 455-462.

Bottoms, C.A., Schuermann, J.P., Agah, S., Henzl, M.T., and Tanner, J.J. Crystal structure of rat α-parvalbumin at 1.05 Å resolution. *Protein Sci.* (2004) 13, 7, 1724-1734.

Brandt, P.W., Cox, R.N., and Kawai, M. Can the binding of Ca^{2+} to two regulatory sites on troponin C determine the steep pCa/tension relationship of skeletal muscle? *Proc. Nat. Acad. Sci. USA* (1980) 77, 8, 4717-4720.

Brandts, J.F., Brennan, M., and Lung-Nam Lin, Unfolding and refolding occur much faster for a proline-free protein than for most proline-containing proteins. *Proc. Nat. Acad. Sci. USA* (1977) 74, 10, 4178-4181.

Breen, P.J., Hild, E.K., and Horrocks, W.DeW., Jr. Spectroscopic studies of metal ion binding to tryptophan-containing parvalbumin. *Biochemistry* (1985a) 24, 19, 4991-1997.

Breen, P.J., Johnson, K.A., Horrocks, W.DeW., Jr. Stopped-flow kinetic studies of metal ion dissociation or exchange in a tryptophan-containing parvalbumin. *Biochemistry* (1985b) 24, 19, 4997-5004.

Bremel, R.D. Myosin linked calcium regulation in vertebrate smooth muscle. *Nature* (1974) 252, 5482, 405-407.

Brewer, J.M., Arnold, J., Beach, G.G., Ragland, W.L., and Wunderlich, J.K. Comparison of the amino acid sequences of tissue-specific parvalbumins from chicken muscle and thymus and possible evolutionary significance. *Biochem. Biophys. Res. Commun.* (1991) 181, 1, 226-231.

Brewer, L.M., Durkin, J.P., and MacManus, J.P. Immunocytochemical detection of oncomodulin in tumor tissue. J. Histochem. Cytochem. (1984) 32, 10, 1009-1016.

Brewer, L.M., and MacManus, J.P. Localization and synthesis of the tumor protein oncomodulin in extraembryonic tissues of the fetal rat. *Dev. Biol.* (1985) 112, 1, 49-58.

Brewer, L.M., and MacManus, J.P. Detection of oncomodulin, an oncodevelopmental protein in human placenta and choriocarcinoma cell lines. *Placenta* (1987) 8, 4, 351-363.

Brewer, J.M., Wunderlich, J.K., and Ragland, W.L. The amino acid sequence of avian thymic hormone, a parvalbumin. *Biochimie* (1990) 72, 653-660.

Briggs, N. Identification of the soluble relaxing factor as a parvalbumin. Fed. Proc. (1975) 34, 3, 540.

Brinley, F.J., Scarpe, F., and Tiffert, J.F. The concentration of ionized magnesium in barnacle muscle fibers. *J. Physiol.* (1977) 266, 3, 545-565.

Brodersen, D.E., Etzerodt, M., Madsen, P., Celis, J.E., Thogersen, H.C., Nyborg, J. EF-hands at atomic resolution: the structure of human psoriasin (S100A7) solved by MAD phasing. *Structure* (1998) 6, 477-489.

Bu, J., Sathyendra, V., Nagykery, N., and Geula, C. Age-related changes in calbindin-D28k, calretinin, and parvalbumin-immunoreactive neurons in the human cerebral cortex. *Exp. Neurol.* (2003) 182, 1, 220-231.

Bugajska-Schretter, A., Elfman, L., Fuchs, T., et al. Parvalbumin, a crossreactive fish allergen, contains IgE binding epitopes sensitive to periodate treatment and Ca^{2+} depletion. *J. All. Clin. Immunol.* (1998) 101, 67–74.

Bugajska-Schretter, A., Grote, M., Vangelista, L., Valent, P., Sperr, W.R., Rumpold, H., Pastore, A., Reichelt, R., Valenta, R., and Spitzauer, S. Purification, biochemical, and immunological characterisation of a major food allergen: different immunoglobulin E recognition of the apo- and calcium-bound forms of carp parvalbumin. *Gut* (2000) 46, 5, 661-669.

Buller, A.J., Eceles, J.C., and Eceles, R.M. Differentiation of fast and slow muscles in the cat hind limb. *J. Physiol.* (1960) 150, 2, 399-416.

Burgoyne, R.D., and Weiss, J.L. The neuronal calcium sensor family of Ca^{2+}-binding proteins. *Biochem. J.* (2001) 353, 1-12.

Burns, K., and Michalak, M. Interactions of calreticulin with proteins of the endoplasmic and sarcoplasmic reticulum membranes. *FEBS Letters* (1993) 318, 2, 181-185.

Burstein, E.A., Permyakov, E.A., Yashin, V.A., Burkhanov, S.A., and Finazzi-Agro, A. The fine structure of luminescence spectra of azurin. *Biochim. Biophys. Acta* (1977) 491, 1, 155-159.

Burtnick, L.D., Koepf, E.K., Grimes, J.M., Jones, E.Y., Stuart, D.I., McLaughlin, P.J., and Robinson, R.C. The crystal structure of plasma gelsolin: implications for actin severing, capping and nucleation. *Cell* (1997) 90, 661-670.

Butters, C., Tobacman, J.B., and Tobacman, L.S. Cooperative effect of calcium binding to ajacent troponin molecules on the thin filament-myosin subfragment 1 MgATPase rate. *J. Biol. Chem.* (1997) 272, 20, 13196-13202.

Cai, D. Q., Li, M., Lee, K. K. H., Lee, K. M., Qin, L., and Chan, K.M. Parvalbumin expression is downregulated in rat fast-twitch skeletal muscles during aging. *Arch. Biochem. Biophys.* (2001) 387, 202–208.

Caillard, O., Moreno, H., Schwaller, B., Llano. I., Celio, M.R., and Marty, A. Role of the calcium-binding protein parvalbumin in short-term synaptic plasticity. *Proc. Natl. Acad. Sci. USA.* (2000) 97, 24, 13372-13377.

Calabretta, B., Battini, R., Kaczmarek, L. et al. Molecular cloning of the cDNA for growth factor-inducible gene with strong homology to S-100, a calcium-binding protein. *J. Biol. Chem.* (1986) 261, 27, 12628-12632.

Cannell, M.B., and Allen, D.G. Model of calcium movements during activation in the sarcomere of frog skeletal muscle. Biophys. J. (1984) 45, 913-925.

Capony, J.-P., Demaille, J.G., Pina, C., and Pechére, J.-F. The amino-acid sequence of the most acidic major parvalbumin from frog muscle. *Eur. J. Biochem.* (1975) 56, 215-227.

Capony, J.-P., Ryden, L., Demaille, J.G., and Pechére, J.-F. The primary structure of the major parvalbumin from hake muscle. Overlapping peptides obtained with chemical and enzymatic methods. The complete amino-acid sequence. *Eur. J. Biochem.* (1973) 32, 97-108.

Capper-Loup, C., Burgunder, J.M., and Kaelin-Lang, A. Modulation of parvalbumin expression in the motor cortex of parkinsonian rats. *Exp Neurol.* (2005) 193, 1, 234-237.

Caputo, C., Gerday, C., Lopez, J.R., Taylor, S.R., and Bolanos, P. Opposite effects of cooling on twitch contractions of skeletal muscle isolated from tropical toads (*Leptodactylidae*) and northern frogs (*Ranidae*). *J. Comp. Physiol. [B]* (1998) 168, 8, 600-610.

Carafoli, E. Calcium signaling: a tale for all seasons. *Proc. Natnl. Acad. Sci. USA* (2002) 99, 3, 1115-1122.

Carafoli, E. Calcium-mediated cellular signals: a story of failures. *Trends Biol. Sci.* (2004) 29, 7, 371-379.

Carlier, M.R. Actin: protein structure and filament dynamics. *J. Biol. Chem.* (1991) 266, 1, 1-4.

Carroll, S.L., Klein, M.G., and Schneider, M.F. Decay of calcium transients after electrical stimulation in rat fast- and slow-twitch skeletal muscle fibres. *J. Physiol.* (1997) 501, Pt 3, 573-588.

Castillo, M.B., Berchtold, M.W., Rulicke, T., Schwaller, B., Gotzos, V., Pinzani, M., Reichen, J., and Celio, M.R. Ectopic expression of the calcium-binding protein parvalbumin in mouse liver endothelial cells. *Hepatology* (1997) 25, 5, 1154-1159.

Castillo, M.B., Celio, M.R., Andressen, C., Gotzos, V., Rulicke, T., Berger, M.C., Weber, J., and Berchtold, M.W. Production and analysis of transgenic mice with ectopic expression of parvalbumin. *Arch. Biochem. Biophys.* (1995) 317, 1, 292-298.

Cates, M.S., Berry, M.B., Ho, E.L., Li, Q., Potter, J.D., and Phillips, G.N. Metal-ion affinity and specificity in EF-hand proteins: coordination geometry and domain plasticity in parvalbumin. *Structure* (1999a) 7, 1269-1278.

Cates, M.S., Berry, M.B., Ho, E.L., Li, Q., Potter, J.D., and Phillips, G.N. Metal-ion affinity and specificity in EF-hand proteins: coordination geometry and domain plasticity in parvalbumin. *Structure Fold. Des.* (1999) 7, 10, 1269-1278.

Cates, M.S., Teodoro, M.L., and Phillips, G.N. Jr. Molecular mechanisms of calcium and magnesium binding to parvalbumin. *Biophys. J.* (2002) 82, 1133-1146.

Cave, A., Daures, M.F., Parello, J., Saint-Ives, A., and Sempre, R. NMR studies of primary and secondary sites of parvalbumins using the two paramagnetic probes Gd(III) and Mn(II). *Biochimie* (1979a) 61, 7, 755-765.

Cave, A., Dobson, C.M., Parello, J., and Williams, R.J.P. Conformational mobility within the structure of muscular parvalbumins. An NMR study of the aromatic resonances of phenylalanine residues. *FEBS Letters* (1976) 65, 2, 190-194.

Cave, A., Pages, M., and Morin, P. Conformational studies on muscular parvalbumins: cooperative binding of calcium (II) to parvalbumins. *Biochimie* (1979b), 61, 7, 607-613.

Cave, A., Parello, J., Drakenberg, T., Thulin, E., and Lindman, B. Mg^{2+} binding to parvalbumins studied by ^{25}Mg and ^{113}Cd NMR spectroscopy. *FEBS Letters* (1979c), 100, 1, 148-152.

Celio, M.R. Parvalbumin in most γ-aminobutiric acid-containing neurons of the rat cerebral cortex. *Science*, 231, 995-997.

Celio, M.R., and Heizmann, C.W. Calcium-binding protein parvalbumin is associated with fast contracting muscle fibers. Nature (1982) 297, 504-506.

Chard, P.S., Bleakman, D., Christakos, S., Fullmer, C.S., and Miller, R.J. Calcium buffering properties of calbindin D_{28k} and parvalbumin in rat sensory neurones. *J. Physiol.* (1993) 472, 341-357.

Chen, G., Carroll, S., Racay, P., Dick, J., Pette, D., Traub, I., Vrbova, G., Eggli, P., Celio, M., and Schwaller, B. Deficiency in parvalbumin increases fatigue resistance in fast-twitch muscle and upregulates mitochondria. *Am. J. Physiol. Cell. Physiol.* (2001) 281, C114–C122.

Chikou, A., Huriaux, F., Laleye, P., Vandewalle, P., and Focant, B. Isoform distribution of parvalbumins and of some myofibrillar proteins in adult and developing *Chrysichthys auratus* (Geoffroy St. Hilaire, 1808) (Pisces, Claroteidae). *Arch. Physiol. Biochem.* (1997) 105, 6, 611-617.

Chin, E.R., Grange, R.W., Viau, F., Simard, A.R., Humphries, C., Shelton, J., Bassel-Duby, R., Williams, R.S., and Michel, R.N. Alterations in slow-twitch muscle phenotype in transgenic mice overexpressing the Ca^{2+} buffering protein parvalbumin. *J. Physiol.* (2003) 547, Pt 2, 649-663.

Chiquet, C., Dkhissi-Benyahya, O., and Cooper, H.M. Calcium-binding protein distribution in the retina of strepsirhine and haplorhine primates. *Brain Res. Bull.* (2005) 68, 3, 185-194.

Choe, H., Burtnick, L.D., Mejillano, M., Yin, H.L., Robinson, R.C., and Choe, S. The calcium activation of gelsolin: insights from the 3 Å structure of the G4-G6/actin complex. *J. Mol. Biol.* (2002) 324, 691-702.

Christianson, D.W. Structural biology of zinc. *Adv. Protein Chem.* (1991) 42, 281-355.

Clapham, D.C. Calcium sinaling. *Cell* (1995) 80, 259-268.

Clarke, T.E., and Vogel, H.J. Cadmium-113 and Lead-207 NMR Spectroscopic Studies of Calcium-Binding Proteins. *Calcium-Binding Protein Protocols.* Volume 2: Methods and Techniques. (2002) 205-215.

Closset, J., and Gerday, C., Conformational studies on parvalbumins by circular dichroism. *Biochim. Biophys. Acta* (1975) 205, 2, 228-235.

Coffee, C.J., and Bradshaw, R.A. Carp muscle calcium-binding protein. I. Characterization of the tryptic peptides and the complete amino acid sequence of component *B. J. Biol. Chem.* (1973) 248, 3302-3312.

Coffee, C.J., Bradshaw, R.A., and Kretsinger, R.H. The coordination of calcium ions by carp muscle calcium binding proteins A, B and C. *Adv. Exp. Med. Biol.* (1974) 48, 211-233.

Coffee, C.J., and Solano, C. Preparation and properties of carp muscle parvalbumin fragments A (residues 1→75) and B (76→108). *Biochim. Biophys. Acta* (1976) 453, 67-80.

Collin, T., Chat, M., Lucas, M.G., Moreno, H., Racay, P., Schwaller, B., Marty, A., and Llano, I. Developmental changes in parvalbumin regulate presynaptic Ca^{2+} signaling. *J. Neurosci.* (2005) 25, 1, 96-107.

Collins, J.H., Cox, J.A., and Theibert, J.L. Amino acid sequence of a sarcoplasmic calcium-binding protein from the sandworm *Nereis diversicolor. J. Biol. Chem.* (1988) 263, 30, 15378-15385.

Collins, J.H., Johnson, J.D., and Szent-Gyorgyi, A.G. Purification and characterization of a scallop sarcoplasmic calcium-binding protein. *Biochemistry* (1983) 22, 2, 341-345.

Conde, F., Lund, J.S., Jacobowitz, D.M., Baimbridge, K.G., Lewis, D.A. Local circuit neurons immunoreactive for calretinin, calbindin D-28k or parvalbumin in monkey prefrontal cortex: distribution and morphology. *J. Comp. Neurol.* (1994) 341, 95-116.

Contreras-Rodriguez, J., Gonzalez-Soriano, J., Martinez-Sainz, P., and Rodriguez-Veiga, E. The thalamic reticular and perireticular nuclei in developing rabbits: patterns of parvalbumin expression. *Brain Res. Dev. Brain Res.* (2002) 136, 2, 123-133.

Corson, D.C., Lee, L., McQuaid, G.A., and Sykes, B.D. An optical stopped-flow and ^{1}H and ^{113}Cd nuclear magnetic resonance study of the kinetics and stoichiometry of the interaction of the lanthanide Yb^{3+} with carp parvalbumin. *Can. J. Biochem. and Cell Biol.* (1983) 61, 8, 860-867.

Corson, D.C., Williams, T.C., Kay, L.E., and Sykes, B.D. Proton NMR spectroscopic studies of calcium-binding proteins. 1. Stepwise proteolysis of the carboxyl-terminal alpha-helix of a helix- loop-helix metal-binding domain. *Biochemistry* (1986) 25, 7, 1817-1826.

Coutu, P., Bennett, C.N., Favre, E.G., Day, S.M., and Metzger, J.M. Parvalbumin corrects slowed relaxation in adult cardiac myocytes expressing hypertrophic cardiomyopathy-linked α-tropomyosin mutations. *Circ Res.* (2004) 94, 9, 1235-1241.

Coutu, P., Hirsch, J.C., Szatkowski, M.L., and Metzger, J.M. Targeting diastolic dysfunction by genetic engineering of calcium handling proteins. *Trends Cardiovasc. Med.* (2003) 13, 2, 63-67.

Coutu, P., and Metzger, J.M. Genetic manipulation of calcium-handling proteins in cardiac myocytes. I. Experimental studies. *Am. J. Physiol. Heart. Circ. Physiol.* (2005a) 288, 2, H601-H612.

Coutu, P., and Metzger, J.M. Genetic manipulation of calcium-handling proteins in cardiac myocytes. I. Mathematical modeling studies. *Am. J. Physiol. Heart. Circ. Physiol.* (2005b) 288, 2, H613-H631.

Coutu, P., and Metzger, J.M. Optimal range for parvalbumin as relaxing agent in adult cardiac myocytes: gene transfer and mathematical modeling. *Biophys J.* (2002) 82, 5, 2565-2579.

Cox, J.A., Durussel, I., Scott, D.J., and Berchtold, M.W. Remodeling of the AB site of rat parvalbumin and oncomodulin into a canonical EF-hand. *Eur. J. Biochem.* (1999) 264, 3, 790-799.

Cox, J.A., Milos, M., and MacManus, J.P. Calcium- and magnesium-binding properties of oncomodulin. Direct binding studies and microcalorimetry. *J. Biol. Chem.* (1990) 265, 12, 6633-6637.

Cox, J.A., and Stein, E.A. Characterization of a new sarcoplasmic calcium-binding protein with magnesium-induced cooperativity in the binding of calcium. *Biochemistry* (1981) 20, 19, 5430-5436.

Cox, J.A., Wingle, D.R., and Stein, E.A. Calcium, magnesium and conformation of parvalbumin during muscular activity. *Biochimie* (1979) 61, 7, 601-605.

Cox, J.A., Wnuk, W., and Stein, E.A. Isolation and properties of a sarcoplasmic calcium-binding protein from crayfish. *Biochemistry* (1976) 15, 12, 2613-2618.

Cox, J.A., Wnuk, W., and Stein, E.A. Regulation of calcium binding by magnesium. In: *Calcium Binding Proteins and Calcium Function.* Wasserman, R.H. et al. eds, North Holland, N.Y. (1977) 266-269.

Cozens, B., and Reithmeier, R.A.F. Size and shape of rabbit skeletal muscle calsequestrin. *J. Biol. Chem.* (1984) 259, 10, 6248-6252.

Cronce, D.T., and Horrocks, W.D. Jr. Probing the metal-binding sites of cod parvalbumin using europium(III) ion luminescence and diffusion-enhanced energy transfer. *Biochemistry* (1992) 31, 34, 7963-7969.

Czurylo, E.A., Emelyanenko, V.I., Permyakov, E.A., and Dabrowska, R. Spectrofluorimetric studies on C-terminal 34 kDa fragment of caldesmon. *Biophys. Chem.* (1991) 40, 2, 181-188.

Dalgarno, D.C., Levine, B.A., Williams, R.J., Fullmer, C.S., and Wasserman, R.H. Proton-NMR studies of the solution conformations of vitamin-D-induced bovine intestinal calcium-binding protein. *Eur. J. Biochem.* (1983) 137, 3, 523-529.

Davideau, J.L., Celio, M.R., Hotton, D., and Berdal A. Developmental pattern and subcellular localization of parvalbumin in the rat tooth germ. *Arch. Oral. Biol.* (1993) 38, 8, 707-715.

Daviss, S.R., and Lewis, D.A. Local circuit neurons of the prefrontal cortex in schizophrenia: selective increase in the density of calbindin-immunoreactive neurons. *Psychiatry Res.* (1995) 59, 81-96.

Declercq, J.P., Evrard, C., Lamzin, V., and Parello, J. Crystal structure of the EF-hand parvalbumin at atomic resolution (0.91 Å) and at low temperature (100 K). Evidence for conformational multistates within the hydrophobic core. *Protein Sci.* (1999) 8, 10, 2194-2204.

Declercq, J.P., Tinant, B., Parello, J., Etienne, G., and Huber, R. Crystal structure determination and refinement of pike 4.10 parvalbumin (minor component from *Esox lucius*). *J. Mol. Biol.* (1988) 202, 2, 349-353.

Declercq, J. P., Tinant, B., Parello, J. and Rambaud, J. Ionic interactions of parvalbumin. Crystal structure of pike 4.10 parvalbumin in four different ionic environments. *J. Mol. Biol.* (1991) 220, 1017-1039.

Dedman, J.R. Mediation of intracellular calcium: variation on a common theme. *Cell Calcium* (1986) 7, 2, 297-307.

DeFelipe, J. Types of neurons, synaptic connections and chemical characteristics of cells immunoreactive for calbindin-D_{28K}, parvalbumin and calretinin in the neocortex. *J. Chem. Neuroanat.* (1997) 14, 1-19.

Dekkers, J., Bayley, P., Dick, J.R., Schwaller, B., Berchtold, M.W., and Greensmith, L. Over-expression of parvalbumin in transgenic mice rescues motoneurons from injury-induced cell death. *Neuroscience* (2004) 123, 2, 459-466.

de Lecea, L., del Rio, J.A., Soriano, E. Developmental expression of parvalbumin mRNA in the cerebral cortex and hippocampus of the rat. *Brain Res. Mol. Brain Res.* (1995) 32, 1–13.

Derancourt, J., Haiech, J., and Pechére, J.F. Binding of calcium by parvalbumin fragments. *Biochim. Biophys. Acta* (1978) 532, 2, 373-375.

Dizhoor, A.M., Ray, S., Kumar, S., Niemi, G., Spencer, M., Brolley, D., Walsh, K.A., Philipov, P.P., Hurley, J.B., and Stryer, L. Recoverin: a calcium sensitive activator of retinal rod guanylate cyclase. *Science* (1991) 251, 915-918.

Dobrovolski, Z., Xu, G.Q., Chen, W., and Hitchcock-De Gregory, S.E. Analysis of the regulatory and structural defects of troponin C central helix mutants. *Biochemistry* (1991) 30, 29, 7089-7096.

Donato, R. S100: a multigenic family of calcium-modulated proteins of the EF-hand type with intracellular and extracellular functional roles. *Int. J. Biochem. Cell Biol.* (2001) 33, 637-668.

Donato, H., and Martin, R.B. Conformation of carp muscle calcium binding parvalbumin. *Biochemistry* (1974) 13, 22, 4575-4579.

D'Orlando, C., Celio, M.R., and Schwaller, B. Calretinin and calbindin D-28k, but not parvalbumin protect against glutamate-induced delayed excitotoxicity in transfected N18-RE 105 neuroblastoma-retina hybrid cells. Brain Res. (2002) 945, 2, 181-190.

Dowd, D.R., Calcium regulation of apoptosis. In: Means, A.R. (Ed.), *Advances in Second Messenger and Phosphoprotein Research,* (1995) vol. 30, Raven Press, New York, 255–280.

Dowd, D.R., MacDonald, P.N., Komm, B.S., Haussler, M.R., and Miesfeld, R.L., Stable expression of the calbindin-D28K complementary DNA interferes with the apoptotic pathway in lymphocytes. *Mol. Endocrinol.* (1992) 6, 1843–1848.

Drakenberg, T., Hofman, T., and Chazin, W.J. ^{1}H NMR studies of porcine calbindin 9kD in solution: sequential resonance assignment, secondary structure, and global fold. *Biochemistry* (1989) 28, 14, 5946-5954.

Dreessen, J., Lutum, C., Schafer, B.W., Heizmann, C.W., and Knopfel, T. α-Parvalbumin reduces depolarization-induced elevations of cytosolic free calcium in human neuroblastoma cells. *Cell Calcium* (1996) 19, 6, 527-533.

Drohat, A.C., Amburgey, J.C., Abildgaard, F., Starich, M.R., Baldisseri, D., and Weber, D.J. Solution structure of rat apo-S100B (ββ) as determined by NMR spectroscopy. *Biochemistry* (1996) 35, 11577-11588.

Dudev, T., and Lim, C. Principles governing Mg, Ca, and Zn binding and selectivity in proteins. *Chem. Rev.* (2003) 103, 773-787.

Durussel, I., Luan-Rilliet, Y., Petrova, T., takagi, T., and Cox, J.A. Cation binding and conformation of tryptic fragments of *Neris* sarcoplasmic calcium-binding protein: calcium-induced homo- and heterodimerization. *Biochemistry* (1993) 32, 9, 2394-2400.

Durussel, I., Pauls, T.L., Cox, J.A., and Berchtold, M.W. Chimeras of parvalbumin and oncomodulin involving exchange of the complete CD site show that the Ca^{2+}/Mg^{2+} specificity is an intrinsic property of the site. *Eur. J. Biochem.* (1996) 242, 2, 256-263.

Eberhard, M., and Erne, P. Calcium and magnesium binding to rat parvalbumin. (1994) *Eur. J. Biochem.* 222, 21-26.

Eberspach, J., Strasburger, W., Glatter, U., Gerday, C., and Wollmer, A. Interaction of parvalbumin of pike II with calcium and terbium ion. *Biochim. Biophys. Acta.* (1988) 952, 1, 67-76.

Eftink, M.R., and Wasylewski, Z. Fluorescence lifetime and solute quenching studies with single tryptophan containing protein parvalbumin from codfish. *Biochemistry* (1989) 28, 1, 382-391.

Elkins, K.M., Gatzeva-Topalova, P.Z., and Nelson, D.J. Molecular dynamic study of Ca^{2+} binding loop variants of parvalbumin with modifications at the 'gateway' position. *Protein Eng.* (2001) 14, 2, 115-126.

Elsayed, S., and Aas, K. Characterization of a major allergen (cod). Observations on effect of denaturation on the allergenic activity. *J. Allergy* (1971) 47, 283–291.

Elsayed, S., and Bennich, H. The primary structure of allergen M from cod. *Scand. J. Immunol.* (1975) 4, 203-208.

Elsayed, S., and Bennich, H. The primary structure of allergen M from cod. Scand J. Immunol. (1974) 3, 683–686.

Enfield, D.L., Ericsson, L.H., Blum, H.E., Fischer, E.H., and Neurath, H. Amino-acid sequence of parvalbumin from rabbit skeletal muscle. *Proc. Natl. Acad. Sci. USA.* (1975) 72, 1309-1313.

Endo, T., Takazawa, K., and Onaya, T. Parvalbumin exists in rat endocrine glands. *Endocrinology* (1985) 117, 527–531.

Engelkamp, D., Schafer, B.W., Erne, P., and Heizmann, C.W. S100 , CAPL, and CACY: molecular cloning and expression analysis of three calcium-binding proteins from human heart. *Biochemistry* (1992) 31, 42, 10258-10264.

Epstein, P., Means, A.R., and Berchtold, M.W. Isolation and characterization of a rat parvalbumin gene and full length cDNA. *J. Biol. Chem.* (1986) 261, 5886-5891.

Erickson, J.R., and Moerland, T.S. Functional characterization of parvalbumin from the Arctic cod (*Boreogadus saida*): Similarity in calcium affinity among parvalbumins from polar teleosts. *Comp. Biochem. Physiol. A Mol. Integr. Physiol.* (2006) 143, 2, 228-233.

Erickson, J.R., Sidell, B.D., and Moerland, T.S. Temperature sensitivity of calcium binding for parvalbumins from Antarctic and temperate zone teleost fishes. *Comp. Biochem. Physiol. A Mol. Integr. Physiol.* (2005) 140, 2, 179-185.

Estes, J.E., Selden, L.A., and Gershman, L.C. Tight binding of divalent cations to monomeric actin. *J. Biol. Chem.* (1987) 262, 11, 4952-4957.

Esteve-Romero, J.S., Yman, I.M., Bossi, A., and Righetti, P.G. Fish species identification by isoelectric focusing of parvalbumins in immobilized pH gradients. *Electrophoresis* (1996) 17, 8, 1380-1385.

Eybalin, M., and Ripoll, C. Immunolocalization of parvalbumin in two glutamatergic cell types of the guinea pig cochlea: inner hair cells and spinal ganglion neurons. *C.R. Acad. Sci. Paris* (1990) 310, 13, 639-644.

Fahie, K., Pitts, R., Elkins, K., and Nelson, D.J. Molecular dynamic study of the Ca^{2+} binding loop variants of silver hake parvalbumin with aspartic acid at the "gateway" position. *J. Biomol. Struct. Dyn.* (2002) 5, 821-837.

Falke, J.J., Drake, S.K., Hazard, A.L., and Peersen, O.B. Molecular tuning of ion binding to calcium signaling proteins. *Quart. Rev. Biophys.* (1994) 27, 219-290.

Fallon, J.L., and Quiocho, F.A. A closed compact structure of native Ca^{2+}-calmodulin. *Structure* (2003) 11, 1303-1307.

Feinstein, E., Deikus, G., Rusinova, E., Rachofsky, E.L., Ross, J.B., and Laws, W.R. Constrained analysis of fluorescence anisotropy decay: application to experimental protein dynamics. *Biophys J.* (2003) 84, 1, 599-611.

Ferreira, S.T. Fluorescence studies of the conformational dynamics of parvalbumin in solution. Lifetime and rotational motions of the single tryptophan residue. *Biochemistry* (1989) 28, 26, 10066-10072.

Filimonov, V.V., Pfeil, W., Tsalkova, T.N., and Privalov, P.L. Thermodynamic investigations of proteins. IV. Calcium binding protein parvalbumin. *Biophys. Chem.* (1978) 8, 1, 117-122.

Flaherty, K.M., Zozulya, S., Stryer, L., and McKay, D.B. Three-dimensional structure of recoverin, a calcium sensor in vision. *Cell* (1993) 75, 4, 709-716.

Fleron, J. Etude du calcium lié à des albumines musculaires de faible poids moléculaire de poisson. Mémoire de licence sciences chimiques. Université de Liége. (1968).

Focant, B., and Pechére, J.F. Contribution a l'etude des proteins de faible poids moleculaires des myogenes de vertebras inferieurs. *Arch. Int. Physiol. Biochem.* (1965) 73, 334-354.

Focant, B., Vandewalle, P., and Huriaux, F. Expression of myofibrillar proteins and parvalbumin isoforms during the development of a flatfish, the common sole *Solea solea*: comparison with the turbot *Scophthalmus maximus*. *Comp. Biochem. Physiol. B Biochem. Mol. Biol.* (2003) 135, 3, 493-502.

Foehr, U.G., Weber, B.R., Muentener, M., Staudenmann, W., Hughes, G.J., Frutiger, S., Banville, D., Schaefer, B.W., and Heizmann,C.W., Human alpha and beta parvalbumins. Structure and tissue-specific expression. *Eur. J. Biochem.* (1993) 215, 719-727.

Fohr, U.G., Weber, B.R., Muntener, M., Staudenmann, W., Hughes, G.J., Frutiger, S., Banville, D., Schafer, B.W., and Heizmann, C.W. Human α and β parvalbumins. Structure and tissue-specific expression, *Eur. J. Biochem.* (1993) 215, 719-727.

Franchini, P.L., and Reid, R.E. A model for circular dichroism monitored dimerization and calcium binding in an EF-hand synthetic peptide. *J. Theor. Biol.* (1999) 199, 2, 199-211.

Frankenne, F., Joassin, L., and Gerday, C. The amino acid sequence of the pike (*Esox lucius*) parvalbumin 3. *FEBS Lett.* (1973) 35, 145-147.

Fraser, I.D.C., and Marston, S.B. *In vitro* motility analysis of actin-tropomyosin regulation by troponin and calcium. The thin filament is switched as a single cooperative unit. *J. Biol. Chem.* (1995) 270, 14, 7836-7841.

Freund, T.F., Buzsaki, G., Leon, A., Baimbridge, K.G., and Somogyi, P. Relationship of neuronal vulnerability and calcium binding protein immunoreactivity in ischemia. *Exp. Brain Res.* (1990) 83, 55-66.

Friedberg, F. Parvalbumin isoforms in zebrafish. Mol. Biol. Rep. (2005) 32, 3, 167-175.

Fritz-Niggli, H., Nievergelt-Egido, C., and Heizmann, C.W. Calcium binding parvalbumin in Drosophila testis in connection with in vivo irradiation. *Radiat. Environ. Biophys.* (1988) 27, 59-65.

Fromherz, S., and Scent-Gyorgyi, A.G. Role of essential light chain EF hand domains in calcium binding and regulation of scallop myosin. *Proc. Natl. Acad. Sci. USA* (1995) 92, 7652-7656.

Füchtbauer, E.M., Rowlerson, A.M., Gotz, K., Friedrich, G., Mabuchi, K., Gergely, J., and Jokusch, H. Didect correlation of parvalbumin levels with myosin isoforms and succinate dehydrogenase activity on frozen sections of rodent muscle. *J. Histochem. Cytochem.* (1991) 39, 355-361.

Fujii, T., Machino, K., Andoh, H., Satoh, T., and Kondo, Y. Calcium-dependent control of caldesmon-actin interaction by S100 protein. *J. Biochem.* (1990) 107, 1, 133-137.

Gailly, P. New aspects of calcium signaling in skeletal muscle cells: implications in Duchenne muscular dystrophy. *Biochim. Biophys. Acta.* (2002) 1600, 1-2, 38-44.

Gailly, P., Hermans, E., Octave, J.N., and Gillis, J.M. Specific increase of genetic expression of parvalbumin in fast skeletal muscles of mdx mice. *FEBS Lett.* (1993) 326, 1-3, 272-274.

Galarreta, M., and Hestrin, S. Electrical and chemical synapses among parvalbumin fast-spiking GABAergic interneurons in adult mouse neocortex. *Proc. Natl. Acad. Sci. USA* (2002) 99, 19, 12438-12443.

Gerday, C. The primary structure of the parvalbumin II of pike (*Esox lucius*). *Eur. J. Biochem.* (1976) 70, 305-318.

Gerday, C. Soluble calcium-binding proteins from fish and invertebrate muscle. *Mol. Physiol.* (1982) 2, 1, 63-87.

Gerday, C. Soluble calcium binding proteins in vertebrate and invertebrate muscles. In: *Calcium and Calcium Binding Proteins.* Gerday, C. et al. eds., Springer-Verlag, Berlin, Heidelberg (1988) 23-39.

Gerday, C., and Bhushana Rao, K.S.P. Tryptic peptide maps and terminal amino acid residues of low molecular weight proteins from the white muscles of *Cyprinus carpio*, *Gadus callarias* and *Tilapia macrochir*. *Comp. Biochem. Physiol.* (1970) 36, 229-240.

Gerday, C., Collins, S., and Gerardin-Otthiers, N. The amino acid sequence of the parvalbumin from the very fast swimbladder muscle of the toadfish (*Opsanus tau*). *Comp. Biochem. Physiol.* (1989) 93B, 49-55.

Gerday, C., Collins, S., and Gerardin-Otthiers, N. The soluble calcium-binding protein from muscle of the sandworm, *Nereis virens*. *J. Muscle Res. Cell Motility* (1981) 2, 225-238.

Gerday, C., Collins, S., and Piront, A. Phylogenetic relationships between Cyprinidae Parvalbumins-II. The amino acid sequence of the Parvalbumin V of Chub (*Leuciscus cephalus L.*). *Comp. Biochem. Physiol.* (1978) 61B, 451-457.

Gerday, C., and Gillis, J.M. The possible role of parvalbumins in the control of contraction. *J. Physiol.* (1976) 258, 96P-97P.

Gerday, C., Goffard, P., and Taylor, S.R. Isolation and characterization of parvalbumins from skeletal muscles of a tropical amphibian, *Leptodactylus insularis*. *J. Comp. Physiol.* [B]. (1991) 161, 5, 475-481.

Gerday, C., Joris, B., Gerardin-Otthiers, N., Collin, S., and Hamoir, G. Parvalbumins from lungfish (*Protopterus dolloi*). *Biochimie* (1980) 61, 589-599.

Gerke, V., and Moss, S.E., Annexins: from structure to function. Physiol. Rev. (2002) 82, 331–371.

Gershman, L.C., Selden, L.A., and Estes, J.E. High affinity binding of divalent cation to actin monomer is much stronger than previously reported. *Biochem. Biophys. Res. Commun.* (1986) 135, 2, 607-614.

Gershman, L.C., Selden, L.A., and Estes, J.E. High affinity divalent cation exchange on actin. Association rate measurements support the simple competitive model. *J. Biol. Chem.* (1991) 266, 1, 76-82.

Gillis, J.M. The biological significance of muscle parvalbumins. In: *Calcium Binding Proteins: Structure and Function.* N.Y., Elsevier; North Holland (1980) 251-256.

Gillis, J.M. Relaxation of vertebrate skeletal muscle. A synthesis of the biochemical and physiological approaches. *Biochim. Biophys. Acta* (1985) 811, 2, 97-146.

Gillis, J.M. and Gerday, C. Calcium movement between myofibrils, parvalbumin and sarcoplasmic reticulum in muscle. In: *Calcium Binding Proteins and Calcium Function.* N.Y., Elsevier; North Holland (1977) 193-196.

Gillis, J.M., Piront, A., and Gosselin-Rey, C. Parvalbumins distribution and physical state inside the muscle cell. *Biochim. Biophys. Acta* (1979) 585, 444-450.

Gillis, J.M., Thomason, D., Lefevre, J., and Kretsinger, R. Parvalbumins and muscle relaxation: a computer simulation study. *J. Muscle Res. Cell Mot.* (1982) 3, 4, 377-398.

Goll, D.E., Thompson, V.F., Taylor, R.G., and Christiansen, J.A. Role of the calpain system in muscle growth. *Biochimie* (1992) 74, 225-337.

Goncalves, L.R., Yamanouye, N., Nunez-Burgos, G.B., Furtado, M.F., Britto, L.R., and Nicolau, J. Detection of calcium-binding proteins in venom and Duvernoy's glands of South American snakes and their secretions. *Comp. Biochem. Physiol. C Pharmacol. Toxicol. Endocrinol.* (1997) 118, 2, 207-211.

Gonzalez-Soriano, J., Gonzalez-Flores, M.L., Contreras-Rodriguez, J., Rodriguez-Veiga, E., and Martinez-Sainz, P. Calbindin D_{28k} and parvalbumin immunoreactivity in the rabbit superior colliculus: an anatomical study. *Anat Rec.* (2000) 259, 3, 334-346.

Goodman, M., and Pechére, J.-F. The evolution of muscular parvalbumins investigated by the maximum parsimony method, *J. Mol. Evol.* (1977) 9, 131-158.

Gosselin-Rey, C. Fish parvalbumins: immunochemical reactivity and biological distribution. In: *Calcium Binding Proteins.* Drabikowski, W. et al. eds. PSP Warsaw (1974) 697-701.

Gosselin-Rey, C., Bernard, N., and Gerday, C. Conformation and immunochemistry of parvalbumin III from pike white muscle: modification of the arginine with 1,2 cyclohexandione. *Biochim. Biophys. Acta* (1973) 303, 90-104.

Gosselin-Rey, C., and Gerday, C. Parvalbumins from frog skeletal muscle (*Rana temporaria L.*). Isolation and characterization. Structural modifications associated with calcium binding. *Biochim. Biophys. Acta,* (1977) 492, 53–63.

Grandjean, J., Laszlo, P., and Gerday, C. Sodium complexation by the calcium binding site of parvalbumin. *FEBS Letters* (1977) 81, 2, 376-380.

Grateron, L., Cebada-Sanchez, S., Marcos, P., Mohedano-Moriano, A., Insausti, A.M., Munoz, M., Arroyo-Jimenez, M.M., Martinez-Marcos, A., Artacho-Perula, E., Blaizot, X., and Insausti, R. Postnatal development of calcium-binding proteins immunoreactivity (parvalbumin, calbindin, calretinin) in the human entorhinal cortex. *J. Chem. Neuroanat.* (2003) 26, 4, 311-316.

Gryczynski, I., Lakovich, J.R., and Steiner, R.F. Frequency-domain measurements of the rotational dynamics of the tyrosine groups of calmodulin. *Biophys. Chem.* (1988) 30, 1, 49-59.

Gulati, J., Akella, A.B., Su, H., Mehler, E.L., and Weinstein, H. Functional role of arginine-11 in the N-terminal helix of skeletal troponin C: combined mutagenesis and molecular dynamics investigation. *Biochemistry* (1995) 34, 22, 7348-7355.

Gulati, J., Babu, A., Su, H., and Zhang, Y.F. Identification of the regions conferring calmodulin-like properties to troponin C. *J. Biol. Chem.* (1993) 268, 16, 11685-11690.

Hackney, C.M., Mahendrasingam, S., Jones, E.M., and Fettiplace, R. The distribution of calcium buffering proteins in the turtle cochlea. *J. Neurosci.* (2003) 23, 11, 4577-4589.

Hackney, C.M., Mahendrasingam, S., Penn, A., and Fettiplace, R. The concentrations of calcium buffering proteins in mammalian cochlear hair cells. *J. Neurosci.* (2005) 25, 34, 7867-7875.

Haeseleer, F., Imanishi, Y., Sokal, I., Filipek, S., and Palczewski, K. Calcium-binding proteins: intracellular sensors from the calmodulin superfamily. *Biochem. Biophys. Res. Commun.* (2002) 290, 615-623.

Haiech, J., Derancourt, J., Pechére, J.F., and Demaille, J.G. Magnesium and calcium binding to parvalbumins: evidence for differences between parvalbumins and an explanation of their relaxing function. *Biochemistry* (1979a) 18, 13, 2752-2758.

Haiech, J., Derancourt, J., Pechére, J.F., and Demaille, J.G. A new large-scale purification procedure for muscular parvalbumins. *Biochimie* (1979b) 61, 6, 583-587.

Hamada, Y., Tanaka, H., Ishizaki, S., Ishida, M., Nagashima, Y., and Shiomi, K. Purification, reactivity with IgE and cDNA cloning of parvalbumin as the major allergen of mackerels. Submitted SEP-2002 to the EMBL GenBank DDBJ databases.

Hamada, Y., Tanaka, H., Ishizaki, S., Ishida, M., Nagashima, Y., Shiomi, K. Purification, reactivity with IgE and cDNA cloning of parvalbumin as the major allergen of mackerels. *Food Chem. Toxicol.* (2003) 41, 8, 1149-1156.

Hamoir, G., Gerardin-Otthiers, N., and Focant, B. Protein differentiation of the superfast swimbladder muscle of the toadfish *Opsanus tau*. *J. Mol. Biol.* (1980) 143, 155-160.

Hamoir, G., Gerardin-Otthiers, N., Grodent, V., and Vandewalle, P. Sarcoplasmic differentiation of head muscles of carp (*Cyprinus carpio*). *Mol. Physiol.* (1981) 1, 45-58.

Hamoir, G., and Konosu, S. Carp myogens of white and red muscles. General composition and isolation of low molecular weight components of abnormal amino-acid composition. *Biochem. J.* (1965) 96, 85-97.

Hapak, R.C., Lammers, P.J., Palmisano, W.A., Birnbaum, E.R., and Henzl, M.T. Site-specific substitution of glutamate for aspartate at position 59 of rat oncomodulin. *J. Biol. Chem.* (1989) 264, 18751-18760.

Hapak, R.C., Stanley, C.M., and Henzl, M.T. Intrathymic distribution of the two avian thymic parvalbumins. *Exp. Cell Res.* (1996) 222, 234-245.

Hapak, R.C., Zhao, H., Boschi, J.M., and Henzl, M.T. Novel avian thymic parvalbumin displays high degree of sequence homology to oncomodulin. *J. Biol. Chem.* (1994) 269, 5288-5296.

Hapak, R.C., Zhao, H., and Henzl, M.T. Oligomerization of an avian thymic parvalbumin. Chemical evidence for a Ca^{2+}-specific conformation. *FEBS Lett.* (1994) 349, 2, 295-300.

Hartmann, J., and Konnerth, A. Determinants of postsynaptic Ca^{2+} signaling in Purkinje neurons. *Cell Calcium.* (2005) 37, 5, 459-466.

Hasebe, T., Umezawa, K., Sugita, M., Iwata, T., Yamamoto, K., Obinata, T., and Kikuyama, S. Postmetamorphic changes in parvalbumin expression in the hindlimb skeletal muscle of the bullfrog, *Rana catesbeiana. Biochim. Biophys. Acta.* (2003) 1646, 1-2, 42-48.

Hauer, C.R., Staudenmann, W., Kuster, T., Neuheiser, F., Hughes, G.J., Seto-Ohshima, A., Tanokura, M., and Heizmann, C.W. Protein sequence determination by ESI-MS and LSI-MS tandem mass spectrometry: parvalbumin primary structures from cat, gerbil and monkey skeletal muscle. *Biochim. Biophys. Acta* (1992) 1160, 1-7.

Hazama, M., Watanabe, D., Suzuki, M., Mizoguchi, A., Pastan, I., and Nakanishi, S. Different regulatory sequences are required for parvalbumin gene expression in skeletal muscles and neuronal cells of transgenic mice. *Brain Res. Mol. Brain Res.* (2002) 100, 1-2, 53-66.

Heidorn, D.B., and Trewhella, J. Comparison of the crystal and solution structures of calmodulin and troponin C. *Biochemistry* (1988) 27, 3, 909-915.

Heierhorst, J., Kobe, B., Feil, S.C. Parker, M.W., Benian, G.M., Weiss K.R., and Kemp, B.E. Ca^{2+}/S100 regulation of giant protein kinases. *Nature* (1996) 380, 636-639.

Heilmann, C., and Spamer, C. *Guidebook to the calcium-binding proteins.* Ed. M.R. Celio, Oxford, UK, Oxford Univ. Press, 1996.

Heizmann, C.W. Calcium signaling in the brain. *Acta Neurobiol Exp (Wars).* (1993) 53, 1, 15-23.

Heizmann, C.W. Parvalbumin, an intracellular calcium-binding protein; distribution, properties and possible roles in mammalian cells. *Experientia* (1984) 40, 9, 910-921.

Heizmann, C.W. Parvalbumin in non-muscle cells. In: *Calcium and Calcium Binding Proteins.* Gerday, C., Bolis, L., and Gilles, R. eds., Springer-Verlag, Berlin, Heidelberg (1988) 93-101.

Heizmann, C.W., and Berchtold, M.W. Expression of parvalbumin and other Ca^{2+}-binding proteins in normal and tumor cells: a topical review. *Cell Calcium* (1987) 8, 1-41.

Heizmann, C.W., Berchtold, M.W., and Rowlerson, A.M. Correlation of parvalbumin concentration with relaxation speed in mammalian muscles. *Proc. Nat. Acad. Sci. USA, Biol. Sci.* (1982) 79, 23, 7243-7247.

Heizmann, C.W., and Braun, K. Changes in Ca^{2+}-binding proteins in human neurodegenerative disorders. *TINS* (1992) 15, 259-264.

Heizmann, C.W., and Celio, M.R. Immunolocalization of parvalbumin. *Methods Enzymol.* (1987) 139, 552-570.

Heizmann, C.W., and Cox, J.A. New perspectives on S100 proteins: a multi-functional Ca^{2+}-, Zn^{2+}- and Cu^{2+}-binding protein family. *Biometals* (1998) 11, 4, 383-397.

Heizmann, C.W., and Strehler, E.E. Chicken parvalbumin: comparison with parvalbumin-like protein and three other components (M_r=8,000 to 13,000). *J. Biol. Chem.* (1979) 254, 10, 4296-4303.

Heller, S., Bell, A.M., Denis, C.S., Choe, Y., and Hudspeth, A.J. Parvalbumin 3 is an abundant Ca^{2+} buffer in hair cells. *J. Assoc. Res. Otolaryngol.* (2002) 3, 4, 488-498.

Hemric, M.E., Lu, F.W.M., Shrager, R., Carey, J., and Chalovich, J.M. Reversal of caldesmon binding to myosin with calcium-calmodulin or by phosphorilating caldesmon. *J. Biol. Chem.* (1993) 268, 20, 15305-15311.

Henrotte, J.G. A crystalline component of carp myogen precipitating at high ionic strength. *Nature* (1955) 176, 1221-1223.

Henzl, M.T., and Agah, S. Divalent ion-binding properties of the two avian β-parvalbumins. *Proteins* (2006) 62, 1, 270-278.

Henzl, M.T., Agah, S., and Larson, J.D. Characterization of the metal ion-binding domains from rat α- and β-parvalbumins. *Biochemistry* (2003a) 42, 12, 3594-3607.

Henzl, M.T., Agah, S., and Larson, J.D. Association of the AB and CD-EF domains from rat α- and β-parvalbumins. *Biochemistry* (2004a) 43, 34, 10906-10917.

Henzl, M.T., Agah, S., and Larson, J.D. Rat α- and β-parvalbumins: comparison of their pentacarboxylate and site-interconversion variants. *Biochemistry* (2004) 43, 9307-9319.

Henzl, M.T., and Graham, J.S. Conformational stabilities of the rat α- and β-parvalbumins. *FEBS Lett.* (1999) 442, 2-3, 241-245.

Henzl, M.T., Hapak, R.C., and Likos, J.J. Interconversion of the ligand arrays in the CD and EF sites of oncomodulin. Influence on Ca^{2+}-binding affinity. *Biochemistry* (1998) 37, 25, 9101-9111.

Henzl, M.T., Larson, J.D., and Agah, S. Influence of monovalent cation on rat α- and β-parvalbumin stabilities. *Biochemistry* (2000) 39, 5859-5867.

Henzl, M.T., Larson, J.D., and Agah, S. Estimation of parvalbumin Ca^{2+}- and Mg^{2+}-bindng constants by global least-squares analysis of isothermal titration calorimetry data. *Anal. Biochem.* (2003b) 319, 216-233.

Henzl, M.T., McCubbin, W.D., Kay, C.M., and Birnbaum, E.R. Luminescence studies of lanthanide ion binding to parvalbumin. *J. Biol. Chem.* (1985) 260, 8447-8455.

Henzl, M.T., Shibasaki, O., Comegys, T.H., Thalmann, I., and Thalmann, R., Oncomodulin is abundant in the organ of Corti. *Hear. Res.* (1997) 106, 105-111.

Henzl, M.T., Wycoff, W.G., Larson, J.D., and Likos, J.J. ^{15}N nuclear magnetic resonance relaxation studies on rat β-parvalbumin and the pantacarboxylate variants, S55D and G98D. *Protein Science* (2002) 11, 158-173.

Henzl, M.T., Zhao, H., and Saez, C.T. Self-association of CPV3, an avian thymic parvalbumin. *FEBS Lett.* (1995) 375, 1-2, 137-142.

Herzberg, O., and James, M.N.G. Structure of the calcium regulatory muscle protein troponin C at 2.8 Å resolution. *Nature* (1985) 313, 6004, 653-659.

Hilger, C., Grigioni, F., Thill, L., Mertens, L., and Hentges, F. Severe IgE-mediated anaphylaxis following consumption of fried frog legs: definition of α-parvalbumin as the allergen in cause. *Allergy* (2002) 57, 1053–1058.

Hilger, C., Thill, L., Grigioni, F., Lehners, C., Falagiani, P., Ferrara, A., Romano, C., Stevens, W., and Hentges, F. IgE antibodies of fish allergic patients cross-react with frog parvalbumin. *Allergy* (2004) 59, 6, 653-660.

Hirsch, J.C., Borton, A.R., Albayya, F.P., Russell, M.W., Ohye, R.G., and Metzger, J.M. Comparative analysis of parvalbumin and SERCA2a cardiac myocyte gene transfer in a large animal model of diastolic dysfunction. *Am. J. Physiol. Heart Circ. Physiol.* (2004) 286, 6, H2314-H2321.

Ho, B-K., Alexianu, M.E., Colom, L.V., Mohamed, A.H., Serrano, F., and Appel, S.H., Expression of calbindin-D28K in motoneuron hybrid cells after retroviral infection with calbindin-D28K cDNA prevents amyotrophic lateral sclerosis IgG mediated cytotoxicity. *Proc. Natl. Acad. Sci. USA* (1996) 93, 6796– 6801.

Hoeflich, K.P., and Ikura, M. Calmodulin in action: diversity in target recognition and activation mechanismas. *Cell* (2002) 108, 739-742.

Hof, P.R., Glezer, I.I., Flagg, R.A., Rubin, M.B., Nimchinsky, E.A., and Vogt Weisenhorn, D.M. Cellular distribution of the calcium-binding proteins parvalbumin, calbindin, and calretinin in the neurocortex of mammals: phylogenic and developmental patterns. *J. Chem. Neuroanat.* (1999) 16, 2, 77-116.

Hofman, T., Eng, S., Lilja, H., Drakenberg, T., Vogel, H.J., and Forsen, S. Site-site interactions in EF-hand calcium-binding proteins. Laser-exited europium luminescence studies of 9 kDa calbindin, the pig intestinal calcium-binding protein. *Eur. J. Biochem.* (1988) 172, 2, 307-313.

Hogue, C.W., MacManus, J.P., Banville, D., and Szabo, A.G. Comparison of terbium (III) luminescence enhancement in mutants of EF hand calcium binding proteins. *J. Biol. Chem.* (1992) 267, 19, 13340-13347.

Holroyde, M.J., Robertson, S.P., Johnson, J.D., Solaro, R.J., and Potter, J. The calcium and magnesium binding sites on cardiac troponin and their role in the regulation of myofibrillar adenosine triphosphatase. *J. Biol. Chem.* (1980) 255, 24, 11688-11693.

Hou, T.T., Johnson, J.D., and Rall, J.A. Parvalbumin content and Ca^{2+} and Mg^{2+} dissociation rates correlated with changes in relaxation rate of frog muscle fibres. *J. Physiol.* (1991) 441, 285-304.

Hou, T.T., Johnson, J.D., and Rall, J.A. Effect of temperature on relaxation rate and Ca^{2+}, Mg^{2+} dissociation rates from parvalbumin of frog muscle fibres. *J. Physiol.* (1992) 449, 399-410.

Hou, T.T., Johnson, J.D., and Rall, J.A. Role of parvalbumin in relaxation of frog skeletal muscle. *Adv. Exp. Med. Biol.* (1993) 332, 141-151.

Hsiao, C.-D., Tsai, W.-Y., and Tsai, H.-J. Molecular cloning and developmental expression of parvalbumin genes in zebrafish. Submitted JAN-2002 to the EMBL GenBank DDBJ databases.

Hsiao, C.D., Tsai, W.Y., and Tsai, H.J. Isolation and expression of two zebrafish homologues of parvalbumin genes related to chicken CPV3 and mammalian oncomodulin. *Mech Dev.* (2002) 119, Suppl 1, S161-S166.

Hubbard, M.J. Abundant calcium homeostasis machinery in rat dental enamel cells. Up-regulation of calcium store proteins during enamel mineralization implicates the endoplasmic reticulum in calcium transcytosis. *Eur. J. Biochem.* (1996) 239, 3, 611-623.

Hubbard, S.B., Hedgson, K.O., and Doniach, S. Small-angle X-ray scattering investigation of the solution structure of troponin C. *J. Biol. Chem.* (1988) 263, 9, 4151-4158.

Huber, B., and Pette, D. Dynamics of parvalbumin expression in low-frequency-stimulated fast-twitch rat muscle. *Eur. J. Biochem.* (1996) 236, 3, 814-819.

Hughes, R.E., Brzovic, P.S., Klevit, R.E., and Hurley, J.B. Calcium-dependent solvation of the myristoyl group of recoverin. *Biochemistry* (1995) 34, 36, 11410-11416.

Huq, F., Lebeche, D., Iyer, V., Liao, R., and Hajjar, R.J. Gene transfer of parvalbumin improves diastolic dysfunction in senescent myocytes. *Circulation* (2004) 109, 22, 2780-2785.

Huriaux, F., Vandewalle, P., and Focant, B. Polymorphism of white muscle myosin and parvalbumins in the genus *Barbus* (Teleostei: Cyprinidae). *J. Fish Biol.* (1992) 41, 873–882.

Huriaux, F., Vandewalle, P., and Focant, B. Immunological study of muscle parvalbumin isotypes in three African catfish during development. *Comp. Biochem. Physiol. B Biochem. Mol. Biol.* (2002) 132, 3, 579-584.

Hutnik, C.M.L., MacManus, J.P., Banville, D., and Szabo, A.G. Comparison of metal ion-induced conformational changes in parvalbumin and oncomodulin as probed by the intrinsic fluorescence of tryptophan 102. *J. Biol. Chem.* (1990a) 265, 20, 11456-11464.

Hutnik, C.M.L., MacManus, J.P., and Szabo, A.G. A calcium-specific conformational response of parvalbumin. *Biochemistry* (1990b) 29, 31, 7318-7328.

Iaizzo, P.A. The effect of temperature on relaxation in frog skeletal muscle: the role of parvalbumin. *Pflügers Arch.* (1988) 412, 195-202.

Ichikawa, H., Deguchi, T., Mitani, S., Nakago, T., Jacobowitz, D.M., Yamaai, T., and Sugimoto, T. Neural parvalbumin and calretinin in the tooth pulp. *Brain Res.* (1994) 647, 1, 124-130.

Iio, T., and Hoshihara, Y. Static and kinetic studies on carp muscle parvalbumins. *J. Biochem.* (1984) 96, 321-328.

Iio, T., and Kondo, H. Comparison of the kinetic properties of troponin C and dansylaziridine-labelled troponin C. *J. Biochem.* (1980a) 88, 2, 547-556.

Iio, T., and Kondo, H. Fluorescence stopped-flow study of N'-(Z- dimethylamino-4-methyl-3-coumaryl)-maleimide-labelled troponin C. *J. Biochem.* (1980b) 88, 4, 1087-1092.

Imajoh, S., Aoki, K., Ohno, S., Emori, Y., Kawasaki, H., Sugihara, H., and Suzuki, K. Molecular cloning of the cDNA for the large subunit of the high-Ca^{2+}-requiring form of human Ca^{2+}-activated neural protease. *Biochemistry* (1988) 27, 21, 8122-8128.

Ince, P., Stout, N., Shaw, P., Slade, J., Hunziker, W., Heizmann, C.W., Parvalbumin and calbindin D-28k in the human motor system and in motor neuron disease. *Neuropathol. Appl. Neurobiol.* (1993) 19, 291–299.

Iritani, S., Kuroki, N., Ikeda, K., and Kazamatsuri, H. Calbindin immunoreactivity in the hippocampal formation and neocortex of schizophrenics. *Prog. Neuro-Psychopharmacol. Biol. Psychiatry* (1999) 23, 409-421.

Ishikawa, K., Nakagawa, A., Tanaka, I., Suzuki, M., and Nishihira, J. The structure of human MRP8, a member of the S100 calcium-binding protein family, by MAD phasing at 1.9 Å resolution. *Acta Crystallog. Sect. D* (2000) 56, 559-566.

Ishikawa, T., and Wakabayashi, T. Calcium-induced changes in the location and conformation of troponin in skeletal muscle thin filaments. *J. Biochem.* (1999) 126, 1, 200-211.

Itou, H., Yao, M., Fujita, I., Watanabe, N., Suzuki, M., Nishihira, J., and Tanaka, I. The crystal structure of human MRP14 (S100A9), a Ca^{2+}-dependent regulator protein in inflammatory process. *J. Mol. Biol.* (2002) 316, 265-276.

Jacquemond, V., and Schneider, M.F. Effects of low myoplasmic Mg^{2+} on calcium binding by parvalbumin and calcium uptake by the sarcoplasmic reticulum in frog skeletal muscle. *J. Gen. Physiol.* (1992) 100, 1, 115-135.

Jakab, R.L., and Goldman-Rakic, P.S. Segregation of serotonin 5-HT2A and 5-HT3 receptors in inhibitory circuits of the primate cerebral cortex. *J. Comp. Neurol.* (2000) 417, 337-348.

Janmey, P.A. Phosphoinositides and calcium as regulators of cellular actin assembly and disassembly. *Annu. Rev. Physiol.* (1994) 56, 169-191.

Jardetzky, O. The study of molecular switching mechanism. *Life Science* (1978) 22, 1245-1252.

Jauregui-Adell, J., and Pechére, J.-F. Parvalbumins from coelacanth muscle. III. Amino acid sequence of the major component. *Biochim. Biophys. Acta* . (1978) 536, 275-282.

Jauregui-Adell, J., Pechére, J.-F., Briand, G., Richet, C., and Demaille, J.G. Amino-acid sequence of an α-parvalbumin, pI = 4.88, from frog skeletal muscle. *Eür. J. Biochem.* (1982) 123, 337-345.

Jiang, Y., Johnson, J.D., and Rall, J.A. Parvalbumin relaxes frog skeletal muscle when sarcoplasmic reticulum Ca^{2+}-ATPase is inhibited. *Am. J. Physiol.* (1996) 270, 2 Pt 1, C411-C417.

Jin, J.P., and Smillie, L.B. An unusual metal-binding cluster found exclusively in the avian breast muscle troponin T of *Galliformers* and *Graciformers*. *FEBS Letters* (1994) 341, 135-140.

Jinno, S., and Kosaka, T. Parvalbumin is expressed in glutamatergic and GABAergic corticostriatal pathway in mice. *J. Comp. Neurol.* (2004) 477, 2, 188-201.

Joassin, L., and Gerday, C. The amino acid sequence of the major parvalbumin of the whiting (*Gadus merlangus*).*Comp. Biochem. Physiol.* (1977) 57B, 159-161.

Jockusch, H., Friedrich, G., and Zippel, M. Serum parvalbumin, an indicator of muscle disease in murine dystrophy and myotonia. *Muscle Nerve* (1990) 13, 6, 551-555.

Johnson, J.D., Jiang, Y., and Rall, J.A. Intracellular EDTA mimics parvalbumin in the promotion of skeletal muscle relaxation. *Biophys. J.* (1999) 76, 3, 1514-1522.

Julenius, K., Thulin, E., Linse, S., and Finn, B.E. Hydrophobic core substitutions in calbindin D_{9k}: effects on stability and structure. *Biochemistry* (1998) 37, 25, 8915-8925.

Jurado, L.A., Chockalingam, P.S., and Jarrett, H.W. Apocalmodulin. *Physiol. Rev.* (1999) 79, 3, 661-682.

Kalus, P., Senitz, D., and Beckmann, H. Altered distribution of parvalbumin-immunoreactive local circuit neurons in the anterior cingulate cortex of schizophrenic patients. *Psychiatry Res.* (1997) 75, 49-59.

Kasprzak, A.A. Myosin subframent 1 inhibits dissociation of nucleotide and calcium from G-actin. *J. Biol. Chem.* (1993) 268, 18, 13261-13266.

Katz, A.K., Glusker, J.P., Beebe, S.A., and Bock C.W. Calcium ion coordination: a comparison with that of beryllium, magnesium, and zinc. *J. Am. Chem. Soc.* (1996) 118, 24, 5752-5763.

Kauffman, J.F., Hapak, R.C., and Henzl, M.T. Interconversion of the CD and EF sites in oncomodulin. Influence on the Eu^{3+} $^7F_0 \rightarrow {}^5D_0$ excitation spectrum. *Biochemistry* (1995) 34, 3, 991-1000.

Kay, B.K., Shah, A.J., and Halstead, W.E. Expression of the $Ca2^+$-binding protein, parvalbumin, during embryonic development of the frog, *Xenopus laevis*. *J. Cell Biol.* (1987) 104, 841-847.

Kendrick-Jones, J., and Jakes, R. Regulatory light chains in myosin. In: *Excitation-Contraction Coupling in Smooth Muscle*. N.Y. North Holland (1977) 343-367.

Kerrick, W.G.L., Hoar, P.E., and Cassidy, P.S. Calcium-activated tension: the role of myosin light chain phosphorylation. *Fed. Proc.* (1980) 39, 5, 1558-1563.

Kerschbaum, H.H., Singh, S.K., Hermann, A., Parvalbumin-immunoreactive material in the kidney of *Xenopus laevis*. *Tissue Cell* (1994) 26, 75–81.

Kobayashi, T., Sano, M., Tsukagoshi, H., and Kamo, I. Muscle inactivity reduces the content of parvalbumin in rat thymic myoid cells in vitro. *Neurosci. Lett.* (1991) 131, 2, 221-224.

Konishi, M. Cytoplasmic free concentrations of Ca^{2+} and Mg^{2+} in skeletal muscle fibers at rest and during contraction. *Jap. J. Physiol.* (1998) 48, 421-438.

Konosu, S., Hamoir, G., and Pechére, J.F. Carp myogens of white and red muscles. Properties and amino acid composition of the main low molecular weight components of white muscle. *Biochem. J.* (1965) 96, 98-112.

Kordel, J., Forsen, S., and Chazin, W.J. Proton NMR sequential resonance assignments secondary structure and global fold in solution of the major trans Pro 43 form of bovine calbindin D-9K. *Biochemistry* (1989) 28, 17, 7065-7074.

Kortvely, E., and Gulya, K. Calmodulin, and various ways to regulate its activity. *Life Sci.* (2004) 74, 1065-1070.

Kragelund, B.B., Jonsson, M., Bifulco, G., Chazin, W.J., Nilsson, H., Finn, B.E., and Linse, S. Hydrophobic core substitutions in calbindin D_{9k}: effects on Ca^{2+} binding and dissociation. *Biochemistry* (1998) 37, 25, 8926-8937.

Kraulis, P.J. MOLSCRIPT: A program to produce both detailed and schematic plots of protein structures. *J. Appl. Crystallography* (1991) 24, 946-950.

Krause, K.H., Milos, M., Luan-Rilliet, Y., Lew, D.P., and Cox, J.A. Thermodynamics of cation binding to rabbit skeletal muscle calsequestrin. Evidence for distinct Ca^{2+}- and Mg^{2+}-binding sites. *J. Biol. Chem.* (1991) 266, 15, 9453-9459.

Krestinger, R.H. Calcium coordination and the calmodulin fold. Divergent versus convergent evolution. Cold Spring Harbor Symposium *Quant Biol.* (1987) 52, 499-510.

Kretsinger, R.H. Structure and evolution of calcium-modulated proteins. *CRC Crit. Rev. Biochem.* (1980) 8, 119-174.

Kretsinger, R.H., Moncrief, D., and Persechini, A. The EF-hand family of calcium modulated proteins. *Trends Neurosci.* (1989) 12, 462– 467.

Kretsinger, R.H., and Nockolds, C.E. Carp muscle calcium binding protein. II. Structural determination and general description. *J. Biol. Chem.* (1973) 248, 9, 3313-3326.

Kretsinger, R.H., Rudnick, S.E., and Weissman, L.J. Crystal structure of calmodulin. *J. Inorg. Biochem.* (1986) 28, 2-3, 289-302.

Kumar, V.D., Lee, L., and Edwards, B.F. Refined crystal structure of ytterbium-substituted carp parvalbumin 4.25 at 1.5 Å, and its comparison with the native and cadmium-substituted structures. *FEBS Lett.* (1991) 283, 2, 311-316.

Kushmerick, M. J. and Podolsky, R. J. Ionic mobility in muscle cells. *Science* (1969) 166, 1297-1298.

Kuster, T., Staudenmann, W., Hughes, G.J., and Heizmann, C.W. Parvalbumin isoforms in chicken muscle and thymus. Amino acid sequence analysis of muscle parvalbumin by tandem mass spectrometry. *Biochemistry* (1991) 30, 8812-8816.

Kuznicki, J., Filipek, A., Heimann, P., Kaczmarek, L., and Kaminska, B. Tissue specific distribution of calcyclin - 10.5 kDa Ca^{2+}-binding protein. *FEBS Letters* (1989) 254, 1-2, 141-144.

Laberge, M., Wright, W.W., Sudhakar, K., Liebman, P.A., and Venderkooi, J.M. Conformational effects of calcium release from parvalbumin: comparison of computational simulations with spectroscopic investigations. *Biochemistry* (1997) 36, 18, 5363-5371.

Laforet, C., Feller, G., Narinx, E., and Gerday, C. Parvalbumin in the cardiac muscle of normal and haemoglobin-myoglobin-free antarctic fish. *J. Muscle Res. Cell. Motil.* (1991) 12, 5, 472-478.

Lambooy, P.K., Steiner, R.F., and Sternberg, H. Molecular dynamics of calmodulin as monitored by fluorescence anisotropy. *Arch. Biochem. Biophys.* (1982) 217, 2, 517-528.

Laney, E.L., Shabanowitz, J., King, G., Hant, D.F., and Nelson, D.J. The isolation of parvalbumin isoforms from the tail muscle of the American alligator (*Alligator mississipiensis*). *J. Inorg. Biochem.* (1997) 66, 1, 67-76.

Leberer, E., and Pette, D. Immunochemical quantitation of sarcoplasmic reticulum Ca-ATPase, a calsequestrin and of parvalbumin in rabbit skeletal muscles of defined fiber composition. *Eur. J. Biochem.* (1986) 156, 489-496.

Lech, J.J., Lewis, S.K., Friedman, M.A., Johnson, L.A., and Mende-Mueller, L.M. Binding of acrylonitrile to parvalbumin. Fundam. *Appl. Toxicol.* (1996) 29, 2, 260-266.

Lech, J.J., Waddell, W.J., Friedman, M.A., and Johnson, L.A. Uptake, disposition, and persistence of acrylonitrile in rainbow trout. Fundam. *Appl. Toxicol.* (1995) 27, 291–294.

Leclerc, E., Leclerc, L., Cassoly, R., der Terrossian, E., Wajcman, H., Poyard, C., and Marden, M.C. Heme binding to calmodulin, troponin C, and parvalbumin, as a probe of calcium-dependent conformational changes. *Arch. Biochem. Biophys.* (1993) 306, 1, 163-168.

LeDonne, N.C., and Coffee, C.J. Inability of parvalbumin to function as a calcium-dependent activator of cyclic nucleotide phosphodiesterase activity. *J. Biol. Chem.* (1979) 254, 11, 4317-4320.

Lee, J., Park, K., Lee, S., Whang, K., Kang, M., Park, C., and Huh, Y. Differential changes of calcium binding proteins in the rat striatum after kainic acid-induced seizure. *Neurosci. Lett.* (2002) 333, 2, 87-90.

Lee, L., and Sykes, B.D. Nuclear magnetic resonance determination of metal-proton distances in the EF site of carp parvalbumin using the susceptibility contribution to the line broadening of lanthanide-shifted resonances. *Biochemistry* (1980a) 19, 14, 3208-3214.

Lee, L., and Sykes, B.D. Strategies for the use of lanthanide NMR shift probes in the determination of protein structure ion solution: application to EF calcium binding site of carp parvalbumin. *Biophys. J.* (1980b) 32, 1, 193-210.

Lee, L., and Sykes, B.D. The use of lanthanide NMR shift probes in the determination of the structure of calcium binding proteins in solution: application to the EF calcium binding site of carp parvalbumin. In: *Calcium Binding Proteins: Structure and Function.* N.Y., Elsevier, North Holland (1980c) 323-326.

Lee, L., and Sykes, B.D. Proton nuclear magnetic resonance determination of the sequential ytterbium replacement of calcium in carp parvalbumin. *Biochemistry* (1981) 20, 5, 1156-1162.

Lee, L., and Sykes, B.D. Use of lanthanide-induced NMR shifts for determination of protein structure in solution: EF calcium binding site of carp parvalbumin. *Biochemistry* (1983) 22, 19, 4366-4373.

Lee, L., Sykes, B.D., and Birnbaum, E.R. A determination of the relative compactness of the Ca^{2+}-binding sites of a Ca^{2+}-binding fragment of troponin C and parvalbumin using lanthanide-induced 1H NMR shifts. *FEBS Letters* (1979) 98, 1, 169-172.

Lee, S.H., Rosenmund, C., Schwaller, B., and Neher, E. Differences in Ca^{2+} buffering properties between excitatory and inhibitory hippocampal neurons from the rat. *J. Physiol. (London)* (2000a) 525, 405-418.

Lee, S.H., Schwaller, B., and Neher, E. Kinetics of Ca^{2+} binding to parvalbumin in bovine chromaffin cells: implications for $[Ca^{2+}]$ transients of neuronal dendrites. *J. Physiology* (2000b), 525, 2, 419-432.

Lee, Y.H., Tanner, J.J., Larson, J.D., and Henzl, M.T. Crystal structure of a high-affinity variant of rat α-parvalbumin. *Biochemistry* (2004) 43, 31, 10008-10017.

Lehky, P., Blum, H.E., Stein, E.A., and Fisher, E.H. Isolation and characterization of parvalbumins from the skeletal muscle of higher vertebrates. *J. Biol. Chem.* (1974) 249, 13, 4332-4334.

Lehky, P., and Stein, E.A. Perch muscle parvalbumin: general characterization and magnesium-binding properties. *Comp. Biochem. Physiol. B* (1979) 63, 2, 253-259.

Le Moine, C., Gaspar, P. Subpopulations of cortical GABAergic interneurons differ by their expression of D1 and D2 dopamine receptor subtypes. *Brain Res. Mol. Brain Res.* (1998) 58, 231-236.

Le Peuch, C.J., Demaille, J.G., and Pechére, J.F. Radioelectrophoresis: a specific microassay for parvalbumins. Application to muscle biopsies from man and other vertebrates. *Biochim. Biophys. Acta* (1978) 537, 153-159.

Leuba, G., Kraftsik, R., and Saini, K. Quantitative distribution of parvalbumin, calretinin, and calbindin D-28k immunoreactive neurons in the visual cortex of normal and Alzheimer cases. *Exp Neurol.* (1998) 152, 2, 278-291.

Lewis, D.A. Chandelier cells: shedding light on altered cortical circuitry in schizophrenia. *Mol. Psychiatry* (1998) 3, 468-471.

Li, H.C., and Fajer, P.G. Orientational changes of troponin C associated with thin filament activation. *Biochemistry* (1994) 33, 47, 14324-14332.

Li, H.C., and Fajer, P.G. Structural coupling of troponin C and actomyosin in muscle fibers. *Biochemistry* (1998) 37, 19, 6628-6635.

Likic, V.A., Strehler, E.E., and Gooley, P.R. Dynamics of Ca^{2+}-saturated calmodulin D129N mutant studied by multiple molecular dynamics simulations. *Protein Sci.* (2003) 12, 2215-2229.

Lin, L.N., and Brandts, J.F. Further evidence suggesting that the slow phase in protein unfolding is due to proline isomerization: a kinetic study of carp parvalbumins. *Biochemistry* (1978) 17, 19, 4102-4110.

Lindstroem, C.D.-V., van Do, T., Hordvik, I., Endresen, C., and Elsayed, S. Cloning of two distinct cDNAs encoding parvalbumin, the major allergen of Atlantic salmon (*Salmo salar*). *Scand. J. Immunol.* (1996) 44, 335-344.

Liu, Y., Carroll, S.L., Klein, M.G., and Schneider, M.F. Calcium transients and calcium homeostasis in adult mouse fast-twitch skeletal muscle fibers in culture. *Am. J. Physiol.* (1997) 272, 6 Pt 1, C1919-C1927.

Loffing, J., Loffing-Cueni, D., Valderrabano, V., Klausli, L., Hebert, S.C., Rossier, B.C., Hoenderop, J.G., Bindels, R.J., and Kaissling, B. Distribution of transcellular calcium and sodium transport pathways along mouse distal nephron. *Am. J. Physiol. Renal. Physiol.* (2001) 281, F1021–F1027.

Luby-Phelps, K., Hori, M., Phelps, J.M., and Won, D. Ca^{2+}-regulated dynamic compartmentalization of calmodulin in living smooth muscle cells. *J. Biol. Chem.* (1995) 270, 37, 21532-21538.

Lukas, W., and Jones, K.A., Cortical neurons containing calretinin are selectively resistant to calcium overload and excitotoxicity *in vitro*. *Neuroscience* (1994) 61, 307– 316.

Lund, J.S., and Lewis, D.A. Local circuit neurons of developing and mature macaque prefrontal cortex: golgi and immunocytochemical characteristics. *J. Comp. Neurol.* (1993) 328, 282- 312.

Lynch, W.P., Riseman, V.M., and Bretscher, A. Smooth muscle caldesmon is an extended flexible monomeric protein in solution that can readily undergo reversible intra- and intermolecular sulfhydryl cross-linking. A mechanism for caldesmon's F-actin binding activity. *J. Biol. Chem.* (1987) 262, 15, 7429-7437.

MacManus, J.P. Occurrence of a low-molecular-weight calcium-binding protein in neoplastic liver. *Cancer Res.* (1979) 39, 8, 3000-3005.

MacManus, J.P., and Whitfield, J.F. Oncomodulin: a calcium-binding protein from hepatoma. In *Calcium and Cell Function,* Cheung W.Y. ed, New York, Academic Press, 411-440.

MacManus, J.P., Whitfield, J.F., Boynton, A.L., Durkin, J.P., and Swierenga, S.H. Oncomodulin - a widely distributed, tumour-specific, calcium-binding protein. *Oncodev. Biol. Med.* (1982) 3, 2-3, 79-90.

Maeda, N., Zhu, D., and Fitch, W.M. Amino acid sequences of lower vertebrate parvalbumins and their evolution: parvalbumins of boa, turtle, and salamander. *Mol. Biol. Evol.* (1984) 1, 473-488.

Maetzler, W., Nitsch, C., Bendfeldt, K., Racay, P., Vollenweider, F., and Schwaller, B. Ectopic parvalbumin expression in mouse forebrain neurons increases excitotoxic injury provoked by ibotenic acid injection into the striatum. *Exp. Neurol.* (2004) 186, 1, 78-88.

Mani, R.S., McCubbin, W.D., Kay, C.M. Calcium-dependent regulation of caldesmon by an 11-kDa smooth muscle calcium-binding protein, caltropin. *Biochemistry* (1992) 31, 47, 11896-11901.

Marston, S., Lehman, W., Woody, C., Pritchard, K., and Smith, C. Caldesmon and Ca^{2+} regulation in smooth muscles. In: *Calcium and Calcium Binding Proteins.* Springer-Verlag, Berlin, Heidelberg (1988) 69-81.

Marston, S.B., and Redwood, C.S. The molecular anatomy of caldesmon. *Biochem. J.* (1991) 279, 1, 1-16.

Martin, R. B. Bioinorganic Chemistry of Magnesium. Marcel Dekker, Inc., New York (1990).

Martignoni, G., Pea, M., Chilosi, M., Brunelli, M., Scarpa, A., Colato, C., Tardanico, R., Zamboni, G., and Bonetti, F. Parvalbumin is constantly expressed in chromophobe renal carcinoma. *Mod. Pathol.* (2001) 14, 8, 760-767.

Matsumura, H., Shiba, T., Inoue, T., Harada, S., and Kai, Y. A novel mode of target recognition suggested by the 2.0 Å structure of holo S100B from bovine brain. *Structure* (1998) 6, 2, 233-241.

Matsuura, I., Ishihara, K., Nakai, Y., Yazawa, M., Toda, H., and Yagi, K. A site-directed mutagenesis study of yeast calmodulin. *J. Biochem.* (1991) 109, 1, 190-197.

Matsuura, I., Kimura, E., Tai, K., and Yazawa, M. Mutagenesis of the fourth calcium-binding domain of yeast calmodulin. *J. Biol. Chem.* (1993) 268, 18, 13267-13273.

Mattson, M.P., Rychlik, B., Chu, C., and Christakos, S., Evidence for calcium-reducing and excitoprotective roles for the calciumbinding protein calbindin-D28K in cultured hippocampal neurons. *Neuron* (1991) 6, 41– 51.

Maughan, D.W., and Godt, R.E. Parvalbumin concentration and diffusion coefficient in frog myoplasm. *J. Muscle Res. Cell Motil.* (1999) 20, 199–209.

Maughan, D.W., and Godt, R.E. Protein osmotic pressure and the state of water in frog myoplasm. *Biophys J.* (2001) 80, 1, 435-442.

Maximov, E.E., Zapevalova, N.P., and Mitin, Y.V. On the calcium-binding ability of the synthetic evolutionary ancestor of calcium-binding proteins. *FEBS Letters* (1978) 88, 80-82.

McManus, J.P., Watson, D.C., and Yaguchi, M. The complete amino acid sequence of oncomodulin - a parvalbumin-like calcium-binding protein from Morris hepatoma 5123tc. *Eur. J. Biochem.* (1983) 136, 9-17.

McPhalen, C.A., Sielecki, A.R., Santarsiero, B.D., and James, M.N. Refined crystal structure of rat parvalbumin, a mammalian α-lineage parvalbumin, at 2.0Å resolution. *J. Mol. Biol.* (1994) 235, 2, 718-732.

Medvedkin, V.N., Mitin, Y.V., and Permyakov, E.A. Influence of Arg-74 on formation of structure of the C-terminal domain of pike parvalbumin III. *Bioorg. Khimija (Moscow)* (1987) 13, 2, 177-182.

Meyers, M.B. Sorcin. In *Guidebook to the Calcium-Binding Proteins*. Edited by M.R. Celio. Oxford, UK; Oxford University Press (1996) 167-168.

Michele, D.E., Szatkowski, M.L., Albayya, F.P., and Metzger, J.M. Parvalbumin gene delivery improves diastolic function in the aged myocardium *in vivo*. *Mol. Ther.* (2004) 10, 2, 399-403.

Miki, M., Kobayashi, T., Kimura, H., Hagiwara, A., Hai, H., and Maeda, Y. Ca^{2+}-induced distance change between points on actin and troponin in skeletal muscle thin filaments estimated by fluorescence energy transfer spectroscopy. *J. Biochem.* (1998) 123, 324-331.

Mikkonen, M., Alafuzoff, I., Tapiola, T., Soininen, H., and Miettinen, R. Subfield- and layer-specific changes in parvalbumin, calretinin and calbindin-D_{28K} immunoreactivity in the entorhinal cortex in Alzheimer's disease. *Neuroscience* (1999) 92, 2, 515-532.

Mitchell, R.D., Simmerman, H.K.B., and Jones, L.R. Ca^{2+} binding effects on protein conformation and protein interactions of canine cardiac calsequestrin. *J. Biol. Chem.* (1988) 263, 3, 1376-1381.

Mize RR. Calbindin 28kD and parvalbumin immunoreactive neurons receive different patterns of synaptic input in the cat superior colliculus. *Brain Res.* (1999) 843, 1-2, 25-35.

Mockel, V., and Fischer, G.I., Vulnerability to excitotoxic stimuli of cultured rat hippocampal neurons containing the calciumbinding proteins calretinin and calbindin D28K. *Brain Res.* (1994) 648, 109–120.

Moeshler, H.J., and Schaer, J.J. Thermodynamics of Ca- and Mg-binding to parvalbumin. *Experientia* (1979) 35, 7, 940.

Moeschler, H.J., Schaer, J.J., and Cox, J.C. A thermodynamic analysis of the binding of calcium and magnesium ions to parvalbumin. *Eur. J. Biochem.* (1980) 11, 1, 73-78.

Moews, P.C., and Kretsinger, R.H. Refinement of the structure of carp muscle calcium binding parvalbumin by model building and difference Fourier analysis. *J. Mol. Biol.* (1975) 91, 2, 201-228.

Moisejev, P.A., Asisova, N.A., and Kuranova, I.N. *Ichthyology* Moscow, Legkaya Promyshlennost (1981).

Moncrief, N.D., Kretsinger, R.H., and Goodman, M. Evolution of EF-hand calcium-modulated proteins. I. Relatioships based on amino acid sequences. *J. Mol. Evol.* (1990) 30, 522-562.

Moncrieffe, M.C., Juranic, N., Kemple, M.D., Potter, J.D., Macura, S., and Prendergast, F.G. Structure-fluorescence correlations in a single tryptophan mutant of carp parvalbumin: solution structure, backbone and side-chain dynamics. *J. Mol. Biol.* (2000) 297, 1, 147-163.

Moncrieffe, M.C., Venyaminov, S.Y., Miller, T.E., Guzman, G., Potter, J.D., and Prendergast, F.G. Optical spectroscopic characterization of single tryptophan mutants of chicken skeletal troponin C: evidence for interdomain interaction. *Biochemistry* (1999) 38, 37, 11973-11983.

Mooren, F.C., and Kinne, R.K.H. Cellular calcium in health and disease. *Biochim. Biophys. Acta* (1998) 127-151.

Morimoto, S., and Ohtsuki, I. Ca^{2+} binding to cardiac troponin C in the myofilament lattice and its relation to the myofibrillar ATPase activity. *Eur. J. Biochem.* (1994) 226, 597-602.

Moroz, O.E., Antson, A.A., Murshudov, G.N., Maitland, N.J., Dodson, G.G., and Wilson, K.S. The three-dimensional structure of human S100A12. *Acta Crystallog. Sect. D* (2001) 57, 20-29.

Muller, B.K., Kabos, P., Belhage, B., Neumann, T., and Kater, S.B. Transfected parvalbumin alters calcium homeostasis in teratocarcinoma PCC7 cells. *Eur. J. Cell Biol.* (1996) 69, 4, 360-367.

Müntener, M., Kaser, L., Weber, J., and Berchtold, M.W. Increase of skeletal muscle relaxation speed by direct injection of parvalbumin cDNA. *Proc. Natl. Acad. Sci. USA* (1995) 92, 6504-6508.

Müntener, M., Rowlerson, A.M., Berchtold, M.W., and Heizmann, C.W. Changes in the concentration of the calcium-binding parvalbumin in cross-reinnervated rat muscles. Caomparison of biochemical with physiological and histochemical parameters. *J. Biol. Chem.* (1989) 262, 1, 465-469.

Müntener, M., van Hardeveld, C., Everts, M.E., and Heizmann, C.W. Analysis of the Ca^{2+}-binding parvalbumin in rat skeletal muscles of different thyroid states. *Exp. Neurol.* (1987) 98, 3, 529-541.

Nakayama, S., Moncrief, N.D., and Kretsinger, R.H. Evolution of EF-hand calcium-modulated proteins. II. Domains of several subfamilies have diverse evolutionary histories. *J. Mol. Evol.* (1992) 34, 416-448.

Nara, M., Tasumi, M., Tanokura, M., Hiraoki, T., Yazawa, M., and Tsutsumi, A. Infrared studies of interaction between metal ions and Ca^{2+}-binding proteins. Marker bands for identifying the types of coordination of the side-chain COO^- groups to metal ions in pike parvalbumin (pI = 4.10). *FEBS Letters* (1994) 349, 1, 84-88.

Nelson, D.J., Opella, S.J., and Jardetsky, O. ^{13}C nuclear magnetic resonance study of molecular motions and conformational transitions in muscle calcium binding proteins. *Biochemistry* (1976) 15, 25, 5552-5560.

Nishida, J., Machida, N.W., Tagome, M., and Kasugai, Y. Existence of parvalbumin and biochemical characterization in quail and pigeon skeletal muscles with different fiber type compositions. *J. Exp. Zool.* (1997) 277, 4, 283-292.

Nishida, J., Machida, N.W., Tagome, M., and Kasugai, Y. Distribution of parvalbumin in specific fibre types of chicken skeletal muscles. *Br. Poult. Sci.* (1995) 36, 4, 585-597.

Nishimura, T., and Goll, D.E. Binding of calpain fragments to calpastatin. *J. Biol. Chem.* (1991) 266, 18, 11842-11850.

Nockolds, C.E., Kretsinger, R.H., Coffee, C.J., and Bradshaw, R.H. Structure of a calcium binding carp myogen. *Proc. Nat. Acad. Sci. USA* (1972) 69, 3, 581-584.

Noegel, A.A. Alpha-actinin. In *Guidebook to the Calcium-Binding Proteins,* edited by M.R. Celio. Oxford, UK, Oxford Iniv. Press (1996) 21-23.

Novak, E., Strzelecka-Golaszewska, H., and Goody, R.S. Kinetics of nucleotide and metal ion interaction with G-actin. *Biochemistry* (1988) 27, 5, 1785-1792.

O'Day, D.H., and Myre, M.A. Calmodulin-binding domains in Alzheimer's disease proteins: extending the calcium hypothesis. *Biochem. Biophys. Res. Commun.* (2004) 320, 1051-1054.

Ogawa, Y., Tanokura, M. Steady-state properties of calcium binding to parvalbumin from bullfrog skeletal muscle: effects of Mg^{2+}, pH, ionic strength, and temperature. *J. Biochem.* (1986a) 99, 1, 73-80.

Ogawa, Y., Tanokura, M. Kinetic studies of calcium binding to parvalbumins from bullfrog skeletal muscle. *J. Biochem.* (1986b) 99, 1, 81-89.

Olive, M., and Ferrer, I. Parvalbumin immunohistochemistry in denervated skeletal muscle. *Neuropathol. Appl. Neurobiol.* (1994) 20, 5, 495-500.

Opella, S.J., Nelson, D.J., and Jardetsky, O. Carbon magnetic resonance study of the conformational changes in carp muscle calcium-binding parvalbumin. *J. Amer. Chem. Soc.* (1974) 96, 22, 7157-7158.

Otterbein, L.R., Kordowska, J., Witte-Hoffman, C., Wang, C.L., and Dominguez, R. Crystal structures of S100A6 in the Ca^{2+}-free and Ca^{2+}-bound states. The calcium sensor mechanism of S100 proteins revealed at atomic resolution. *Structure* (2002) 10, 557-567.

Ou, W.J., Bergeron, J.J.M., Li, Y., Kang, C.Y., and Thomas, D.Y. Conformational changes induced in the endoplasmic reticulum luminal domain of calnexin by Mg-ATP and Ca^{2+}. *J. Biol. Chem.* (1995) 270, 30, 18051-18059.

Ozawa, M. Structure of the gene encoding mouse reticulocalbin, a novel endoplasmic reticulum-resident Ca^{2+}-binding protein with multiple Efhand motifs. *J. Biochem.* (1995) 118, 1, 154-160.

Ozawa, M., and Muramatsu, T., Reticulocalbin, a novel endoplasmic reticulum resident Ca^{2+}-binding protein with multiple EF-hand motifs and a carboxyl-terminal HDEL sequence. *J. Biol. Chem.* (1993) 268, 1, 699-705.

Pack, A.K., and Slepecky, N.B. Cytoskeletal and calcium-binding proteins in the mammalian organ of Corti: cell type-specific proteins displaying longitudinal and radial gradients. *Hear Res.* (1995) 91, 1-2, 119-135.

Palmisano, W.A., Trevino, C.L., and Henzl, M.T. Site-specific replacement of amino acid residues within the CD binding loop of rat oncomodulin. *J. Biol. Chem.* (1990) 265, 24, 14450-14456.

Parello, J., Cave, A., Puigdomenech, P., Maury, C., Capony, J.P., and Pechére, J.F. Conformational studies on muscular parvalbumins. II Nuclear magnetic resonance analysis. *Biochimie* (1974) 56, 1, 61-76.

Parello, J., Lilia, H., Cave, A., and Lindman, B. A ^{43}Ca NMR study of the binding of calcium to parvalbumins. *FEBS Letters* (1978) 87, 2, 191-195.

Parello, J., Reimarssen, P., Thulin, E., and Lindman, B. Na^+ binding to parvalbumin studied by ^{23}Na NMR. *FEBS Letters* (1979) 100, 1, 153-156.

Pauls, T.L., Cox, J.A., and Berchtold, M.W. The Ca^{2+}-binding proteins parvalbumin and oncomodulin and their genes: new structural and functional findings. *Biochim. Biophys. Acta* (1996a) 1306, 39-54.

Pauls, T.L., Durussel, I., Berchtold, M.W., and Cox, J.A. Inactivation of individual Ca^{2+}-binding sites in the paired EF-hand sites of parvalbumin reveals asymmetrical metal-binding properties. *Biochemistry* (1994) 33, 34, 10393-10400.

Pauls, T.L., Durussel, I., Clark, I.D., Szabo, A.G., Berchtold, M.W., and Cox, J.A. Site-specific replacement of amino acid residues in the CD site of rat parvalbumin changes the metal specificity of this Ca^{2+}/Mg^{2+}-mixed site toward a Ca^{2+}-specific site. *Eur. J. Biochem.* (1996b) 242, 2, 249-255.

Pauls, T.L., Durussel, I., Cox, J.A., Clark, I.D., Szabo, A.G., Gagne, S.M., Sykes, B.D., and Berchtold, M.W. Metal binding properties of recombinant rat parvalbumin wild-type and F102W mutant. (1993) *J. Biol. Chem.* 268, 20897-20903.

Pauls, T.L., Portis, F., Macri, E., Belser, B., Heitz, P., Doglioni, C., and Celio M.R. Parvalbumin is expressed in normal and pathological human parathyroid glands. *J. Histochem. Cytochem.* (2000) 48, 1, 105-111.

Pechére, J.F. Muscular parvalbumins as homologous proteins. *Comp. Biochem. Physiol.* (1968) 24, 289-295.

Pechére, J.F. Isolement d'une parvalbumine du muscle de lapin. *C.R. Acad. Sc. Paris* (1974) 278, 2577-2579.

Pechére, J.F. The significance of parvalbumins among muscular calciproteins. In: *Calcium Binding Proteins and Calcium Function.* N.Y., North Holland (1977) 213-221.

Pechére, J.F., Capony, J.P., and Ryden, L. The primary structure of the major parvalbumin from hake muscle. Isolation and general properties of the protein. *Eur. J. Biochem.* (1971a) 23, 421-428.

Pechére, J.F., Capony, J.P., Ryden, L., and Demaille, J. The amino acid sequence of the major parvalbumin from hake muscle. *Biochem. Biophys. Res. Commun.* (1971b) 43, 1106-1111.

Pechére, J.F., Demaille, J., and Capony, J.P. Muscular parvalbumins: preparative and analyticalmethods of general applicability. *Biochim. Biophys. Acta* (1971c) 236, 391-408.

Pechére, J.F., Derancourt, J., and Haiech, J. The participation of parvalbumins in the activation-relaxation cycle of vertebrate fast skeletal muscle. *FEBS Letters* (1976) 75, 1, 111-114.

Pechére, J.-F., Rochat, H., and Ferraz, C. Parvalbumins from coelacanth muscle. II. Amino acid sequence of the two less acidic components. *Biochim. Biophys. Acta* (1978) 536, 269-274.

Permyakov, E.A. *Parvalbumin and Related Calcium Binding Proteins.* Nauka, Moscow (1985).

Permyakov, E.A. *Calcium Binding Proteins.* Nauka, Moscow (1993a).

Permyakov, E.A. *Luminescent spectroscopy of Proteins.* CRC Press, Boca Raton, Ann Arbor, London, Tokyo (1993b).

Permyakov, E.A., Burstein, E.A., Emelyanenko, V.I., Alexandrov, Y.M., Glagolev, K.V., Makhov, V.N., Syreishchikova, T.I., and Yakimenko, M.N. Effects of calcium binding on the decay time of intrinsic fluorescence of calcium binding proteins. *Biofizika (Moscow)* (1983c) 28, 3, 393-398.

Permyakov, E.A., Kalinichenko, L.P., Yarmolenko, V.V., Burstein, E.A., and Gerday, C. Binding of nucleotides to parvalbumins. Biochem. Biophys. *Res. Commun.* (1982a) 105, 4, 1059-1065.

Permyakov, E.A., Kreimer, D.I., Kalinichenko, L.P., Orlova, A.A. and Shnyrov, V.L. Interaction of parvalbumins with model phospholipid vesicles. *Cell Calcium.* (1989a) 10, 2, 71-79.

Permyakov, E.A., Kreimer, D.I., Kalinichenko, L.P., and Shnyrov, V.L. Interaction of calcium binding proteins, parvalbumin and α-lactalbumin with dipalmitoylphosphatidylcholine vesicles. *Gen. Physiol. Biophys.* (1988b) 7, 1, 95-107.

Permyakov, E.A., Medvedkin, V.N., Kalinichenko, L.P., and Burstein, E.A. Comparative study of physico-chemical properties of two pike parvalbumins by means of their intrinsic tyrosine and phenylalanine fluorescence. *Arch. Biochem. Biophys.* (1983a) 227, 1, 9-20.

Permyakov, E.A., Medvedkin, V.N., Kalinichenko, L.P., Burstein, E.A., and Gerday, C. Sodium and potassium binding to parvalbumins measured by means of intrinsic protein fluorescence. *Biochim. Biophys. Acta* (1983b) 749, 1, 185-191.

Permyakov, E.A., Medvedkin, V.N., Korneichuk, G.A., Kostrzhevskaia, E.G., and Murzin, A.G. Binding of melittin to parvalbumin inhibited by Ca^{2+} ions. New function for parvalbumin. *Mol. Biologia (Moscow)* (1989b) 23, 3, 693-698.

Permyakov, E.A., Medvedkin, V.N., Mitin, Y.V., and Kretsinger, R.H., Noncovalent complex between domain AB and domains CD*EF of parvalbumin. Biochim. *Biophys. Acta* (1991b) 1076, 67-70.

Permyakov, E.A., Morozova, L.A., Kalinichenko, L.P., and Derezhkov, V.Y., Interaction of α-lactalbumin with Cu(II). *Biophys. Chem.* (1988c) 32, 1, 37-42.

Permyakov, E.A., Ostrovsky, A.V., and Kalinichenko, L.P. Stopped flow kinetic studies of Ca^{2+} and Mg^{2+} dissociation in cod parvalbumin and bovine α-lactalbumin. *Biophys. Chem.* (1987a) 28, 2, 225-233.

Permyakov, E.A., Ostrovsky, A.V., Kalinichenko, L.P., and Deikus, G.Y. Kinetics of dissociation of complexes of parvalbumin with calcium and magnesium ions. *Mol. Biol. (Moscow)* (1987b) 2, 4, 1017-1022.

Permyakov, E.A., Ostrovsky, A.V., Burstein, E.A., Pleshanov, P.G., Gerday, C. Parvalbumin conformers revealed by steady-state and time-resolved fluorescence spectroscopy. *Arch. Biochem. Biophys.* (1985) 240, 2, 781-791.

Permyakov, E.A., and Yarmolenko, V.V. Possible artifacts during the work with glass pH electrode on water solutions of Ca^{2+} binding systems. *Biofizika (Moscow)* (1980) 26, 1, 134-135.

Permyakov, E.A., Yarmolenko, V.V., Burstein, E.A., and Gerday, C. Intrinsic fluorescence spectra of a tryptophan containing parvalbumin as a function of thermal, pH and urea denaturation. *Biophys. Chem.* (1982) 15, 1, 19-26.

Permyakov, E.A., Yarmolenko, V.V., Emelyanenko, V.I., Burstein, E.A., Gerday, C., and Closset, J. Fluorescence studies of the calcium binding to whiting (*Gadus merlangus*) parvalbumin. *Eur. J. Biochem.* (1980) 109, 2, 307-315.

Permyakov, E.A., Yarmolenko, V.V., Kalinichenko, L.P., and Burstein, E.A. Evaluation of Ca^{2+} binding constants of proteins using changes of tyrosine and phenylalanine fluorescence. *Bioorg. Khimija (Moscow)* (1981) 7, 11, 1660-1668.

Permyakov, S.E., Millett, I.S., Doniach, S., Permyakov, E.A., and Uversky, V.N. Natively unfolded C-terminal domain of caldesmon remains substantially unstructered after the effective binding to calmodulin. *Proteins: Structure, Function, and Genetics* (2003) 53, 855-862.

Piront, A., and Gosselin-Rey, C. Immunological cross-reactions among *Cyprinidae* parvalbumins. *Biochem. Syst. Ecol.* (1974) 2, 103-107.

Piront, A., and Gosselin-Rey, C. Immunological cross-reactions among *Gadidae* parvalbumins. *Biochem. Syst. Ecol.* (1975) 2, 251-255.

Plogmann, D., and Celio, M.R. Intracellular concentration of parvalbumin in nerve cells. *Brain Res.* (1993) 600, 273-279.

Pohl, V., Pattyn, G., and Berchtold, M. Parvalbumin expression during developmental differentiation of the rat ovary. *Differentiation* (1995) 59, 4, 235-242.

Pontremoli, S., Melloni, E. Extralysosomal protein degradation. *Ann. Rev. Biochem.* (1986) 55, 455-481.

Pontremoli, S., Sparatore, B., Melloni, E., Michetti, M., and Horecker, B.L. Activation by hemoglobin of the Ca^{2+}-requiring neutral protease of human erythrocytes: structural requirements. *Biochem. Biophys. Res. Comm.* (1984) 123, 1, 331-337.

Por, A., Pocsai, K., Rusznak, Z., and Szucs, G. Presence and distribution of three calcium binding proteins in projection neurons of the adult rat cochlear nucleus. *Brain Res.* (2005) 1039, 1-2, 63-74.

Porter, L.L., Rizzo, E., and Hornung, J.P. Dopamine affects parvalbumin expression during cortical development *in vitro*. J. *Neurosci.* (1999) 19, 20, 8990-9003.

Porteros, A., Arevalo, R., Brinon, J.G., Crespo, C., Aijon, J., and Alonso, J.R. Parvalbumin immunoreactivity during the development of the cerebellum of the rainbow trout. *Brain Res. Dev. Brain Res.* (1998) 109, 2, 221-227.

Potter, J.D., Dedman, J.R., and Means, A.R. Ca^{2+}-dependent regulation of cyclic-AMP phosphodiesterase by parvalbumin. *J. Biol. Chem.* (1977) 252, 16, 5609-5611.

Potter, J.D., and Gergely, J. The calcium and magnesium binding sites on troponin and their role in the regulation of myofibrillar adenine triphosphatase. *J. Biol. Chem.* (1975) 250, 12, 4628-4633.

Potter, J.D., Johnson, J.D., and Mandel, F. Fluorescence stopped flow measurements of Ca^{2+} and Mg^{2+} binding to parvalbumin. *Fed. Proc.* (1978) 37, 8, 1608.

Potter, J.D., Robertson, S.P., Collins, J.H., and Johnson, J.D. The role of the Ca^{2+} and Mg^{2+} binding sites on troponin and other myofibrillar proteins in regulation of muscle contraction. In: *Calcium Binding Proteins: Structure and Function.* N.Y., Elsevier, North Holland (1980) 301-302.

Potter, J.D., Robertson, S.P., and Johnson, J.D. Magnesium and the regulation of muscle contraction. *Fed. Proc.* (1981) 40, 12, 2653-2656.

Potter, J.D., Sheng, Z., Pan, B.S., and Zhao, J. A direct regulatory role for troponin T and a dual role for troponin C in the Ca^{2+} regulation of muscle contraction. *J. Biol. Chem.* (1995) 270, 6, 2557-2562.

Potts, B., Smith, J., Akke, M., Macke, T., Okazaki, K., Hidaka, H. The structure of calcyclin reveals a novel homodimeric fold for S100 Ca^{2+}-binding proteins. *Nature Struct. Biol.* (1995) 2, 790-796.

Pusch, M., and Neher, E. Rates of diffusional exchange between small cells and a measuring patch pipette. *Pflügers Arch.* (1988) 411, 204–211.

Putkey, J.A., Liu, W., and Sweeney, H.L. Function of the N-terminal calcium-binding sites in cardiac/slow troponin C assessed in fast skeletal muscle fibers. *J. Biol. Chem.* (1991) 266, 23, 14881-14884.

Putkey, J.A., Ono, T., Vanberkum, M.F.A., and Means, A.R. Functional significance of the central helix in calmodulin. *J. Biol. Chem.* (1988) 263, 23, 11242-11249.

Qin, Z.P., Ye, S.M., Du, J.Z., and Shen, G.Y. Postnatal developmental expression of calbindin, calretinin and parvalbumin in mouse main olfactory bulb. *Acta Biochim. Biophys. Sin. (Shanghai)* (2005) 37, 4, 276-282.

Racay, P., Gregory, P., and Schwaller, B. Parvalbumin deficiency in fast-twitch muscles leads to increased 'slow-twitch type' mitochondria, but does not affect the expression of fiber specific proteins. *FEBS J.* (2006) 273, 1, 96-108.

Ramakrishnan, S., and Hitchcock-DeGregori, S.E. Investigation of the structural requirements of the troponin C central helix for function. *Biochemistry* (1995) 34, 51, 16789-16796.

Raymackers, J.M., Debaix, H., Colson-Van Schoor, M., De Backer, F., Tajeddine, N., Schwaller, B., Gailly, P., and Gillis, J.M. Consequence of parvalbumin deficiency in the *mdx* mouse: histological, biochemical and mechanical phenotype of a new double mutant. *Neuromuscul. Disord.* (2003) 13, 5, 376-387.

Raymackers, J.M., Gailly, P., Colson-Van Schoor, M., Pette, D., Schwaller, B., Hunziker, W., Celio, M.R., and Gillis, J.M., Tetanus relaxation of fast skeletal muscles of the mouse made parvalbumin deficient by gene inactivation. *J. Physiol.* (2000) 527, 355– 364.

Real, M.A., Davila, J.C., and Guirado, S. Expression of calcium-binding proteins in the mouse claustrum. *J. Chem. Neuroanat.* (2003) 25, 3, 151-160.

Rehbein, H., Kundiger, R., Pineiro, C., and Perez-Martin, R.I. Fish muscle parvalbumins as marker proteins for native and urea isoelectric focusing. *Electrophoresis* (2000) 21, 8, 1458-1463.

Reid, R.E., and Hodges, R.S. Co-operativity and calcium/magnesium binding to troponin C and muscle calcium binding parvalbumin: an hypothesis. *J. Theor. Biol.* (1980) 84, 401-444.

Rety, S., Osterloh, D., Arie, J.P., Tabaries, S., Seeman, J., Russo-Marie, F. Structural basis of the Ca^{2+}-dependent association betwee S100C (S100A11) and its target, the N-terminal part of annexin I. *Struct. Fold. Des.* (2000) 8, 175-184.

Rety, S., Sopkova, J., Renouard, M., Osterloh, D., Gerke, V., and Tabaries, S. The crystal structure of the complex of p11 with the annexin II N-terminal peptide. *Nature Struct. Biol.* (1999) 6, 89-95.

Revett, S.P., King, G., Shabanowitz, J., Hunt, D.F., Hartman, K.L., Laue, T.M., and Nelson, D.J. Characterization of a helix-loop-helix (EF-hand) motif of silver hake parvalbumin isoform B. *Protein Sci.* (1997) 6, 11, 2397-2408.

Reynolds, G.P., and Beasley, C.L. GABAergic neuronal subtypes in the human frontal cortex– development and deficits in schizophrenia. *J. Chem. Neuroanat.* (2001) 22, 95-100.

Rhyner, J.A., Durussel, I., Cox, J.A., Ilg, E.C., Schafer, B.W., and Heizmann, C.W. Human recombinant α-parvalbumin and nine mutants with individually inactivated calcium- and magnesium-binding sites: biochemical and immunological properties. *Biochim. Biophys. Acta* (1996) 1313, 3, 179-186.

Richardson, R.C., King, N.M., Harrington, D.J., Sun, H., Royer, W.E., and Nelson, D.J. X-ray crystal structure and molecular dynamics simulations of silver hake parvalbumin (Isoform B). *Protein Sci.* (2000) 9, 73-82.

Rigden, D.J., and Galperin, M.Y. The DxDxDG motif for calcium binding: multiple structural contexts and implication for evolution. *J. Mol. Biol.* (2004) 343, 971-984.

Ritzler, J.M., Sawhney, R., Geurts van Kessel, A.H., Grzeschik, K.H., Schinzel, A., and Berchtold, M.W. The genes for the highly homologous Ca^{2+}-binding proteins oncomodulin and parvalbumin are not linked in the human genome. *Genomics* (1992) 12, 3, 567-572.

Rizo, J., and Südhof, T.C. C_2-domains, structure and function of universal Ca^{2+}-binding domain. *J. Biol. Chem.* (1998) 273, 26, 15879-15882.

Robertson, S.R., and Kerrick, W.G.L. Use of parvalbumin as a calcium buffer in biological systems. *Fed. Proc.* (1978) 37, 3, 557.

Robertson, S.R., and Potter, J.D. On the role of various myofibrillar binding sites in the Ca^{2+}-regulation of vertebrate striated muscle. In: Intern. Symp. on Calcium Binding in Health and Decease. Madison, Wisc. (1980) 62.

Rome, L.C. Designe and function of superfast muscles: new insights into the physiology of skeletal muscle. *Annu. Rev. Physiol.* (2006) 68, 193-221.

Roquet, F., Declercq, J.-P., Tinant, B., Rambaud, J., and Parello, J. Crystal structure of the unique parvalbumin component from muscle of the leopard shark (*Triakis semifasciata*). The first X-ray study of an α-parvalbumin. *J. Mol. Biol.* (1992) 223, 705-720.

Ross, C., Hevener, S., Clark, R., Hartmann, J.X., and Mari, F. Isolation of parvalbumin isotypes by preparative HPLC techniques. *Prep. Biochem. Biotechnol.* (1998) 28, 1, 49-60.

Ross, N.R., and Porter, L.L. Effects of dopamine and estrogen upon cortical neurons that express parvalbumin *in vitro*. *Brain Res. Dev. Brain Res.* (2002) 137, 1, 23-34.

Sakaguchi, N., Henzl, M.T., Thalmann, I., Thalmann, R., and Schulte, B.A. Oncomodulin is expressed exclusively by outer hair cells in the organ of Corti. *J. Histochem. Cytochem.* (1998) 46, 1, 29-40.

Sasaki, T., Tanokura, M., and Asaoka, K. The complete amino acid sequence of bullfrog (*Rana catesbeiana*) parvalbumin pI 4.97. *FEBS Lett.* (1990) 268, 249-251.

Sastry, M., Ketchem, R.R., Crescenzi, O., Weber, C., Lubienski, M.J., Hidaka, H., and Chazin, W.J. The three-dimensional structure of Ca^{2+}-bound calcyclin: implications for Ca^{2+}-signal transduction by S100 proteins. *Structure* (1998) 6, 223-231.

Satyshur, K.A., Pyzalska, D., Greaser, M., Rao, S.T., and Sundaralingam. M. Structure of chicken skeletal muscle troponin C at 1.78 A resolution. Acta Crystallogr. D Biol. *Crystallogr.* (1994) 50, Pt 1, 40-49.

Sauter, A., Staudenmann, W., Hughes, G.J., and Heizmann, C.W. A novel EF-hand Ca^{2+}-binding protein from abdominal muscle of crustaceans with similarity to calcyphosine from dog thyroidea. *Eur. J. Biochem.* (1995) 227, 1, 97-101.

Scharfman, H.E., and Schwartzkroin, P.A., Protection of dentate hilar cells from prolonged stimulation by intracellular calcium chelation. *Science* (1989) 246, 257– 260.

Schibeci, A., and Martonosi, A. Ca^{2+}-binding proteins in nuclei. *Eur. J. Biochem.* (1980) 113, 1, 5-14.

Schleef, M., Zuhlke, C., Jockusch, H., and Schoffl, F. The structure of the mouse parvalbumin gene. *Mamm Genome* (1992) 3, 4, 217-225.

Schmidt, H., Brown, E.B., Schwaller, B., and Eilers, J. Diffusional mobility of parvalbumin in spiny dendrites of cerebellar Purkinje neurons quantified by fluorescence recovery after photobleaching. *Biophys. J.* (2003a) 84, 4, 2599-2608.

Schmidt, H., Stiefel, K.M., Racay, P., Shwaller, B., and Eilers, J. Mutational analysis of dendritic Ca^{2+} kinetics in rodent Purkinje cells: role of parvalbumin and calbindin D_{28k}. *J. Physiol.* (2003b) 551, Pt 1, 13-32.

Schmidt, U., Zhu, X., Lebeche, D., Huq, F., Guerrero, J.L., and Hajjar, R.J. In vivo gene transfer of parvalbumin improves diastolic function in aged rat hearts. *Cardiovasc Res.* (2005) 66, 2, 318-323.

Schmitt, T.L., and Pette, D. Fiber type-specific distribution of parvalbumin in rabbit skeletal muscle. A quantitative microbiochemical and immunohistochemical study. *Histochemistry* (1991) 96, 6, 459-465.

Schneeberger, P.R., and Heizman, C.W. Parvalbumin in rat kidney. Purification and localization. *FEBS Letters* (1986) 201, 1, 51-56.

Scholey, J.M., Taylor, K.A., and Kendrick-Jones, J. The role of myosin light chains in regulating actin-myosin interaction. *Biochimie* (1981) 63, 4, 255-271.

Schwaller, B., Dick, J., Dhoot, G., Carroll, S., Vbrova, G., Nicotera, P., Pette, D., Wyss, A., Bluethmann, H., Hunziker, W., and Celio, M.R. Prolonged contraction-relaxation cycle of fast-twitch muscles in parvalbumin knockout mice. *Am. J. Physiol. Cell. Physiol.* (1999) 276, C395-C403.

Schwaller, B., Meyer, M., and Schiffmann, S. 'New' functions for 'old' proteins: the role of the calcium-binding proteins calbindin D-28k, calretinin and parvalbumin, in cerebellar physiology. Studies with knockout mice. *Cerebellum* (2002) 1, 4, 241-258.

Schwaller, B., Tetko, I.V., Tandon, P., Silveira, D.C., Vreugdenhil, M., Henzi, T., Potier, M.C., Celio, M.R., and Villa, A.E. Parvalbumin deficiency affects network properties resulting in increased susceptibility to epileptic seizures. *Mol. Cell. Neurosci.* (2004) 25, 4, 650-663.

Schwarzenbach, G., and Flaschka, H. *Die Komplexonometrische Titration*, Ferdinand Enke Verlag, Stuttgart (1965).

Sheng, Z., Strauss, W.L., Francois, J.M., and Potter, J.D. Evidence that both Ca^{2+}-specific sites of skeletal muscle TnC are required for full activity. *J. Biol. Chem.* (1990) 265, 35, 21554-21560.

Shirinsky, V.P., Bushueva, T.L., and Frolova, S.I. Caldesmon-calmodulin interaction. Study by the method of protein intrinsic tryptophan fluorescence. *Biochem. J.* (1988) 255, 2, 203-208.

Schutt, C. Hands on the calcium switch. *Nature* (1985) 15, 6014, 15.

Sia, S.K., Li, M.X., Spyracopoulos, L., Garne, S.M., Liu, W., Putkey, J.A., and Sykes, B.D. Structure of cardiac muscle troponin C unexpectedly reveals a closed regulatory domain. *J. Biol. Chem.* (1997) 272, 29, 18216-18221.

Sloviter, R.S. Calcium-binding protein (calbindin-D28K) and parvalbumin immunocytochemistry: localization in the rat hippocampus with specific reference to the selective vulnerability of hippocampal neurons to seizure activity. *J. Comp. Neurol.* (1989) 280, 183-196.

Slupsky, C.M., and Sykes, B.D. NMR solution structure of calcium-saturated skeletal muscle troponin C. *Biochemistry* (1995) 34, 49, 15953-15964.

Smith, C.W.J., Pritchard, K., and Marston, S.B. The mechanism of Ca^{2+} regulation of vascular smooth muscle thin filaments by caldesmon and calmodulin. *J. Biol. Chem.* (1987) 262, 1, 116-122.

Smith, L., Greenfield, N.J., Hitchcock-DeGregori, S.E. Mutations in the N- and D-helices of the N-domain of troponin C affect the C-domain and regulatory function. *Biophys J.* (1999) 76, 1 Pt 1, 400-408.

Smith, M.J., and Koch, G.L.E. Multiple zones in the sequence of calreticulin (CBP55, calregulin, HACBP), a major calcium binding ER/SR protein. *EMBO Journal* (1989) 8, 12, 3581-3586.

Smith, S.P., and Shaw, G.S. A novel calcium-sensitive switch revealed by the structure of human S100B in the calcium-bound form. *Structure* (1998) 6, 2, 211-222.

Sobue, K., and Sellers, J.R. Caldesmon, a novel regulatory protein in smooth muscle and nonmuscle actomyosin systems. *J. Biol. Chem.* (1991) 266, 19, 12115-12118.

Soderling, T.R. Structure and regulation of calcium/calmodulin-dependent protein kinases II and IV. *Biochim. Biophys. Acta* (1996) 1297, 2, 131-138.

Sommerville, L.E., and Hartshorne, D.J. Intracellular calcium and smooth muscle contraction. *Cell Calcium* (1986) 7, 353-364.

Sorenson, M.M., da Silva, A.C.R., Gouveia, C.S., Sousa, V.P., Oshima, W., Ferro, J.A., and Reinach, F.C. Concerted action of the high affinity calcium binding sites in skeletal muscle troponin C. *J. Biol. Chem.* (1995) 270, 17, 9770-9777.

Sorimachi, H., and Suzuki, K. The structure of calpain. J. Biochem. (2001) 129, 653-664.

Soto-Prior, A., Cluzel, M., Renard, N., Ripoll, C., Lavigne-Rebillard, M., Eybalin, M., and Hamel, C.P. Molecular cloning and expression of α parvalbumin in the guinea pig cochlea. *Brain Res. Mol. Brain Res.* (1995) 34, 2, 337-342.

Stichel, C.C., Singer, W., Heizmann, C.W., and Norman, A.W. Immunochemical localization of calcium-binding proteins, parvalbumin and calbindin D-28 K, in the adult and developing visual cortex of cats: a light and electron microscopic study. *J. Comp. Neurol.* (1987) 262, 2, 563-577.

Strobl, S., Fernandez-Catalan, C., Braun, M., Huber, R., Masumoto, H., Nakagawa, K., Irie, A., Sorimachi, H., Bourenkow, G., Bartunik, H., Suzuki, K., and Bode, W. The crystal structure of calcium-free human m-calpain suggests an electrostatic switch mechanism for activation by calcium. Proc. Natl. Acad. Sci. U S A (2000) 97, 2, 588-592.

Stryer, L. Visual excitation and recovery. *J. Biol. Chem.* (1991) 266, 17, 10711-10714.

Strzelecka-Golaszewska, H., Boguta, G., Zmorzynski, S., and Moraczewska, J. Biochemical and theoretical approach to localization of metal-ion-binding sites in the actin primary structure. *Eur. J. Biochem.* (1989) 182, 2, 299-305.

Stull, J.T., Blumenthal, D.K., and Cooke, R. Regulation of contraction by myosin phosphorylation. A comparison between smooth and skeletal muscles. *Biochem. Pharmacol.* (1980) 29, 2537-2543.

Sudhakar, K., Erecinska, M., and Vanderkooi, J.M. Interaction of polyamines with Ca^{2+}-binding protein parvalbumin. *Eur. J. Biochem.* (1995a) 230, 498-502.

Sudhakar, K., Phillips, C.M., Owen, C.S., and Vandercooi, J.M. Dynamics of parvalbumin studied by fluorescence emission and triplet absorption spectroscopy of tryptophan. *Biochemistry* (1995b) 34, 1355-1363.

Sudhakar, K., Phillips, C.M., Williams, S.A., and Vanderkooi, J.M. Excited states of tryptophan in cod parvalbumin. Identification of a short-lived emitting triplet state at room temperature. *Biophys. J.* (1993) 64, 5, 1503-1511.

Sugita, S., Ho, A., and Sudhof, T.C. NECABs: a family of neuronal Ca^{2+}-binding proteins with an unusual domain structure and a restricted expression pattern. *Neuroscience* (2002) 112, 1, 51-63.

Sundaralingam, M., Drendel, W., and Greaser, M. Stabilization of the long central helix of troponin C by intrahelical salt bridges between charged amino acid side chains. *Proc. Nat. Acad. Sci. USA* (1985) 82, 23, 7944-7947.

Suzuki, K., Sorimachi, H., Yoshizawa, T., Kinbara, K., and Ishiura, S. Calpain: novel family members, activation, and physiologic function. *Biol. Chem. Hoppe-Seyler* (1995) 376, 523-529.

Swoboda, I., Bugajska-Schretter, A., Verdino, P., Keller, W., Sperr, W.R., Valent, P., Valenta, R., and Spitzauer, S., Recombinant carp parvalbumin, the major cross-reactive fish allergen: a tool for diagnosis and therapy of fish allergy. *J. Immunol.* (2002) 168, 4576–4584.

Szatkowski, M.L., Westfall, M.V., Gomez, C.A., Wahr, P.A., Michele, D.E., DelloRusso, C., Turner, I.I., Hong, K.E., Albayya, F.P., and Metzger, J.M. *In vivo* acceleration of heart relaxation performance by parvalbumin gene delivery. *J. Clin. Invest.* (2001) 107, 2, 191-198.

Szebenyi, D.M., and Moffat, K. The refined structure of vitamin D-dependent calcium-binding protein from bovine intestine. Molecular details, ion binding, and implications for the structure of other calcium-binding proteins. *J. Biol. Chem.* (1986) 261, 19, 8761-8777.

Szebenyi, D.M.E., Obendorf, S.K., and Moffat, K. Structure of vitamin D-dependent calcium binding protein from bovine intestine. *Nature* (1981) 294, 327-332.

Tada, M., Yamamoto, T., and Tonomura, Y. Molecuilar mechanism of active calcium transport by sarcoplasmic reticulum. *Physiol. Rev.* (1978) 58, 1, 1-79.

Tanner, J.J., Agah, S., Lee, Y.H., and Henzl, M.T. Crystal structure of the D94S/G98E variant of rat α-parvalbumin. An explanation for the reduced divalent ion affinity. *Biochemistry* (2005) 44, 33, 10966-10976.

Tanaka, T., Ames, J.B., Harvey, T.S., Stryer, L., and Ikura, M. Sequestration of the membrane-targeting myristoyl group of recoverin in the calcium-free state. *Nature* (1995) 376, 6539, 444-447.

Tanaka, T., Ames, J.B., Kainosho, M., Stryer, L., and Ikura, M. Differential isotype labelling strategy for determining the structure of myristoylated recoverin by NMR spectroscopy. *J. Biomol. NMR* (1998) 11, 2, 135-152.

Tanokura, M. Heat capacity and entropy changes of the major isotype of the toad (*Bufo*) parvalbumin induced by calcium binding. *Eur. J. Biochem.* (1990) 188, 1, 23-28.

Tardif, E., Chiry, O., Probst, A., Magistretti, P.J., and Clarke, S. Patterns of calcium-binding proteins in human inferior colliculus: identification of subdivisions and evidence for putative parallel systems. *Neuroscience* (2003) 116, 4, 1111-1121.

Terro, F., Yardin, C., Esclaire, F., Ayer-Lelievre, C., and Hugon, J., Mild kainite toxicity produces selective motoneuron death with marked activation of Ca^{2+}-permeable MPA/kainite receptors. *Brain Res.* (1998) 809, 319–324.

Thatcher, D.R., and Pechére, J.-F. The amino-acid sequence of the major parvalbumin from thornback-ray muscle. *Eur. J. Biochem.* (1977) 75, 121-132.

Thepaut, M., Strub, M.P., Cave, A., Baneres, J.L., Berchtold, M.W., Dumas, C., and Padilla, A. Structure of rat parvalbumin with deleted AB domain: implications for the evolution of EF hand calcium-binding proteins and possible physiological relevance. *Proteins: Structure, Function and Genetics* (2001) 45, 117-128.

Thulin, E., Svard, M., Forsen, S., Drakenberg, T., Anderson, T., Seamon, K.B., and Krebs, J. The interaction between calcium binding proteins and a phenothiazine drug (trifluoperazine) as studied by ^{113}Cd and ^{43}Ca NMR. In: Intern. Symp.on Calcium Binding Proteins and Calcium Function in Health and Decease. Madison, Wisc. (1980) 46.

Thys, T.M., Blank, J.M., and Schachat, F.H. Rostral-caudal variation in troponin T and parvalbumin correlates with differences in relaxation rates of cod axial muscle. *J. Exp. Biol.* (1998) 201, Pt 21, 2993-3001.

Tjoelker, L.W., Sayfried, C.E., Eddy, R.L., Byers, M.G., Shows, T.B., Calderon, J., Schreiber, R.B., and Gray, P.W. Human, mouse, and rat calnexin cDNA cloning: identification of potential calcium binding motif and gene localization to human chromosome 5. *Biochemistry* (1994) 33, 11, 3229-3236.

Todtenkopf, M.S., Stellar, J.R., Williams, E.A., and Zahm, D.S. Differential distribution of parvalbumin immunoreactive neurons in the striatum of cocaine sensitized rats. *Neuroscience* (2004) 127, 1, 35-42.

Toury, R., Belqasmi, F., Hauchecorne, M., Heizmann, C.W., and Balmain, N. Ultrastructural localization of α-parvalbumin in the epiphyseal plate cartilage and bone of growing rats. *Bone* (1996) 19, 3, 245-253.

Toury, R., Belqasmi, F., Hauchecorne, M., Leguellec, D., Heizmann, C.W., and Balmain, N.,. Localization of the Ca^{2+}-binding α-parvalbumin and its mRNA in epiphyseal plate cartilage and bone of growing rats. *Bone* (1995) 17, 121–130.

Treves, S., Scutari, E., Robert, M., Groh, S., Ottolia, M., Prestipino, G., Ronjat, M., Zorzato, F. Interaction of S100A1 with the Ca^{2+} release channel (ryanodine receptor) of skeletal muscle. *Biochemistry* (1997) 36, 11496-11503.

Trevino, C.L., Boschi, J.M., and Henzl, M.T. Interactions between residues in the oncomodulin CD domain influence Ca^{2+} ion-binding affinity. *J. Biol. Chem.* (1991) 266, 17, 11301-11308.

Trevino, C.L., Palmisano, W.A., Birnbaum, E.R., and Henzl, M.T. Eu^{3+} luminescence studies of oncomodulin. The origin of the pH-dependent behavior. *J. Biol. Chem.* (1990) 265, 17, 9694-9700.

Tripet, B., Van Eyk, J.E., and Hodges, R.S. Mapping of a second actin-tropomyosin and a second troponin C binding site within the C terminus of troponin I, and their importance in the Ca^{2+}-dependent regulation of muscle contraction. *J. Mol. Biol.* (1997) 271, 728-750.

Troxler, H., Kuster, T., Rhyner, J.A., Gehrig, P., and Heizmann, C.W. Electrospray ionization mass spectrometry: analysis of the Ca^{2+}-binding properties of human recombinant alpha-parvalbumin and nine mutant proteins. *Anal. Biochem.* (1999) 268, 1, 64-71.

Ushio, H., and Watabe, S. Carp parvalbumin binds to and directly interacts with the sarcoplasmic reticulum for Ca^{2+} translocation. Biochem. *Biophys. Res. Commun.* (1994) 199, 1, 56-62.

Valenta, R., Hayek, B., Seiberler, S., Bugajska-Schretter, A., Niederberger, V., Twardosz, A., Natter, S., Vangelista, L., Pastore, A., Spitzauer, S., and Kraft, D. Calcium-binding allergens: from plants to man. Int. Arch. *Allergy Immunol.* (1998) 117, 3, 160-166.

Van Den Bosch, L., Schwaller, B., Vleminckx, V., Meijers, B., Stork, S., Ruehlicke, T., Van Houtte, E., Klaassen, H., Celio, M.R., Missiaen, L., Robberecht, W., and Berchtold, M.W. Protective effect of parvalbumin on excitotoxic motor neuron death. *Exp. Neurol.* (2002) 174, 2, 150-161.

Van Do, T., Elsayed, S., Florvaag, E., Hordvik, I., and Endresen, C. Allergy to fish parvalbumins: studies on the cross-reactivity of allergens from 9 commonly consumed fish. *J. Allergy Clin. Immunol.* (2005) 116, 6, 1314-1320.

Van Do, T., Hordvik, I., Endresen, C., and Elsayed, S. The major allergen (parvalbumin) in codfish is encoded by two isotypic genes. Submitted MAY-2001 to the EMBL GenBank DDBJ databases.

Van Do, T., Hordvik, I., Endresen, C., and Elsayed, S. Expression and analysis of recombinant salmon parvalbumin, the major allergen in Atlantic salmon (*Salmo salar*). *Scand. J. Immunol.* (1999) 50, 6, 619-625.

Van Do, T., Hordvik, I., Endresen, C., and Elsayed, S. The major allergen (parvalbumin) of codfish is encoded by at least two isotypic genes: cDNA cloning, expression and antibody binding of the recombinant allergens. *Mol. Immunol.* (2003) 39, 10, 595-602.

Van Do, T., Hordvik, I., Endresen, C., and Elsayed, S. Characterization of parvalbumin, the major allergen in Alaska pollack, and comparison with codfish Allergen M. *Mol. Immunol.* (2005) 42, 3, 345-53.

Van Do, T., Hordvik, I., Endresen, C., and Elsayed, S. Parvalbumin isotypes in Alaska pollack. Submitted MAY-2001 to the EMBL GenBank DDBJ databases.

Vassylyev, D.G., Takeda, S., Wakatsuki, S., Maeda, K., and Maeda, Y. Crystal structure of troponin C in complex with troponin I fragment at 2.3-Å resolution. *Proc. Natl. Acad. Sci. USA* (1998) 95, 4847-4852.

Vreugdenhil, M., Jefferys, J.G.R., Celio, M.R., and Schwaller, B. Parvalbumin-deficiency facilitates repetitive IPSCs and related inhibitionbased gamma oscillations in the hippocampus. *J. Neurophysiol.* (2003) 89, 1414-1423.

Wada, I., Rindress, D., Cameron, P.H., Ou, W.J., Doherty, J.J., Louvard, D., Bell, A.W., Dignard, D., Thomas, D.Y., and Bergeron, J.J.M. SSR and associated calnexin are major calcium binding proteins of the endoplasmic reticulum membrane. *J. Biol. Chem.* (1991) 266, 29, 19599-19610.

Wahr, P.A., Michele, D.E., and Metzger, J.M. Parvalbumin gene transfer corrects diastolic dysfunction in diseased cardiac myocytes. *Proc. Natl. Acad. Sci. USA.* (1999) 96, 21, 11982-11985.

Weeds, A.G., and McLachlan, A.D. Structural homology of myosin light chains, troponin C and carp calcium binding protein. *Nature* (1974) 252, 5385, 646-649.

Werber, M.M. Metal binding to myosin and DTNB-light chain. *Experientia* (1978) 34, 5, 575-576.

Weruaga, E., Velasco, A., Brinon, J.G., Arevalo, R., Aijon, J., and Alonso, J.R. Distribution of the calcium-binding proteins parvalbumin, calbindin D-28k and calretinin in the retina of two teleosts. *J. Chem. Neuroanat.* (2000) 19, 1, 1-15.

Westerblad, H., and Allen, D.G. Relaxation, $[Ca^{2+}]_i$ and $[Mg^{2+}]_i$ during prolonged tetanic stimulation of intact, single fibres from mouse skeletal muscle. *J. Physiol.* (1994) 480, Pt 1, 31-43.

Westerblad, H., and Allen, D.G. Slowing of relaxation and $[Ca^{2+}]_i$ during prolonged tetanic stimulation of single fibres from Xenopus skeletal muscle. *J. Physiol.* (1996) 492, Pt 3, 723-736.

White, H.D. Kinetic mechanism of calcium binding to whiting parvalbumin. *Biochemistry* (1988) 27, 9, 3357-3365.

White, H., and Closset, J. Kinetics of calcium binding to whiting parvalbumin. Biophys. J. (1979) 25, 2, 247a.

Wild, J.M., Williams, M.N., and Suthers, R.A. Parvalbumin-positive projection neurons characterise the vocal premotor pathway in male, but not female, zebra finches. *Brain Res.* (2001) 917, 2, 235-252.

Williams, T.C., Corson, D.C., Oikawa, K., McCubbin, W.D., Kay, C.M., and Sykes, B.D. ^{1}H NMR spectroscopic studies of calcium-binding proteins. 3. Solution conformations of rat apo-α-parvalbumin and metal-bound rat α-parvalbumin. *Biochemistry* (1986a) 25, 7, 1835-1846.

Williams, T.C., Corson, D.C., McCubbin, W.D., Oikawa, K., Kay, C.M., and Sykes, B.D. ^{1}H NMR spectroscopic studies of calcium-binding proteins. 2. Histidine microenvironments in α- and β-parvalbumins as determined by protonation and laser photochemically induced dynamic nuclear polarization effects. *Biochemistry* (1986b) 25, 7, 1826-1834.

Williams, T.C., Corson, D.C., and Sykes, B.D. Calcium binding proteins: calcium(II)-lanthanide(III) exchange in carp parvalbumin. *J. Am. Chem. Soc.* (1984) 106, 19, 5698-5702.

Wilson, M.A., and Brunger, A.T. The 1.0 Å crystal structure of Ca^{2+}-bound calmodulin: an analysis of disorder and implications for functionally relevant plasticity. *J Mol Biol.* (2000) 301, 5, 1237-1256.

Wnuk, W., Cox, J.A., Kohler, L.G., and Stein, E.A. Calcium- and magnesium-binding properties of a high affinity calcium-binding protein from crayfish sarcoplasm. *J. Biol. Chem.* (1979) 254, 12, 5284-5289.

Woo, T.-U., Whitehead, R.E., Melchitzky, D.S., and Lewis, D.A. A subclass of prefrontal Gama-aminobutyric acid axon terminals are selectively altered in schizophrenia. *Proc. Natl. Acad. Sci.* (1998) 95, 5341-5346.

Wriggers, W., Mehler, E., Pitici, F., Weinstein, H., and Schulten, K. Structure and dynamics of calmodulin in solution. *Biophys. J.* (1998) 74, 1622-1639.

Wu, Y., and Parent, A. Striatal interneurons expressing calretinin, parvalbumin or NADPH-diaphorase: a comparative study in the rat, monkey and human. *Brain Res.* (2000) 863, 1-2, 182-191.

Xiong, H., Feng, X., Gao, L., Xu, L., Pasek, D.A., Seok, J.H., and Meissner, G. Identification of a two EF-hand Ca^{2+} binding domain in lobster skeletal muscle ryanodine receptor/Ca^{2+} release channel. *Biochemistry* (1998) 37, 14, 4804-4814.

Xu, Y., He, J., Wang, X., Lim, T.M., and Gong, Z. Asynchronous activation of 10 muscle-specific protein (MSP) genes during zebrafish somitogenesis. *Dev. Dyn.* (2000) 219, 201-215.

Yang, W., Lee, H.W., Hellinga, H., and Yang, J.J. Structural analysis, identification, and design of calcium-binding sites in proteins. *Proteins* (2002) 47, 344-356.

Zahm, D.S., Grosu, S., Irving, J.C., and Williams, E.A. Discrimination of striatopallidum and extended amygdala in the rat: a role for parvalbumin immunoreactive neurons? *Brain Res.* (2003) 978, 1-2, 141-154.

Zanotti, J.M., Bellissent-Funel, M.C., and Parello, J. Hydration-coupled dynamics in proteins studied by neutron scattering and NMR: the case of the typical EF-hand calcium-binding parvalbumin. *Biophys J.* (1999) 76, 5, 2390-2411.

Zawadowska, B., and Supikova, I. Parvalbumin in skeletal muscles of teleost (*Tinca tinca L. and Misgurnus fossilis L.*). Histochemical and immunohistochemical study. *Folia Histochem. Cytobiol.* (1992) 30, 2, 63-67.

Zhang, C., Speno, H., Clairmont, C., and Nelson, D. The isolation of an unusual parvalbumin from the white muscle of the silver hake (*Merluccius Bilinearis*). *J. Inorg. Biochem.* (1990) 40, 59-79.

Zhang, H., Wang, G., Ding, Y., Wang, Z., Barraclough, R., Rudland, P.S., Fernig, D.G., and Rao, Z. the crystal structure at 2 Å resolution of the Ca^{2+}-binding protein S100P. *J. Mol. Biol.* (2003) 325, 785-794.

Zhang, Z.J., and Reynolds, G.P. A selective decrease in the relative density of parvalbumin-immunoreactive neurons in the hippocampus in schizophrenia. *Schizophr Res.* (2002) 55, 1-2, 1-10.

Zhao, X., Kobayashi, T., Malak, H., Gryczynski, I., Lakowicz, J., Wade, R., and Collins, J.H. Calcium-induced troponin flexibility revealed by distance distribution measurements between engineered sites. *J. Biol. Chem.* (1995) 270, 26, 15507-15514.

Zimmer, D.B., and Landar, A. Analysis of S100A1 expression during skeletal muscle and neuronal cell differentiation. *J. Neurochem.* (1995) 64, 2727-2736.

Zuehlke, C.H., Schoeffl, F., Jockusch, H., Simon, D., and Guenet, J.-L. cDNA sequence and chromosomal localization of the mouse parvalbumin gene, *Pva. Genet. Res.* (1989) 54, 37-43.

Zuschratter, W., Scheich, H., and Heizmann, C.W. Ultrastructural localization of the calcium-binding protein parvalbumin in neurons of the song system of the zebra finch *Peophila guttata. Cell Tissue Res.* (1985) 241, 1, 77-83.

INDEX

A

B

C

D

E

F

G

H

I

J

K

L

M

N

O

P

Q

R

S

T

U

V

W

X

Y

Z